Future-Oriented Technology Assessment

Future-Oriented Technology Assessment

A Manager's Guide with Case Applications

Edited by

Haydar Yalçın
Ege University
Turkey

Tugrul U. Daim
Portland State University
United States

IEEE Press Series on Technology Management, Innovation, and Leadership

IEEE TEMS
Technology & Engineering
Management Society

IEEE Press

WILEY

Published by John Wiley & Sons, Inc., Hoboken, New Jersey.
Published simultaneously in Canada.

For general information on our other products and services or for technical support, please contact our Customer Care Department within the United States at (800) 762-2974, outside the United States at (317) 572-3993 or fax (317) 572-4002.

Wiley also publishes its books in a variety of electronic formats. Some content that appears in print may not be available in electronic formats. For more information about Wiley products, visit our website at www.wiley.com.

Library of Congress Cataloging-in-Publication Data

Names: Yalçın, Haydar, author. | Daim, Tugrul Unsal, 1967- author.
Title: Future-oriented technology assessment : a manager's guide with case applications / Haydar Yalçın, Ege University, Turkey, Tugrul U. Daim, Portland State University, US.
Description: Hoboken, New Jersey : Wiley, [2025] | Series: IEEE Press series on technology management, innovation, and leadership | Includes bibliographical references and index.
Identifiers: LCCN 2024011177 (print) | LCCN 2024011178 (ebook) | ISBN 9781119909859 (hardback) | ISBN 9781119909866 (adobe pdf) | ISBN 9781119909873 (epub)
Subjects: LCSH: Technological innovations–Economic aspects. | Business planning–Case studies. | Organizational change–Case studies.
Classification: LCC HC79.T4 Y355 2025 (print) | LCC HC79.T4 (ebook) | DDC 658.5/14–dc23/eng/20240407
LC record available at https://lccn.loc.gov/2024011177
LC ebook record available at https://lccn.loc.gov/2024011178

Cover Design: Wiley
Cover Image: © Weiquan Lin/Getty Images

Set in 9.5/12.5pt STIXTwoText by Straive, Chennai, India

We dedicate this book to our Dads, who are both in eternity now.

Contents

A Note from the Series Editor

IEEE Press is well known for its books on technical and engineering topics. The *Technology Management, Innovation, and Leadership* series extends the reach of the imprint, from engineering and scientific deep dives to downstream stages of technology value chains and ultimately to societal impact.

The pathway starts with the crucial activities of basic and applied research, but connecting the dots from invention to innovation to the betterment of humanity and our ecosphere has become increasingly critical. Consider some of the other key developments and processes that are essential: new product and service design, system or ecosystem integration, intellectual property protection, manufacturing and supply chain integration, regulatory and compliance and certification, business model design and analysis, cost and price estimation, environmental sustainability assessment, … and much more. The time, effort, and funding required for realizing real-world impact dwarfs what was expended on the research. The skills required for end-to-end success also extend well beyond what is typically covered in STEM disciplines and include effective communication, cross-functional and global collaboration, leadership acumen, and science and technology policy development.

Big-picture insights and perspectives can be offered at an industry-agnostic level, and considerations also vary by industry sector, technology area, geography, and other factors. Accordingly, the series scope encompasses related topics both writ large – e.g. pragmatic assessments of emerging technologies, intrapreneurship and entrepreneurship frameworks, artificial intelligence and digital transformation, cybersecurity and resilience – and in the context of one or more societal application domains such as power and clean energy, logistics and transportation, smart cities and infrastructure, and global healthcare.

I am pleased that, in its first few years, the series has produced several books addressing key topics in its remit. But it's a vast space to cover and we look forward to publishing more titles that are relevant for researchers, practitioners, policy makers, educators, students, business leaders, and others. For those who are seeking to make a positive difference for themselves, their organizations, and the world, technology management, innovation, and leadership are essential skills to hone.

You can review an up-to-date list of books published and upcoming in the series by navigating to the series page at https://ieee-press.ieee.org.

Tariq Samad
Senior Fellow and Honeywell/W.R. Sweatt Chair in
Technology Management Technological Leadership Institute
University of Minnesota, Minneapolis, Minnesota, USA

IEEE Technology and Engineering Management Society Body of Knowledge (TEMSBOK)
1st Ed. | Sep 2023
Gustavo Giannattasio, Elif Kongar, Marina Dabić, Celia Desmond, Michael Condry, Sudeendra Koushik, Roberto Saracco

Future-Oriented Technology Assessment: A Manager's Guide with Case Applications
1st Ed. | Nov 2024
Haydar Yalçın, Tugrul U. Daim

Intrapreneurship Management: Concepts, Methods, and Software for Managing Technological Innovation in Organizations
1st Ed. | Jun 2024
Rainer Hasenauer, Oliver Yu

Persuasive Communication for Science and Technology Leaders: Writing and Speaking with Confidence
Nov 2022
Stephen Wilbers

The Digital Transformation of Logistics: Demystifying Impacts of the Fourth Industrial Revolution
1st Ed. | Mar 2021
Mac Sullivan, Johannes Kern

ICT Policy, Research, and Innovation: Perspectives and Prospects for EU-US Collaboration
Nov 2020
Svetlana Klessova, Sebastian Engell, Maarten Botterman, Jonathan Cave

The Exponential Era: Strategies to Stay Ahead of the Curve in an Era of Chaotic Changes and Disruptive Forces
Dec 2020
David Espindola, Michael W. Wright

Cyber-Physical-Human Systems: Fundamentals and Applications
1st Ed. | Jun 2023
Anuradha M. Annaswamy, Pramod P. Khargonekar, Françoise Lamnabhi-Lagarrigue, Sarah K. Spurgeon

About the Editors

Haydar Yalçın, PhD, is an Associate Professor on Management Information Systems at Ege University in Izmir, Turkey. He is also an affiliated faculty member at Mark O. Hatfield Cybersecurity and Cyber Defense Policy Center, which is a National Academic Center of Excellence for Cybersecurity at Portland State University.

Tugrul U. Daim, PhD, is a Professor of Engineering and Technology Management, Fulbright Scholar, and the Director of Research at Mark O. Hatfield Cybersecurity and Cyber Defense Policy Center, which is a National Academic Center of Excellence for Cybersecurity at Portland State University.

List of Contributors

Ashfaq AlKhalil
Mark O. Hatfield Cybersecurity and
Cyber Defense Policy Center
Department of Engineering and
Technology Management
Portland State University
Portland, OR 97201, USA

Saeed Alzahrani
King Saud University
Riyadh, 11451 Saudi Arabia

Fayez Alsoubaie
Mark O. Hatfield Cybersecurity and
Cyber Defense Policy Center
Department of Engineering and
Technology Management
Portland State University
Portland, OR 97201, USA

Serhat Burmaoglu
Izmir Katip Celebi University
Izmir, Turkey

Umit Cali
School of Physics, Engineering
and Technology
University of York
Heslington
York, YO10 5DD, UK

Ferhat O. Catak
Electrical Engineering & Computer
Science
University of
Stavanger, Rogaland, Norway

Tugrul U. Daim
Mark O. Hatfield Cybersecurity
and Cyber Defense Policy Center
Department of Engineering
and Technology Management
Portland State University
Portland, OR 97201, USA

Mursel Dogrul
Dational Defense University
Istanbul, Turkey

Ahmet Ergurum
University of Wisconsin-Milwaukee
Milwaukee, WI 53201, USA

Sara Ferdousi
Portland State University
Portland, OR 97201, USA

Tugrul Felipe C. Gelbecke
Hamburg Technical University
Hamburg, Germany

Nathasit Gerdsri
College of Management
Mahidol University
Bangkok, Thailand

Ebru Gökalp
Computer Engineering Department
Hacettepe University
Ankara 06800, Turkey

Mert Onuralp Gökalp
Data Analytics Center
Tüpras, Ankara 06510, Turkey

Kerem Kayabay
HLRS, University of Stuttgart
Stuttgart 70569, Germany

Atilla Kılınç
Energy Institute
Instanbul Technical University
Istanbul 34469, Turkey

Aynur Kirbac
Mark O. Hatfield Cybersecurity and
Cyber Defense Policy Center
Department of Engineering and
Technology Management
Portland State University
Portland, OR 97201, USA

Vimal Kumar
Department of Information
Management
Chaoyang University of Technology
Taichung 41349, Taiwan

Murat Kuzlu
Batten College of Engineering &
Technology
Old Dominion University
VA 23507, USA

Kuei-Kuei Lai
Department of Business
Administration
Chaoyang University of Technology
Taichung 41349, Taiwan

Hermann Lodding
Hamburg Technical University
Hamburg, Germany

Ali B. Naeini
Iran University of Science
and Technology
Tehran, Iran

Alison Nalven
Portland State University
Portland, OR 97201, USA

Ummaraporn Pora
Technopreneurship and Innovation
Management Program
Chulalongkorn University
Bangkok, Thailand

Sudatip Puengrusme
College of Management
Mahidol University
Bangkok, Thailand

Salih Sarp
Electrical & Computer Engineering
Department
Virginia Commonwealth University
VA 23284, USA

Saumya Saxena
Portland State University
Portland, OR 97201, USA

Nagendra K. Sharma
Department of Management Studies
Graphic Era (Deemed to be University)
Dehradun 248002, India

Sine B. Skaarup
FL Smidth
Copenhagen, Denmark

Nolan Thompson
Portland State University
Portland, OR 97201, USA

Pratima Verma
Department of Strategic Management
Indian Institute of Management
Kozhikode
Kozhikode 673570, India

Ronald Vatananan
College of Management
Mahidol University
Bangkok, Thailand

Courtney Wright
Portland State University
Portland, OR 97201, USA

Haydar Yalçın
Department of Business
Administration
Ege University
Izmir 35800, Turkey

Aynur Yarga
Izmir Katip Celebi University
Izmir, Turkey

Mehdi Zamani
Kiel University
Kiel, Germany

Jennifer L. Zeitouni
Portland State University
Portland, OR 97201, USA

Hao Zhang
Chengdu Medical College
Chengdu, Sichuan Province, China

Yanxiao Zhao
Electrical & Computer Engineering
Department
Virginia Commonwealth University
VA 23284, USA

Preface

This book provides a set of tools for managers tasked with assessing technologies for different purposes.

The first four chapters demonstrate technology assessment and forecasting through case studies including smart cities, solar technology, cement production, and energy storage. Case studies provide insight into how to apply the novel methods used. By providing a wide range of industry applications, the book establishes a set of guidelines for established as well as emerging sectors.

The next five chapters use various technology intelligence approaches to identify technological insight. Through analysis of published papers and patents, we provide intelligence on technologies including crypto, geothermal, smart grid, and heat pump water heaters.

Finally we introduce technology roadmapping, which integrates all the prior methods to provide a path for technology development. Cases provided cover a wide range of sectors, ranging from cybersecurity in health care to nanotechnology in construction.

We hope that this book will enable the readers to manage technology more effectively.

1

Technology Assessment: Smart City Development Initiatives and Issues

Nagendra K. Sharma[1], Vimal Kumar[2], Pratima Verma[3], Tugrul U. Daim[4], Haydar Yalçın[5], and Kuei-Kuei Lai[6]

[1] *Department of Management Studies, Graphic Era (Deemed to be University), Dehradun 248002, India*
[2] *Department of Information Management, Chaoyang University of Technology, Taichung 41349, Taiwan*
[3] *Department of Strategic Management, Indian Institute of Management Kozhikode, Kozhikode 673570, India*
[4] *Mark O. Hatfield Cybersecurity and Cyber Defense Policy Center, Department of Engineering and Technology Management, Portland State University, Portland, OR 97201, USA*
[5] *Department of Business Administration, Ege University, Izmir 35800, Turkey*
[6] *Department of Business Administration, Chaoyang University of Technology, Taichung 41349, Taiwan*

1.1 Introduction

"*A smart city is a city well performing built on the 'smart' combination of endowments and activities of self-decisive, independent, and aware citizens*" (Giffinger et al. 2007). Urban population concentration and other environmental challenges are among the key antecedents to the concept of smart cities around the world. Traditional cities are not adequately structured to provide a good quality of life for people in today's fast-moving world. The smart cities concept came into the picture with the advent of high-end technologies and corresponding new policies that make it possible to live smartly (Letaifa 2015).

We are facing several kinds of challenges because of population growth in the old cities, which is straining limited, basic resources for their residents. Therefore, a new planning program must be implemented so that cities can be a better place to live. The challenges are compounded because people continue to migrate to urban areas for employment and a better standard of living. Urban jobs and infrastructure attract people looking for a better quality of life. The United Nations (UN) predicts that by 2050 there may be 6.5 billion people living in cities (Streitz 2015).

Future-Oriented Technology Assessment: A Manager's Guide with Case Applications, First Edition.
Edited by Haydar Yalçın and Tugrul U. Daim.

In some ways, they find this better lifestyle in the cities. However, migration is putting much pressure on the urban established cities and their systems and is not likely to stop in the near future (Okai et al. 2018). The proposal of smart cities is therefore widely accepted as a way to deal with such challenges. Cities play a crucial role in building socioeconomic and ecological counterparts around the world.

Because of the expected and current population growth, cities are suffering from various kinds of challenges as the available resources and the built infrastructure are limited and are under pressure. Thus, there is an urgent need to develop new infrastructure that can have a problem-solving approach to these challenges in the cities. *Smart cities* offer the hope that everyone can survive in a better way with the use of technology and engineering. For a city to be called a smart city, the smart city (SC) projects require several important criteria to be met: smart mobility, safety and security of the people, smart healthcare system, clean water and air, major dependencies on renewable energy systems, prompt disaster response system, economic development, and higher social and cultural values among people (Kosowatz 2020). Innovative technological solutions will be needed to develop all these factors for smart cities – solutions that can be easily adopted and fixed in the current or a new system.

Figure 1.1 shows several key components of smart cities. Smart cities implement technologies to become smart and it leads to better quality of life, economic development, a traffic management system for a robust supply chain in the city, and an effective health care system in the city with a proper network of emergency vehicles and ambulances. Beyond the listed factors, an effective communication system is necessary for appropriate information travel among the public and authorities (Ismagilova et al. 2019). Understanding the importance of smart cities and their development requires understanding the main priorities identified in the researchers' smart city concept agendas (Camero and Alba 2019). These researchers are adding knowledge in all the SC areas, such as architecture and planning, civil engineering, information technology, management, policy, and governance.

In recent studies, it was found that SC development is greatly strengthened with the latest technological developments. The challenges in the cities can be minimized with the appropriate utilization of technologies. Information communication technology (ICT), artificial intelligence (AI), the internet of things (IoT), deep learning, machine learning, neural networking, cognitive computing, and big data analytics are some of the promising technologies used in the development of the smart city projects around the world. The engagement of these technologies is aimed to convert the conventional systems of the city into the autonomous system (Ahad et al. 2020). Inclusion of high-end technologies makes the entire system of the city smart work, such as smart mobility, which connects the vehicles on

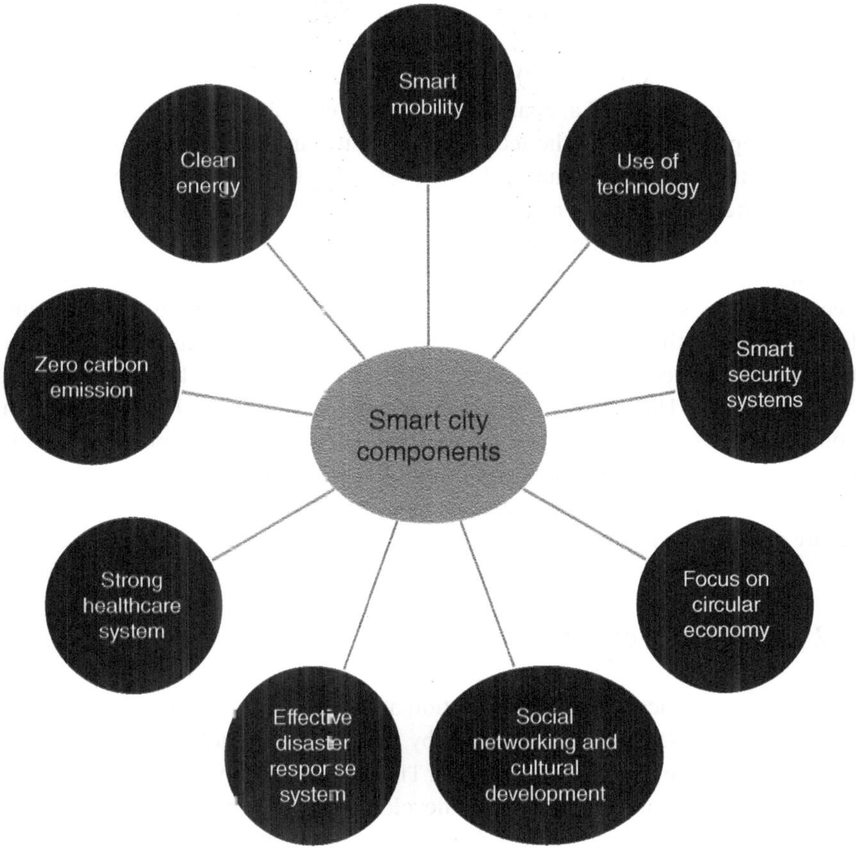

Figure 1.1 Components of smart city. *Source:* Developed by Authors.

board to the traffic system and users can easily understand the traffic situation in the real-time. Smart mobility can also reduce fatalities that result when people injured in traffic accidents cannot get immediate treatment. In this case, smart mobility can be helpful for running ambulances or emergency vehicles with the help of an automatic traffics system linked with these vehicles. Real-time monitoring systems and global positioning systems (GPS) showing hospitals and clinics to the users are some of the smart systems that can save lives. These examples confirm how important the smart city project is for every nation in the world and the significance of the innovation agendas of the government, research agencies, and technology partners toward the development and implementation of smart technologies in the cities (Okai et al. 2018). In today's context, many cities are interested in transforming into smart cities for achieving developmental goals, although the journey is full of challenges and complexities as it includes lots of planning with a public-private partnership (PPP) approach, experts with ICT, government support, and significant funding (Lai et al. 2020; Razmjoo et al. 2021).

This chapter focuses on the technological issues in the development of smart cities around the world and technological engagement in the development process. The major objective is to highlight smart city initiatives adopted by the cities and technology engagement for the same. Another task is to find the issues that are linked to technological aspects. These objectives are important to achieve because it helps in understanding the complexities of adopting the high-end technologies by the countries for smart city development projects. Studies show that most SC projects fail to achieve success because of poor technology adoption or affordability. Hence, this chapter endeavors to highlight these issues with appropriate possible solutions for the successful development of smart cities. This chapter may be helpful for the decision-makers who are engaged with smart city development projects and for researchers who may use this research for exploring future research opportunities.

1.2 Evolution of the Smart City

Urbanization has accelerated the evolution of smart cities. According to De Marco and Mangano (2021), it is expected that by 2050, approximately 66% of the global population will be living in urban areas. The term *smart city* has gained traction around the world, influencing urban development plans and government policies (Berry 2018). Overpopulation of smart cities also has many space challenges. The challenges are traffic congestion, waste management, pollution, and parking allocation – but opportunities exist to solve these problems. One of the biggest challenges or problems is traditional safety and security infrastructure that arises from rapid urbanization (Isafiade and Bagula 2017).

There is no widely accepted definition of the smart city. It varies according to perspective. The SC concept arose a few years ago as a collection of "ideas on how information and communication technology might improve city functioning" (Camero and Alba 2019). In other words, a smart city is a sustainable city that uses fourth-industrial-revolution technology and stakeholder governance to solve urban challenges and improve inhabitants' quality of life (Myeong et al. 2022). Smart cities are also known by different names: Digital City (Ishida 2002); Eco-City (Register 1987); Green City (OECD 2013); Intelligent City (Komninos 2006); Knowledge City (Edvinsson 2006); Sustainable City (Haughton and Hunter 1994); and Wired City (Dutton et al. 1987).

1.3 Need for Smart Cities

The increasing burden of population on the planet and certain other types of environmental challenges led by overexploitation of the limited resources made the cities a poor place to live. The basic utilities such as getting safe drinking water are even a challenge for the common public, especially in densely populated places like Mumbai, Manila, Beijing, and Dhaka. The road traffic in these cities is one of the major problems as it wastes precious time and emits more carbon into the air. Traffic congestion kills many people each year who are delayed while seeking emergency medical help. These issues create extensive pressure on the conventional system of older cities as they lack the communication technologies that could strengthen the information system.

On the other hand, there is a lack of access to the latest devices that can help people in living a good and safe life. The increasing population especially in south Asian countries also leads to unemployment and increased crime. To control crime, the police department must have access to devices that help to monitor and track the situation in real-time with wireless technology. Another law enforcement tool is closed-circuit TVs (CCTVs). Police departments could install CCTVs in prominent locations and track the video feed with the help of fixed stations and patrolling cabs. Such tools utilized at a broader scale can be beneficial for the people living in smart cities. The technological potential of SCs has given new wing that can be promising in achieving the target of smart cities.

There are various factors where smart technologies can be implemented to make the city smart, such as safety and security where the police department may be provided with the latest technological tools to track and control crime. Traffic management can be stronger where the congestion can be reduced with automation and machine learning algorithm. The waste needs to be converted into a value with the circular economy concept. Water sewage treatment is necessary so that water can be reused by the public. Pedestrian and cycling paths must be developed so that citizens can ensure good health. The focus should not be only on nonrenewable energy, but there should also good scope of renewable sources of energy such as wind energy and solar energy. Local community parks provide opportunities for exercise and more interaction among citizens for better social development of the public. Although some of these things are present for the holistic development and the establishment of smart cities, all these factors need to be implemented. The need and development of smart cities are shown in Figure 1.2, which illustrates how population and other challenges have provoked greater effort in pairing the concept of smart cities with the development of smart technologies.

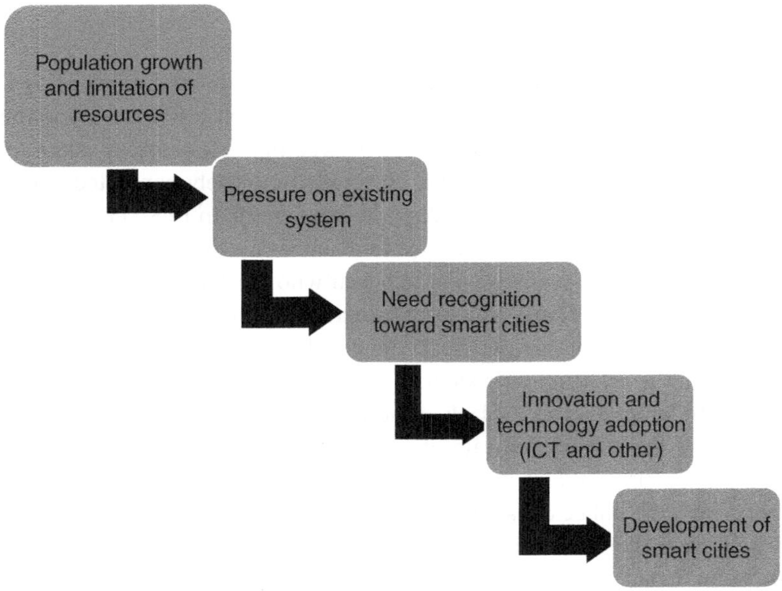

Figure 1.2 Need and development of smart cities. *Source:* Developed by Authors.

1.3.1 Global Smart Cities Initiatives

Development of smart city projects are taking place across the globe; it is one of the key agendas for almost every nation in the world (Table 1.1). Countries have adopted various kinds of initiatives to make their cities smart. A study conducted by Lai et al. (2020) well highlighted the initiatives of the countries toward SC development. In Africa, the city of Konza has been selected for smart city development. Konza is working with ICT toward the development of a strong network used for efficient management of transportation, utilities, and public safety standards. The initiatives also work toward environmental development and local business development. Other cities in Africa are Slavova and Okwechime that are being included in the smart city project under the African Union's Agenda 2063. Power generations with renewable resources, rapid adoption of technologies, establishments of smart mobility labs, automatic parking systems, real-time traffic, transport control system, cycle-sharing systems, smart information desks for visitors, and big data projects in smart cities are some of the major initiatives in Africa. In Asian countries, the major SC projects are carried out by China, primarily in the cloud projects, city brain, collection of data from the video at traffic system, and reduction of traffic congestion that made the emergency vehicles 50% faster than before.

Table 1.1 Global smart city initiatives and challenges.

Name of the smart city	Smart city initiatives	Challenges
London	Working well in human capital and international projection. Better urban planning. It is also well known for its smart mobility and transportation for citizens.	Lack of social cohesion and environmental programs.
New York	Better urban planning and smart mobility.	Lack of social cohesion.
Paris	Strong for its international projection, smart mobility, and transportation. Better human capital inclusion.	Lack of social cohesion and environmental programs.
Tokyo	Great initiatives toward environmental and economic development. Better human capital initiatives. Technology-intensive development of the city. Special initiatives for an aging population.	Lack of social cohesion.
Reykjavik	Great initiatives for environmental development. The city is well known for hydroelectric and geothermal energy sources which make the city a world leader in sustainable energy and smart solutions. Better social cohesion.	Lack of urban planning and economy.
Copenhagen	Initiatives for environmental development that gives the city a low level of pollution and contamination. Good governance.	Poor urban planning.
Berlin	Best smart mobility and transportation. Better in human capital and international projection.	Needs improvements in the economy and environmental development.
Amsterdam	Best for international projection; attracts international tourists. Smart mobility and transportation	Lack of social cohesion.
Singapore	The first city to launch driverless taxis and proposed to launch the same kind of buses too. Incorporation of innovation and technology in the city development. Better international projection and environmental development.	Need to work more on mobility and transportation.
Hong Kong	The city is well known for its use of innovative and technology-intensive programs to increase the quality of life and city management through its remarkable initiatives called the Hong Kong Smart City Blueprint project.	Lack of social cohesion.

Source: Adapted from Forbes 2020.

1.3.2 Role of Technology in Smart City Development

Technology plays a significant role in the development of smart cities as it not only makes the city smart but also makes the city utilities convenient for the public. Most smart cities focus on smart mobility that consists of three main elements: ICT, smart cells, and developmental mechanisms. These elements consist of the internet of things (IoT), cloud computing, AI, big data, etc. Smart cells incorporate, for example, smart and automated vehicles and driverless vehicles. In developmental mechanisms, the system is used for smart traffic management and another mechanism for operations (Yan et al. 2020). In cities, increased traffic congestion affects both safety and the environment, as it plays a crucial role in increasing carbon emissions. The technologies in these areas that are broadly used in the development of smart cities are sensors, controller area network (CAN) bus, GPS, light detection and ranging (LiDAR), and others (Xu and Thakur 2021). These technologies have the potential to make the city smart and sustainable in long run. Smart cities around the world also use lithium batteries, which help in running public vehicles like buses and taxis. On the other hand, cities like Singapore are continuously investing in the automation of vehicles and another self-driving effort for the development of the city. Asian Pacific areas, such as China, Taiwan, Malaysia, Korea, Japan, and Hong Kong, use paperless work and much focus on Radio Frequency Identification (RFID) for payments and other related tasks (Asia Pacific Smart Card Association 2002).

Approximately 30% of the people involved in vehicle accidents die because of they reach the hospital too late due to traffic congestion. In the same way, heart attack and stroke patients are not able to reach the hospital in time. To solve such problems and develop a smooth traffic system during an emergency, England has incorporated the use of AI and added emergency crews to the system. A company called Red Ninja in the United Kingdom works well on the project called Life First Emergency Control (LiFE), which is used for managing traffic and taking the emergency patient to the hospital within a few minutes (Ninja 2022). The project uses much of AI and various other algorithms to make this program successful and save precious lives. These are concrete examples of how technology plays a crucial role in smart city development.

1.3.3 Technology Adoption and Development of Smart Cities

A city with strong, forward-looking performance in its governance, mobility, environment, and quality of life, founded on a clever combination of endowments and citizen initiatives from self-aware, independent, and self-determining individuals (Giffinger et al. 2007). The idea of a "smart city" is regarded as a new paradigm for urban development (Bremser et al. 2019). Everyday operations including government, transportation, agriculture, logistics, maintenance, education, and healthcare are all automated in some fashion because of the widespread use of technology, and smart devices can be used to control, monitor, and access these systems remotely (Ahad et al. 2020). This gave rise to the idea of smart cities, in which ICT are combined with a city's existing traditional infrastructure and then managed and coordinated using digital technology.

A city that inspires its citizens to develop and thrive in their own lives shares culture, information, and life with its citizens (O'Connor and Shaw 2014). The earliest idea was the intelligent city. Top-down strategies are used in the intelligent city, with a concentration on technology (Letaifa 2015). Smart cities are hybrid models that combine centralized city support, coordination, and monitoring with democratized open innovation (Letaifa 2015). Smart people, smart governance, smart mobility, smart environment, and smart lifestyle are the five most frequent markers of the adoption of smart cities (Giffinger et al. 2007).

Planning strategies for smart cities have gathered a lot of traction recently (Kummitha and Crutzen 2017). While supporters of smart cities claim that the adoption of ICTs, improved governance, and human capital among the populace will result in beneficial social transformation, detractors highlight the negative repercussions and flaws in their development and implementation (Kummitha and Crutzen 2017). Transportation, energy, government, and the environment are just a few of the sectors being affected by digital transformation. But it mostly affects people. Consequently, the implementation of the smart city's various components offers tremendous research possibilities (Manfreda et al. 2021). Bremser et al. (2019) begin with a methodical development of technology and data platforms since they perceive new technology and standardized data interchange as a rare opportunity. Research on innovation uptake serves as the theoretical underpinning (Bremser et al. 2019).

1.3.4 Efforts Toward Overcoming Technological Challenges

SC technologies offer the possibility to manage the repercussions of growing urbanization (Ibrahim et al. 2015). Smart city efforts still struggle to take advantage of technological opportunities (Bremser et al. 2019). Policymakers in metropolitan areas around the world are hurriedly implementing IoT devices, sensors, and other contemporary information and communications technologies to address various governance challenges, boost efficiencies, and empower residents in the race to create smart cities (Mondschein et al. 2021). However, there are several obstacles to the development of smart cities, including laws and regulations, funding, infrastructure, and technological issues that are related to sustainability in terms of the environment, the economy, and society (Ibrahim et al. 2015). Numerous studies have emphasized the significance of sustainability ideas in the growth of smart cities. However, little research has been done on the difficulties in creating sustainable smart cities (Ibrahim et al. 2015).

To give towns a comprehensive knowledge of how SC development should be conducted, overcoming the nature of smart city obstacles is essential. Mora et al. (2019) recommended coordinating efforts when examining the strategic principles that underpin the development of smart cities, reaching an agreement on how to conceptualize, analyze, and standardize such developments, and devising creative monitoring and evaluation systems for such strategies by reflecting on the lessons discovered from existing practices. Rapid urbanization is creating chances for cutting-edge applications of developing technology to identify sectors in the issues of city management (Nyberg 2018).

The creation of sustainable smart cities and the provision of more effective and seamlessly connected services can both be improved by the practitioners by anticipating potential problems (Vu and Hartley 2018). Ignoring the truth about these problems and how to solve them will cause delays and result in the failure to build smart cities. ICT infrastructure support (Hashem et al. 2016), rapid urbanization, and a lack of government wisdom (Nyberg 2018), a lack of coordination and externalities (Warwick 2013), PPPs (European Commission 2013), and implementation issues for sustainable smart cities are some of these challenges (Lytras and Visvizi 2018).

1.3.5 Future of Smart Cities with Technology Adoption

In order to create a blueprint for future smart cities, it is necessary to evaluate the opportunities and difficulties that lie ahead (Murthy Nimmagadda and Harish 2022). Therefore, although being viewed as complex, technological advances are recognized as a unique potential for the city's future development (Bremser et al. 2019). SC projects have a technological focus, but the current technological environment is thought to be insufficient for future requirements (Bremser et al. 2019). Smart cities use technology that can help inhabitants live better lives by acting as a foundation for future services, such as autonomous mobility, as well as adapt in the future, providing new technology and a supportive environment (Manfreda et al. 2021). A smart city is an entrepreneurial city (Kummitha 2019). Smart cities and entrepreneurship have a reciprocal relationship (Kummitha 2019). First, entrepreneurs launch technological initiatives that aid in the socio-technical transformation of cities into smart cities. Second, as new technologies are embraced in cities, data is produced that aids businesses in looking for new opportunities.

1.4 Conclusion

In this chapter, the focus has been given largely on the technological aspect of smart city projects. The study has well highlighted the significance and need of the smart city for countries that are getting crowded because of the rising level of the urban population. This increase in the population is happening because of the migration of the people from rural areas to urban areas in search of good education and earning a livelihood. This kind of thing is happening in most places around the world. This incident is making cities under pressure as it was not planned for the excessive population that we are noticing today, and therefore the planner, government, and other stakeholders decided to make the city smart so that it can well handle people. As suggested in our very basic example earlier, although traffic management was once handled by a few cops, now it is not possible without the use of sensors and other technological devices to handle the increased traffic. The chapter also talks about the global initiatives toward the development of smart cities. The major objective of the chapter was to highlight the technological role in the development of smart cities. The chapter has also showcased how the implementation of technology is difficult for some countries because of various reasons such as affordability challenges, lack of technical know-how, missing infrastructure support, and other kinds of policy-related challenges. Technology transfer among the countries is also a major challenge.

These challenges hinder the smart city project as already discussed. Few solutions can be implemented toward adopting these technologies such as PPP, public funding support, technology diffusion strategy between the countries, and training for employees working in smart city projects. These methods can help the smart city projects toward adopting new technologies, and it can help in importing technology from other countries. However, with the intervention of international bodies, some countries are ready to help other developing countries with their technologies in the development of the smart city. Many cities have been established as smart cities, and they can serve as a great example or role model in this domain.

1.5 Implication and Future Research

The chapter is about showcasing the significance of the technological aspect in developing smart cities. The knowledge from this book chapter can be used for understanding the basics of smart cities' concepts and need. Most importantly the technologies that are discussed in the chapter. SC policymakers and experts may get help from this study. The researchers who are working in the field of smart cities may develop some exclusive studies such as how SC technologies can be distributed to developing and underdeveloped countries. The knowledge from the present chapter may be helpful for both researchers and policymakers.

References

Ahad, M.A., Paiva, S., Tripathi, G., and Feroz, N. (2020). Enabling technologies and sustainable smart cities. *Sustainable Cities and Society* 61: 102301.

Asia Pacific Smart Card Association (2002). Contactless smart card schemes in the Asia Pacific region. https://www.securetechalliance.org/secure/reports/Asia_CSC_Report.pdf (accessed June 24, 2022).

Berry, M. (2018). Technology and organised crime in the smart city: an ethnographic study of the illicit drug trade. *City, Territory and Architecture* 5 (1): 16.

Bremser, C., Piller, G., and Helfert, M. (2019). Technology adoption in smart city initiatives: starting points and influence factors. *International Conference on Smart Grids and Green IT Systems* 70–79.

Camero, A. and Alba, E. (2019). Smart city and information technology: a review. *Cities* 93: 84–94.

De Marco, A. and Mangano, G. (2021). Evolutionary trends in smart city initiatives. *Sustainable Futures* 3: 100052. https://doi.org/10.1016/j.sftr.2021.100052.

Dutton, W.H., Kraemer, K.L., and Blumler, J.G. (1987). *Wired Cities: Shaping the Future of Communications*. Macmillan Publishing Co., Inc.

Edvinsson, L. (2006). Aspects on the city as a knowledge tool. *Journal of Knowledge Management* 10 (5): 6–13.

European Commission (2013). Strategic implementation plan. http://www.smartcities.at/assets/Uploads/sip-final-en.pdf (accessed June 26, 2022).

Forbes (2020). These are the 10 smartest cities in the world for 2020. https://www.forbes.com/sites/iese/2020/07/08/these-are-the-10-smartest-cities-in-the-world-for-2020/?sh=223e143e12af (accessed June 25, 2022).

Giffinger, R., Fertner, C., Kramar, H. et al. (2007). *Smart Cities: Ranking of European Medium-sized Cities*. Vienna, Austria: Centre of Regional Science, Vienna University of Technology www.smart-cities.eu/download/smart cities final report.pdf.

Hashem, I.A.T., Chang, V., Anuar, N.B. et al. (2016). The role of big data in smart city. *International Journal of Information Management* 36 (5): 748–758.

Haughton, G. and Hunter, C. (1994). *Sustainable Cities*. J. Kingsley Publishers.

Ibrahim, M., Al-Nasrawi, S., El-Zaart, A., and Adams, C. (2015). Challenges facing e-government and smart sustainable city: An Arab region perspective. In: *15th European Conference on e-Government*, 396–402. Portsmouth, UK: ECEG.

Isafiade, O.E. and Bagula, A.B. (2017). Fostering smart city development in developing nations: A crime series data analytics approach. In: *2017 ITU Kaleidoscope: Challenges for a Data-Driven Society (ITU K)*, 1–8. Nanjing, China: IEEE.

Ishida, T. (2002). Digital city Kyoto. *Communications of the ACM* 45 (7): 76–81.

Ismagilova, E., Hughes, L., Dwivedi, Y.K., and Raman, K.R. (2019). Smart cities: advances in research – an information systems perspective. *International Journal of Information Management* 47: 88–100.

Komninos, N. (2006). The architecture of intelligent cities. In: *2nd International Conference on Intelligent Environments*, 13–20.

Kosowatz, J. (2020). Top 10 growing smart cities American society of mechanical engineers (ASME), https://www.asme.org/topics-resources/content/top-10-growing-smart-cities (accessed July 11, 2022).

Kummitha, R.K.R. (2019). Smart cities and entrepreneurship: an agenda for future research. *Technological Forecasting and Social Change* 149: 119763.

Kummitha, R.K.R. and Crutzen, N. (2017). How do we understand smart cities? An evolutionary perspective. *Cities* 67: 43–52.

Lai, C.S., Jia, Y., Dong, Z. et al. (2020). A review of technical standards for smart cities. *Clean Technologies* 2 (3): 290–310.

Letaifa, S.B. (2015). How to strategize SMART cities: revealing the SMART model. *Journal of Business Research* 68 (7): 1414–1419.

Lytras, M. and Visvizi, A. (2018). Who uses smart city services and what to make of it: towards interdisciplinary smart cities research. *Sustainability* 10 (6): 1998–2013.

Manfreda, A., Ljubi, K., and Groznik, A. (2021). Autonomous vehicles in the smart city era: an empirical study of adoption factors important for millennials. *International Journal of Information Management* 58: 102050.

Mondschein, J., Clark-Ginsberg, A., and Kuehn, A. (2021). Smart cities as large technological systems: overcoming organizational challenges in smart cities through collective action. *Sustainable Cities and Society* 67: 102730.

Mora, L., Deakin, M., Reid, A., and Angelidou, M. (2019). How to overcome the dichotomous nature of smart city research: proposed methodology and results of a pilot study. *Journal of Urban Technology* 26 (2): 89–128.

Murthy Nimmagadda, S. and Harish, K.S. (2022). Review paper on technology adoption and sustainability in India towards smart cities. *Multimedia Tools and Applications* 1–29.

Myeong, S., Park, J., and Lee, M. (2022). Research models and methodologies on the smart city: a systematic literature review. *Sustainability* 14 (3): 1687.

Ninja, R. (2022). The life project, www.redninja.co.uk/design-technology-work/life-project (accessed June 23, 2022).

Nyberg, R.A. (2018). Using 'smartness' to reorganize sectors: energy infrastructure and information engagement. *International Journal of Information Management* 39: 60–68.

O'Connor, J. and Shaw, K. (2014). What next for the creative city? *City, Culture and Society* 5 (3): 165–170.

OECD (2013). *Green Growth in Cities*. OECD Publishing http://doi.org/10.1787/9789264195325.

Okai, E., Feng, X., and Sant, P. (2018). Smart cities survey. In: *2018 IEEE 20th International Conference on High Performance Computing and Communications; IEEE 16th International Conference on Smart City; IEEE 4th International Conference on Data Science and Systems (HPCC/SmartCity/DSS)*, 1726–1730. IEEE, IEEE.

Razmjoo, A., Østergaard, P.A., Denai, M. et al. (2021). Effective policies to overcome barriers in the development of smart cities. *Energy Research & Social Science* 79: 102175.

Register, R. (1987). *Ecocity Berkeley: Building Cities for a Healthy Future*. North Atlantic Books.

Streitz, N.A. (2015). Citizen centered design for humane and sociable hybrid cities. In: *Hybrid City*, 17–20. Athens, Greece.

Vu, K. and Hartley, K. (2018). Promoting smart cities in developing countries: policy insights from Vietnam. *Telecommunications Policy* 42 (10): 845–859.

Warwick, K. (2013).Beyond industrial policy. http://www.enterprise-development.org/wp-content/uploads/Beyond_Industrial_Policy.pdf (July 23, 2022).

Xu, B. and Thakur, G. (2021). Introduction to the special issue on smart transportation. *GeoInformatica* 25 (3): 417–418.

Yan, J., Liu, J., and Tseng, F.M. (2020). An evaluation system based on the self-organizing system framework of smart cities: a case study of smart transportation systems in China. *Technological Forecasting and Social Change* 153: 119371.

2

Technology Assessment: Process Optimization Services in the Cement Industry

Felipe C. Gelbecke[1], *Tugrul U. Daim*[2], *Hermann Lodding*[3], *and Sine B. Skaarup*[4]

[1] *Hamburg Technical University, Hamburg, Germany*
[2] *Mark O. Hatfield Cybersecurity and Cyber Defense Policy Center, Department of Engineering and Technology Management, Portland State University, Portland, OR 97201, USA*
[3] *Hamburg Technical University, Hamburg, Germany*
[4] *FL Smidth, Copenhagen, Denmark*

2.1 Introduction

Cement has allowed the building of modern civilization as we know it, playing a fundamental social and economic role worldwide (Habert et al. 2020). For example, it brings social benefits in the form of sewage systems, housing, and health care facilities, and keeps the economic engine turning by creating jobs globally and enabling the existence of other industries and activities such as real estate, power plants (e.g. nuclear and hydroelectric), and transportation.

Cement is the most used manmade material in construction. Virtually every construction and infra-structure development uses cement – be it in its pure form or as a key ingredient to making concrete. It is estimated that 4.1 Gt of cement were produced globally in 2019 (International Energy Agency 2020), compared to 1.9 Gt of steel in the same year (World Steel Association 2021). Cement is particularly attractive because it is simple to use and shape, it quickly renders an "artificial" rock with numberless applications, it is inexpensive, and it is physically, chemically and mechanically stable (Rodrigues and Joekes 2011).

However, these benefits come at a cost. Cement production is associated with environmental degradation throughout its value chain. For instance, cement production depends on the mining of nonrenewable natural resources – mainly limestone and clay. Arguably, cement's biggest impact refers to air pollution and climate change. For every unit of material produced, roughly a unit of greenhouse gas (GHG) is emitted (Smil 2016), accounting for 5–8% of global CO_2 emissions (Kajaste and Hurme 2016) alongside other pollutants such as nitrogen oxides

Future-Oriented Technology Assessment: A Manager's Guide with Case Applications, First Edition.
Edited by Haydar Yalçın and Tugrul U. Daim.

(NOx), sulfur oxides (SO_x), and particles (Rodrigues and Joekes 2011; Van den Heede and De Belie 2012).

There are five stages in the manufacture of a standard cement: (i) extraction of the main raw materials by quarrying, (ii) crushing and grinding the raw materials, (iii) blending the materials in the appropriate proportions, (iv) burning the prepared mix in a kiln, and (v) grinding the burned product, known as *clinker*, together with gypsum and other additives depending on the type of cement (Britannica Encyclopedia online 2021). To produce clinker, the system requires temperatures above 1600 °C largely sourced from thermal energy. Furthermore, the electrical energy-intensive process of grinding is used in the beginning and in the end of cement production. It was estimated in 2002 that 0.02% of the total world power consumption was used in cement production. Thus, it is not a surprise that energy consumption is another challenge in the industry (International Energy Agency 2021).

There have been many attempts to find environmentally sustainable replacements for cement. None of the alternatives, however, have shown comparable properties and applications. And even if that were the case, large-scale replacement is unattainable within the next decade (Claisse 2015; Davis et al. 2018), thus compromising the fulfillment of the Paris Agreement to curb GHG emissions and cap global temperatures rise. This is aggravated by the expected continued rise of cement production driven by a growing global population and their ongoing urbanization (United Nations n.d.).

Although industry-specific national and international regulations, publications, and actions have increasingly addressed sustainability challenges, especially regarding decarbonization and net-zero emissions (Edwards 2021; ICR Research 2020; Weetch 2021), it remains unclear whether the industry's efforts effectively and timely contribute to keeping global temperatures rise well below 2 °C compared to pre-industrial levels. High competition driving prices down, along with greater scarcity of resources (e.g. energy and raw materials) leading to more complex and costly operations, as well as the aged cement plants, set up conditions in which transition to more sustainable products worldwide has much more potential.

Given the broad use of cement in virtually every construction and infrastructure development, the inexistence – as of now – of a scalable alternative to cement with similar properties and applications, and the expected increasing demand for construction materials driven by a growing and urbanizing global population, it is imperative that the industry finds timely and effective solutions to keep its environmental impact within the Earth's boundaries while ensuring financial viability.

One prominent solution to this issue lies within process optimization. Process optimization uses suitable scientific methodologies to identify improvement opportunities and maximize resource efficiency (e.g. raw materials, energy, equipment), productivity, product quality, etc. (Rodrigues and Joekes, 2011), and at the same time manage and reduce the processes' negative environmental impacts (Benhelal et al. 2013). If coupled with broad stakeholder engagement, process optimization can drive the achievement of "marginal gains throughout the [cement] value chain leading to substantial savings in GHG emissions (…) without heavy investment in new industrial infrastructure" and with the potential to save up to 50% immediate GHG emissions (Habert et al. 2020).

Leading global players currently strive to offer process optimization services to their client portfolio, entailing a wide spectrum of focus areas, from energy and fuel consumption to project management, high performance and quality, to climate change mitigation (FLSmidth 2021a; Krupp 2021; ATech 2021). An important success factor in their endeavor is the balance among potentially conflicting decision-making criteria. For instance, specialized consulting to bring about higher efficiency implies financial expenditures within a cost-sensitive industry. Thus, effective optimization services must be designed according to industry-, region-, and client-specific preferences and constraints.

At ABC, the process optimization service starts with an initial dialogue with the plant staff to understand the plant's individual situation and challenges. Next step is the definition of the service scope, which will describe the main focus, as well as delivery schedules and obligations of the specialist team and the client. Later, the specialist team makes an initial analysis of the equipment and/or operational area based on plant data. Once the plant's specific needs are identified, specialists with the right hands-on experience and appropriate theoretical knowledge are appointed to conduct the specific process optimization service. They identify the bottlenecks and root causes based on specific data collection and visual inspection. Then they produce a service report containing the analyses performed, the main findings and conclusion drawn, and the recommendations for executive decision-makers, while the full extent of the technical evaluation and detailed recommendations are given directly to the plant's engineers. Contrary to the ABC equipment, service performances are not formally documented and there is no official knowledge base that stores and shares lessons learned regarding optimization services. Accordingly, the contracts involving the sale and installation of equipment are accompanied by a binding performance guarantee, which does not happen in the case of optimization services.

ABC is particularly invested in further developing its optimization services, entailing proactive and predictive maintenance, reducing operations complexity, and promoting digitalization and innovation, to mention a few (FLSmidth 2021b).

ABC desires to improve its process optimization services (internal development) to scale and better serve their clients' needs. To support the achievement of these goals, this work informs the further development of ABC's process optimization services by assessing how internal and external expert stakeholders' weigh criteria and barriers to process optimizations using HDM.

The assessment in this work can be divided into three phases:

1. The analytic hierarchy is developed, identifying the main criteria and sub-criteria affecting the client's decision to buy an optimization service, as well as the company's internal barriers and sub-barriers hampering the sale of such service.
2. Primary data are collected through online surveys from internal and external expert stakeholders to determine the most important criteria for the clients and the most troublesome barriers for ABC in the process optimization context.
3. The survey results are analyzed to identify the most important aspects of the process optimization services that the company has to focus on in order to improve.

This research aims at providing science-based evidence and guidance on how the company can enlarge its service optimization outreach, paving the way to consolidating its leading position as a consulting and service provider for the industry's development in technology, performance, financial viability, and sustainability performance.

The remainder of this work is organized as follows. The research design is covered in Section 2.2, including the description of the AHP and how it was applied in this work. Section 2.3 describes the results found. The main discussions are explored in Section 2.4. Finally, Section 2.5 provides conclusions, unique contributions, and limitations.

2.2 Research Design

2.2.1 Model Development

The creation of the hierarchical model for this research was based on literature review as well as on unstructured interviews (see Appendix 2.A.) with internal expert stakeholders from various continents and regions to ensure thorough coverage of ABC's market and to capture important industry-specific strategic and practical concerns (Daim et al. 2018).

The literature review suggested five relevant criteria for investing in process optimization services:

1. *Quality:* Quality is a recurring topic in the literature, often embedded in industry- and service-specific forms. For instance, service performance, management skills, technical skills, innovation, and problem solving refer to the level of quality of different services and products (Chow and Ng 2006; Daim

and Dash 2011; Daim et al 2012; Govindan et al. 2017; Kabir and Sumi 2014; Soh 2010).

2. *Team's capabilities:* Specialist capacities are a special topic of concern when buying a service. The literature reveals that capacities not only refer to technical knowledge but also consider soft skills such as communication, interpersonal skills, relationship management, collaboration and information exchange, and understanding clients' brief (Chow and Ng 2006; Daim et al. 2012; Govindan et al. 2017; Kabir and Sumi 2014; Vayvay et al. 2012).
3. *Financial:* Budgeting and costs are a main area of concern and a decisive decision-criteria in any industry. Accordingly, the literature review highlights the financial aspect, often breaking it down into project-specific topics. A few examples are implementation cost, cost of ownership, adequacy of cost estimate, and overall service cost (Chow and Ng 2006; Daim and Dash 2011; Daim et al. 2012; Kabir and Sumi 2014; Martin and Daim 2012; Vayvay et al. 2012; Soh 2010).
4. *Company's reputation:* Reputation appears in the literature as years of experience, credibility, proven performance record, references, and financial stability (Daim and Dash 2011; Kabir and Sumi 2014; Vayvay et al. 2012; Razi et al. 2019; Soh 2010).
5. *Reliability:* While reputation refers to the company's image in the market, reliability refers to expected results of the service, their competitiveness, and how well they translate the company's brief into actionable recommendations and services (Chow and Ng 2006; Govindan et al. 2017; Kabir and Sumi 2014; Martin and Daim 2012; Soh 2010).

Because these criteria were drawn from academic work focusing on a broad spectrum of fields (e.g. engineering consulting, technology roadmapping, logistics services, total quality management consulting, and resource planning consulting), 15 interviews with internal experts were used to validate their relevance to the cement industry as well as to point out new relevant criteria. The interviews were documented in writing (note-taking) on an interview journal and coded using thematic coding. The main themes that emerged from this were compared to the list of criteria drawn from the literature review. The resulting list of criteria was:

- Quality of the process optimization service:
- Capabilities of the specialist team
- Budget for the process optimization service
- Supplier's reputation
- Resulting benefits of the service

Besides the validation of the criteria, the interviews enriched the creation of the hierarchical model by revealing internal barriers to the execution of process optimization services. This prompted the need to create not one but two HMs: one referring to the criteria and sub-criteria affecting the client's decision to buy a process optimization service and another referring to internal barriers and sub-barriers hampering the sale of such service.

Upon close examination of the interviews' thematic coding, barriers were selected based on their number of appearances and, in most cases, on their correspondence to the defined criteria. For example, the criteria "budget" corresponds to the barrier "financial," while team's capabilities are perceived as a decision criteria by the client, and the lack thereof is perceived as a barrier to the delivery of the service by the staff. The following barriers were selected:

- Internal bureaucracy
- Financial
- Company's vs competitor's capabilities
- Team's capabilities

Once the criteria and barriers were defined, sub-criteria and sub-barriers were added to the HMs, drawn from the literature review and from the interviews. The two complete HMs were then validated by the internal experts. The HMs are illustrated below and accompanied by a brief description of each criterion/barrier and sub-criterion/sub-barrier.

2.2.1.1 Hierarchical Model: Client's Point of View

The *hierarchical model: client's point of view* refers to the criteria relevant to the client when investing in a process optimization service. It entails one objective (hierarchical level 1) and five criteria (hierarchical level 2), with four of these criteria nesting four sub-criteria (hierarchical level 3) and one criteria nesting three sub-criteria. The HM for the client's point of view is shown in Figure 2.1.

Objective: Invest in service optimization

Criterion 1: Quality of the process optimization service

This criterion refers to the quality of the final recommendations, the quality of the service planning including tasks and time to execute them, the digital tools applied and the quality of the final report.

Sub-criterion 1.1: Quality of recommendations

The recommendations were clear and useful for the plant. The client can easily see the recommendation's purposes and how they will affect the production process.

Sub-criterion 1.2: Quality of the planning

The service planning has all the tasks needed to evaluate the process. The tasks are completed in a velocity that fulfill the client's needs.

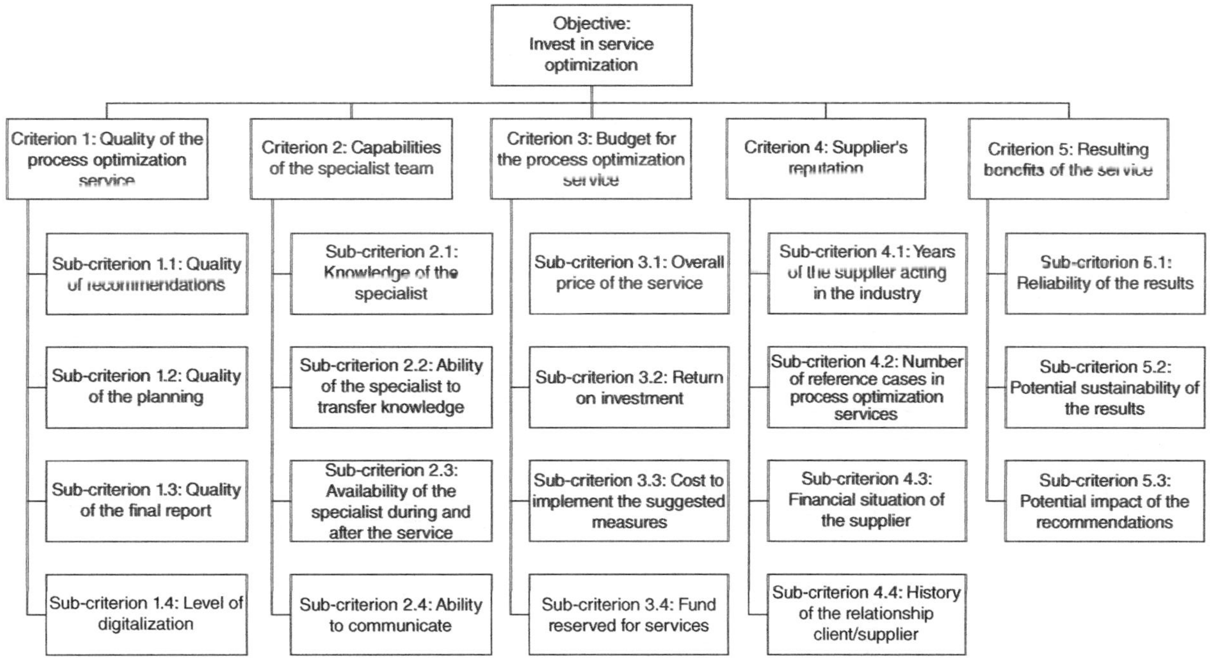

Figure 2.1 The hierarchical model from the client's point of view.

Sub-criterion 1.3: Quality of the final report

The report has a clear structure with clear recommendations, it is well-organized data obtained from the plant and in a language desired by the client.

Sub-criterion 1.4: Level of digitalization

The digital tools used reduced the time of the tasks and improved the quality of the service.

Criterion 2: Capabilities of the specialist team

The capabilities of the team is about their knowledge on the cement production process, their ability to communicate clearly and transfer the knowledge, and their ability to see the issues beyond the part of the process studied. The team showed a genuine desire to improve the clients' process.

Sub-criterion 2.1: Knowledge of the specialist

The team shows solid knowledge about the cement production process and strategies to overcome the issues faced by the client. The team shows the ability to find the root cause of the problem even if it does not belong to the project's scope.

Sub-criterion 2.2: Ability of the specialist to transfer knowledge

The team shows a strong ability to teach the client to continue improving the production process.

Sub-criterion 2.3: Availability of the specialist during and after the service

The specialist team is available for questions and discussions, during and after the service. The team shows a genuine intention to improve the process by listening to the problem and using all the resources to address it. The specialists follow up the implementation of the recommendations.

Sub-criterion 2.4: Ability to communicate

The team shows the ability to clearly communicate the ideas discussed in the project, either in English or in the client's language. No discussion points are missed in the translation.

Criterion 3: Budget for the process optimization service

The budget factor takes into account the overall price of the service, the return of investment of the service, the future cost to implement the suggestions resulting from the service and the budget reserved for these services.

Sub-criterion 3.1: Overall price of the service

The overall price of the service.

Sub-criterion 3.2: Return on investment

The return expected after the implementation of the process optimization service recommendations.

Sub-criterion 3.3: Cost to implement the suggested measures

The future investment to implement the suggestions resulted from the service.

Sub-criterion 3.4: Fund reserved for services

The availability money in the budget reserved for process optimization services.

Criterion 4: Supplier's reputation

The supplier's reputation takes into account the time that the company is acting in the market, the number of references with optimization services, the current financial situation of the supplier, and the history of the client/supplier relationship.

Sub-criterion 4.1: Years of the supplier acting in the industry

Time that the supplier is acting in the market.

Sub-criterion 4.2: Number of reference cases in process optimization services

A list of references from the supplier with the process optimization services and their main results.

Sub-criterion 4.3: Financial situation of the supplier

The financial health of the supplier.

Sub-criterion 4.4: History of the relationship client/supplier

The history of the relationship between the client and supplier. Also the overall relationship of the supplier with other players in the market.

Criteria 5: Resulting benefits of the service

The benefits of the service are about the reliability of the results, the sustainability of the future results, and the potential impact of the recommendations through the production process.

Sub-criterion 5.1: Reliability of the results

The results are aligned with the defined goals for the optimization service.

Sub-criterion 5.2: Potential sustainability of the results

The good results persisted for long periods after the application of the suggestions. The client's team was trained to reply to the suggestion made by the specialist. The specialist followed up the results.

Sub-criterion 5.3: Potential impact of the recommendations

The suggestions resulting from the service have the potential to positively impact the production process beyond the scope of the project. E.g. improvement of the cement mill production after a clinker quality optimization service.

2.2.1.2 Hierarchical Model: Staff's Point of View

The *hierarchical model: staff's point of view* refers to the barriers and sub-barriers hampering the sale of such service from the internal expert stakeholder's point of view. It entails one objective (hierarchical level 1), four barriers (hierarchical level 2), with each of these barriers nesting four sub-barriers (hierarchical level 3). The hierarchical model for the staff point of view is shown in Figure 2.2.

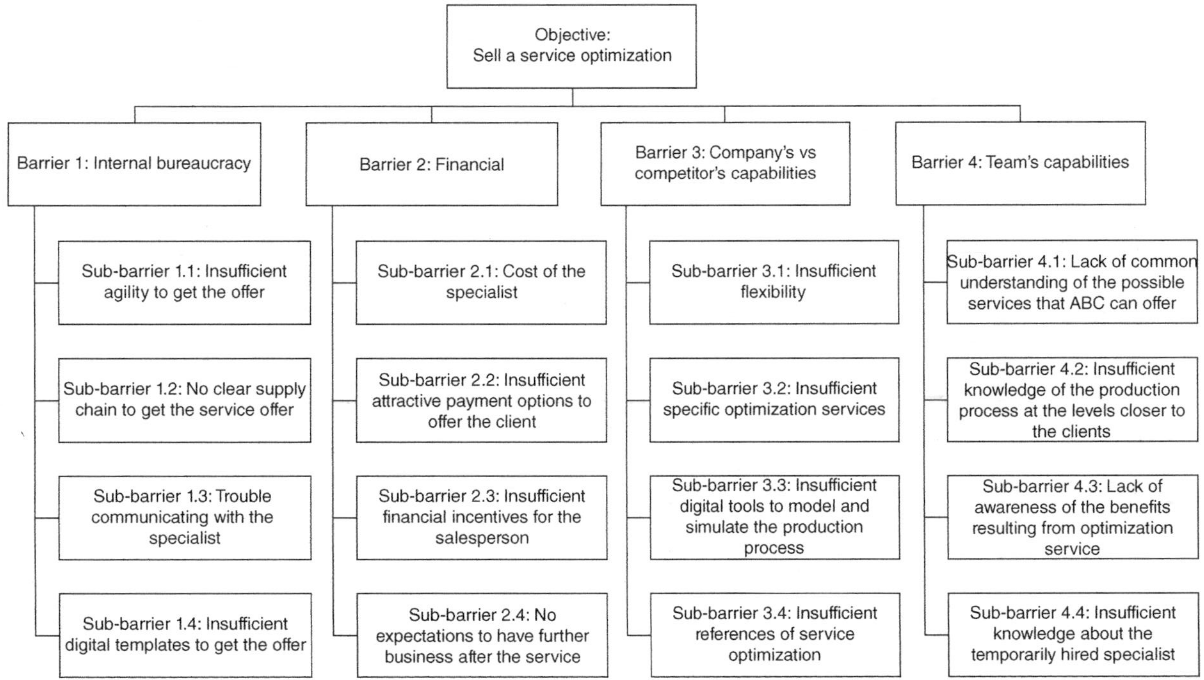

Figure 2.2 The hierarchical model from the staff's point of view.

Objective: Sell a service optimization

Barrier 1: Internal bureaucracy

The internal process and tools do not facilitate the sales of optimization services. This barrier takes into account the insufficient agility to get service offers, the lack of a clear supply chain to get service (who to talk to), trouble to get in touch with the process specialists and insufficient digital templates to facilitate the offer acquisition.

Sub-barrier 1.1: Insufficient agility to get the offer

The offer is not ready in the proper velocity to meet the needs of the clients.

Sub-barrier 1.2: No clear supply chain to get the service offer

It is not clear who to talk to in order to get the offer for optimization service.

Sub-barrier 1.3: Trouble communicating with the specialist

It is complicated to reach the process specialists to better understand the scope/problem of the client and what the company can offer as an optimization service.

Sub-barrier 1.4: Insufficient digital templates to get the offer

The digital templates need improvements in order to facilitate the selection and offer of an optimization service.

Barrier 2: Financial

The price, cost, and incentives are not adequate to promote the optimization services. This barrier refers to the internal cost to get the process specialist, the insufficient payment options to offer the client, the insufficient financial incentives for the salesperson and no expectations to have further business after the optimization service.

Sub-barrier 2.1: Cost of the specialist

The high internal cost of the specialist to analyze the case of the client, especially compared to a competitor local specialist.

Sub-barrier 2.2: Insufficient attractive payment options to offer the client

Insufficient payment options to attract the customer e.g. payment based on performance.

Sub-barrier 2.3: Insufficient financial incentives for the salesperson

Insufficient financial returns to motivate the sales person to sell the optimization services.

Sub-barrier 2.4: No expectations to have further business after the service

No expectations to have to sell products or long-term optimization services after the optimization service.

Barrier 3: Company's vs. competitor's capabilities

The company does not have enough capabilities to differentiate its optimization services from its competitors. This barrier takes into account insufficient flexibility, insufficient optimization services to offer, insufficient digital tools to model or simulate the production process and insufficient references cases of service optimization. All these points in comparison with the competitors (local or not).

Sub-barrier 3.1: Insufficient flexibility

The company does not have enough flexibility to satisfy the customer's technical needs in a proper time frame as the competitors have.

Sub-barrier 3.2: Insufficient specific optimization services

The company does not offer the optimization service requested by the clients.

Sub-barrier 3.3: Insufficient digital tools to model and simulate the production process

The company's digital tools cannot model and simulate the client's production process properly. Digital tools could accelerate the execution of the optimization service.

Sub-barrier 3.4: Insufficient references of service optimization

The company does not have a robust reference list of optimization services.

Barrier 4: Team's capabilities

The team needs to improve its capabilities regarding the optimization services. This barrier takes into account the lack of common understanding of the possible services offered by ABC, insufficient technical knowledge about the production process, and lack of awareness of the benefits that the optimization service can bring to the client. Furthermore the insufficient knowledge about the specialist when he or she is hired temporarily.

Sub-barrier 4.1: Lack of common understanding of the possible services that ABC can offer

The team is not aware of all the company's capabilities regarding optimization services.

Sub-barrier 4.2: Insufficient knowledge of the production process at the levels closer to the clients

The technical level of the team closer to the client can be improved to facilitate the offer of optimization services.

Sub-barrier 4.3: Lack of awareness of the benefits resulting from optimization service

It is not clear for the team what are the benefits that the ABC's optimization services can bring to the client.

Sub-barrier 4.4: Insufficient knowledge about the temporarily hired specialist
Lack of knowledge about the third-party specialist hired for a temporary service since he or she can reach the client to continue the service without the ABC participation.

2.2.1.3 Comparative Judgment of the Variables

Once the HM was built, expert stakeholders performed a comparative judgment of the variables using an online survey. The relative importance of the criteria and the barriers is defined by a series of pairwise comparisons. The same is done for the sub-criteria and sub-barriers within each criterion and barrier. The pairwise comparisons approach facilitates the study of the main objective because it divides the process into several simpler evaluations (Daim et al. 2018). Figure 2.3 illustrates the pairwise combinations among barriers, which was also done for the criteria, while Figure 2.4 refers to pairwise comparisons among sub-barriers, which was also done for the sub-criteria.

The surveys' introduction instructed the respondents to consider each pairwise comparison on its own merit, thus not taking into account the other comparisons, and just assigning the subjective importance scores to the pair being evaluated. In each pairwise comparison, one criterion/barrier was placed on one side of a slide bar and another on the opposite side. A sliding button was placed in the middle of the bar (point zero). The respondent weighted the importance of each element by moving the button closer to the criterion/barrier that is most affecting the main goal. Figure 2.5 illustrates the pairwise comparison between "Quality of process optimization service" and "Capabilities of the specialist team" from the client's point of view survey. In this example the expert moved the button closer to the

Figure 2.3 Illustration on how the main barrier were compared – each line is a comparison.

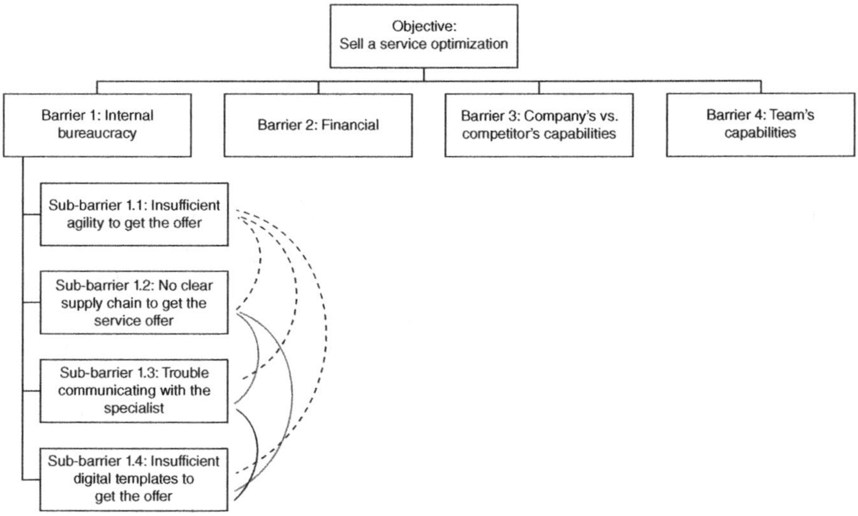

Figure 2.4 Illustration on how the sub-barrier were compared – each line is a comparison.

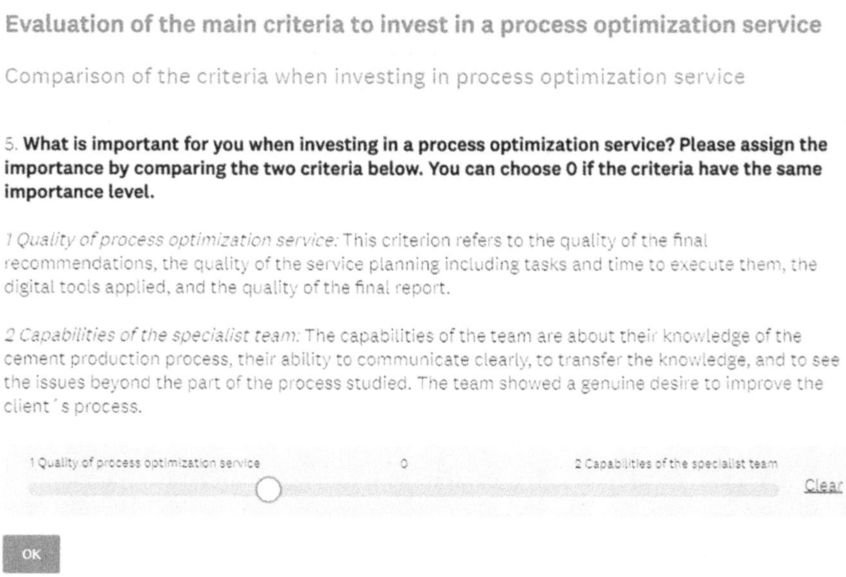

Figure 2.5 Example of a pairwise comparison in the online survey.

criterion "Quality of process optimization" showing that this criterion has a higher importance than "Capabilities of the specialist team."

Survey respondents were purposely selected. For the comparison of barriers to selling the service, personnel from the sales and technical departments were invited to participate, resulting in a total of 12 responses. For the comparison of criteria to invest in service optimization, clients and internal sales personnel with vast knowledge and solid relationships with clients were invited to participate, as their experience enables them to wear the clients' shoes. The initial low number of responses to the survey from the client's point of view, potentially due to the length of the survey, led to the need to create a shortened version where only the main criteria were judged in pairs. As a result, six responses for the long survey and 19 responses to the long survey were obtained. This number of responses is considered satisfactory to render a serious pairwise comparison when using HDM methodology because it surpasses the six-response threshold necessary to stabilize the aggregate weights, as previously reported by Daim et al. (2012).

The last step is the synthesis of priorities. The goal is to provide numerical priorities for every element, representing the relative ability of each element to affect the main goal (Govindan et al. 2017). The weight of each element used in the pairwise comparisons is registered in slide bars (Figure 2.5). The bar corresponds to 100 points. The conversion of the subjective evaluation takes into account the proximity and distance of the sliding button to each element. When the sliding button touches the edge of the bar, the corresponding element takes all the 100 points. Consequently, in each comparison, the expert allocates a total of 100 points to the two elements in the proportion of their relative importance. For example, if one element is three times as important as the other, the first gets 75 points and the second 25 points.

The prioritization of the criteria/barrier is done by averaging the individual evaluation, which reduces the bias of the method. Furthermore, knowing that human judgment has inconsistencies, the logical coherence of each specialist's judgment must be evaluated.

2.2.2 Supporting Qualitative Data

In addition to collecting quantitative data from the pairwise comparisons, qualitative data were collected by allowing respondents to freely insert comments at the bottom of each pairwise comparison. The supporting qualitative data allows for deeper understanding of the reasoning behind the comparisons, thus enriching the data analysis and the development of market-oriented optimization services. The comments were coded using thematic coding and juxtaposed to the corresponding pairwise comparisons to inform the results and the conclusion.

2.2.3 Data Analysis

The Hierarchical Decision Model software version 2.0 was used to analyze the responses recorded in the online survey. Developed by the Portland State University (n.d.), this program averages the response data and runs it through a sequence of matrix calculations and normalizations to arrive at average weight allocations for the criteria and barriers considered in this work. The pairwise responses for both HMs were manually fed into the Hierarchical Decision Model software.

The software can also calculate the inconsistency score of each expert. The lower the score, the more accurate the result will be. The threshold for inconsistency score is 0.10. Ideally, inconsistency scores higher than 0.10 require the specialist to revise the pairwise comparison (Daim et al. 2012). However, comparison reviews were not possible in this work. Therefore, we have not considered inconsistent comparisons in the analysis.

2.3 Results of the Survey

The external survey rendered a total of 25 responses. Six refer to the complete version entailing the criteria and sub-criteria, with an inconsistency level below 0.10; thus, all responses could be used in the analysis. Nineteen responses refer to the simplified version entailing the criteria only, of which one response presented an inconsistency level higher than 0.10. As it was not possible for the respondent to review the answers, this input was discarded from the analysis, and 18 responses were considered valid.

The internal survey received a total of 12 responses. One response had to be discarded because it presented an inconsistency level higher than 0.10. Therefore, 11 responses were considered valid.

Accordingly, the results are divided into two subsections: results of the external survey and results of the internal survey. The results of the external survey are broken down into the two versions of the external survey – the complete survey comparing the criteria and sub-criteria, and the simplified version comparing only the criteria. Other relevant analyses and comparisons are presented, such as regarding the area of operation (local or global companies).

2.3.1 Simplified External Survey

The respondents to the simplified external survey consider the "resulting benefit of the service" to be the most important criterion to hire an optimization service, with a score of 0.403, almost double of the second highest weight criterion "capabilities of the specialist team" with a score of 0.206, followed by the "budget of the process

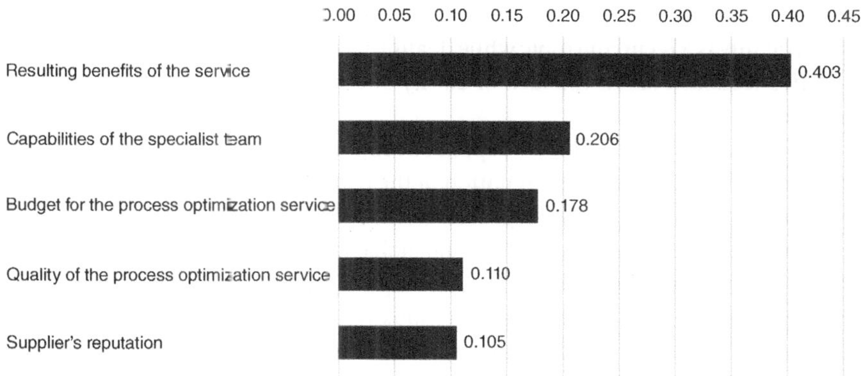

Figure 2.6 Scores of the main criteria to invest in a process optimization service.

optimization service" with 0.178, "quality of the process optimization service" and "supplier's reputation" with 0.110 and 0.105 respectively (Figure 2.6).

The high score of the criteria "resulting benefits of the service" suggests that the clients are strongly focused on the results independently of the other criteria. Many of the respondents stated that they are willing to pay a premium to assure the effectiveness of the process optimization service.

The criterion in second place was the "capabilities of the specialist team." Considering the knowledge gap in the organizations, mediating entities such as specialist consultants can play an important role in successful process optimization. Furthermore, the specialists' capabilities can be linked to the client's confidence in the service to be provided, as stated in one of the survey comments as follows:

> The capabilities of the team provide a good level of confidence in the assertiveness of the future modifications in the process.

Contrasting with previous research that placed the monetary aspect at the top of the criteria ranking for specialized service hiring (Daim et al. 2012; Chow and Ng 2006), "budget for the process optimization service" appears in the third place in the results. This suggests that the clients consider investing in optimization services as long as there is an assurance of the resulting benefits and the specialist team is highly qualified. One respondent uses a *cost* vs. *benefit* logic to justify their choice:

> (...) the budget is always important, however the cost savings obtained from the process optimization can return the investment and save money in the future.

"Quality of the process optimization service" comes in fourth place, contrasting with several internal experts interviewed prior to the development of the HMs who appointed quality as a major factor in investment decision-making. According to them, quality is a recurring topic while dealing with clients, mainly related to some negative feedback about the lack of clarity and quality of the recommendations and the reports.

The lowest score was given to "supplier's reputation," indicating that the clients are not concerned about the reputation of the company as long as they have clear potential results from a knowledgeable specialist at a reasonable price.

Companies operating on a regional and global level are represented in this study through their employee respondents. Although there are some similarities in their priorities and preferred criteria, a closer look into the data indicates relevant differences among them. Table 2.1 presents the scores of each criteria as per global and regional players.

The analysis of the responses according to the company's area of operation suggests that both regional and global players give a higher importance to the "resulting benefits of the service." The difference, however, lies in the importance they give to the "capabilities of the specialist team." Global companies give a relatively low importance to capabilities, placing it in third place, while "budget for the process optimization service" assumes the second position, regional players rate the "capabilities of the specialist team" closely to the "resulting benefits of the service," with scores of 0.333 and 0.388, respectively.

Beyond the area of operation, an analysis based on geolocation was performed. Taking into consideration the amount of respondents per country, it was possible to establish a comparison between Brazil and South Africa. Table 2.2 presents the scores given by respondents from these two countries.

In general, the scores given by the respondents from Brazil and South Africa are fairly similar. Yet, Brazilian customers give less importance to the "Supplier's reputation" and more importance to the "Capabilities of the specialist team" than the South African customers. One comment by a Brazilian customer adequately illustrates this finding.

Table 2.1 Scores of the main criteria for global and regional players respondents.

Criterion	Global player	Regional player
Quality of the process optimization service	0.121	0.053
Capabilities of the specialist team	0.176	0.333
Budget for the process optimization service	0.179	0.135
Supplier's reputation	0.099	0.093
Resulting benefits of the service	0.425	0.388

Table 2.2 Scores of the main criteria for Brazilian and South African respondents.

Criterion	Brazil	South Africa
Quality of the process optimization service	0.108	0.123
Capabilities of the specialist team	0.215	0.177
Budget for the process optimization service	0.175	0.183
Supplier's reputation	0.102	0.143
Resulting benefits of the service	0.402	0.373

> The reputation of the supplier is important, but not as important as the other criteria. We use the reputation as a tiebreaker when two proposals are similarly interesting.

On the other hand, a South African respondent stated:

> The supplier's reputation is confirmation that the company has a good team of specialists. Therefore, the reputation functions as a triage criteria to select potential suppliers.

The respondents' seniority level allows for conclusions regarding importance of the criteria in the service optimization decision-making. Table 2.3 presents the criteria scores according to professional, supervisory and management levels.

The supervision and management levels ranked the criteria similarly, while the professional level presented a different understanding. While supervisors and managers ranked the "capabilities of the specialist team" in second place and "budget for the process optimization service" in third, the professional level respondents did the opposite. In fact, two professional-level respondents stated that budget is a critical criterion, as production plants deal with strict budgeting

Table 2.3 Scores of the main criteria for three different levels of seniority.

Criterion	Professional	Supervisor	Manager
Quality of the process optimization service	0.115/4	0.098/5	0.113/5
Capabilities of the specialist team	0.161/3	0.212/2	0.235/2
Budget for the process optimization service	0.210/2	0.207/3	0.138/3
Supplier's reputation	0.091/5	0.103/4	0.115/4
Resulting benefits of the service	0.423/1	0.380/1	0.402/1

control. Furthermore, professional-level respondents gave less importance to the "supplier's reputation" (fifth place) compared to the other two levels, while supervisors and managers prioritized the "supplier's reputation" in detriment to the quality of the process optimization, placing them in the fourth and fifth places, respectively. This potential internal misalignment was mentioned by one of the internal specialists interviewed during the development of the HMs as follows:

> It happens often that the specialists come to the plant with a scope that is predefined by the management, but then they learn from the operations personnel that the scenario is completely different. This results in misleading expectations regarding the service results, with the management expecting one thing and the operations expecting another, and these expectations cannot be met simultaneously.

2.3.2 Complete External Survey

The complete external survey analyzed the criteria and sub-criteria, allowing a more detailed analysis of the results compared to the simplified external survey where only the criteria were present. The weight given to the main criteria in both surveys were similar. In this way it is assumed that the importance level given for the sub-criteria by the 6 respondents can also describe the opinion of the 19 respondents who compared only the criteria.

For the complete version of the external survey all the six responders have an inconsistency level below 0.10, thus no responses were discarded.

Following the results obtained from the simplified external survey, the "resulting benefits of the service" was considered the most important criterion (Figure 2.7). This was driven by the sub-criterion with the highest impact: "Potential impact of the recommendations" with a score of 0.17. The other two sub-criteria of this first criterion, namely "reliability of the results" and "potential sustainability of the results," got the same importance level of 0.11. These two sub-criteria had a better performance than the other sub-criteria from all other criteria, reinforcing the weight of the "resulting benefits of the service" criterion. Comments from respondents state that process optimization services often do not follow the scope defined or that the results were not sustainable because implementations' benefits were lost in less than one year.

The second most important criterion was the "capabilities of the specialist team," having the sub-criterion "ability to transfer knowledge" as the most important, followed closely by the "knowledge of the specialist" and "ability to communicate." This indicates that a very knowledgeable specialist will not create meaningful results if he or she cannot transfer the knowledge to the client.

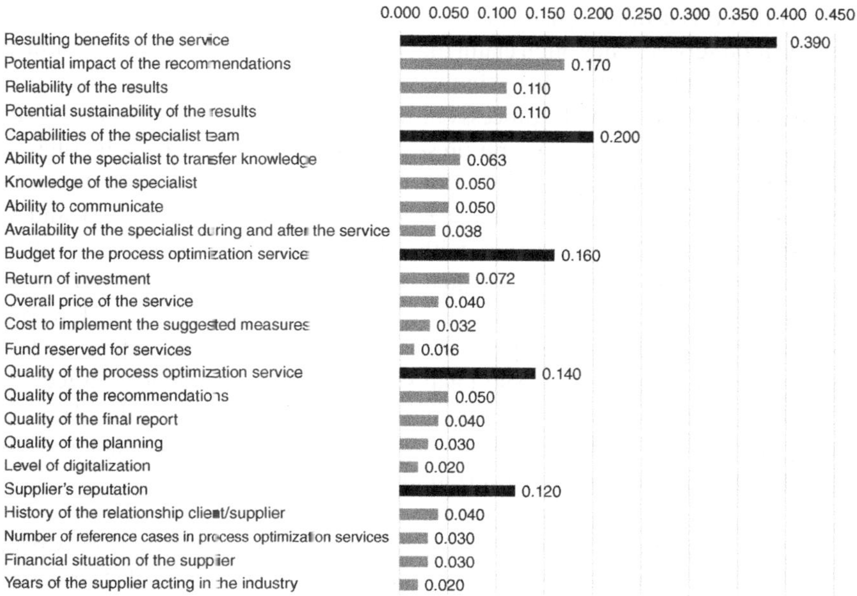

Figure 2.7 Scores of the main criteria and sub-criteria to invest in a process optimization service.

Two comments illustrate particularly well the clients' points of view: one of the respondents wrote:

> (...) if the specialist team is good at transferring knowledge, the results will last long after the optimization service is finished.
> The field engineer is not always trained enough to interact with the client.

In third comes the criterion "budget for the process optimization service," with the most important sub-criterion being the "return on investment." The client's view is well documented in one of the comments as follows:

> The available budget is always a constraint. It would be great if the solution provider could help with the ROI (return on investment) or TCO (total cost of ownership) calculation.

This agrees with the *cost* vs. *benefit* logic expressed by one respondent in the simplified external survey and documented in Section 2.3.1, as both indicate that an assurance of when the investment will pay off facilitates and spurs the hiring process even if the service or implementation costs are high. The sub-criterion with the lowest score for the budget criterion was "fund reserved for services," despite the fact that this is often a justification from the client for not hiring the process optimization service. They often claim that there are no resources to invest in such a service since it was not predicted in the year's budget.

The criterion "quality of the process optimization service" comes in fourth place of importance, with its sub-criteria presenting similar scores. The sub-criterion with the highest score is "quality of the recommendations" with 0.05, followed by "quality of the final report" with 0.04, "quality of planning" with 0.03, and in the last place "level of digitalization" with 0.02. It is understandable that the quality of recommendations exceeds the quality of the report as a sub-criterion, since a poorly structured report can still provide valuable recommendations. However, the report needs to have an acceptable level of quality because, according to a few respondents' comments, the operational staff often uses the report as a technical base to ask for more resources from the management. Regarding the planning quality, its low score can be explained by one of the respondents' comments about the difficulties of planning an optimization service:

> (…) the planning of the tasks can be, always, only approximated. A plan that tries to cover all the possibilities is a plan almost impossible to execute.

"Supplier's reputation" comes in the fifth and last position. Among its sub-criteria, the respondents selected the "history of the relationship client/supplier" as the most important with a score of 0.04. The second place in this rank was a tie between the "number of reference cases in process optimization services" and "financial situation of the supplier" with 0.03. The sub-criterion with the lowest score was the "level of digitalization" with 0.02. This indicates that respondents value direct relationship with the supplier more than the references, financial situation or the history of the supplier in the market.

2.3.3 Internal Survey

The 11 valid responses to the internal survey indicate that the most impactful barriers to process optimization service sales are financial with a score of 0.304 followed by the "team's capabilities" with 0.278, "company's vs. competitor's capabilities" with 0.228 and lastly "internal bureaucracy" with 0.195. Contrasting with the scores assigned to the criteria, the scores assigned to the barriers are well balanced, as shown in Figure 2.8, indicating that all the barriers investigated in this study are worthy of consideration by the company.

Contrary to the external surveys, the financial aspect attained the highest score in the internal survey. The sub-barrier with the highest weight was "cost of the specialist" with 0.093. This was further explained in the comments as a reflection of recurring client complaints regarding the high service costs, especially the cost of

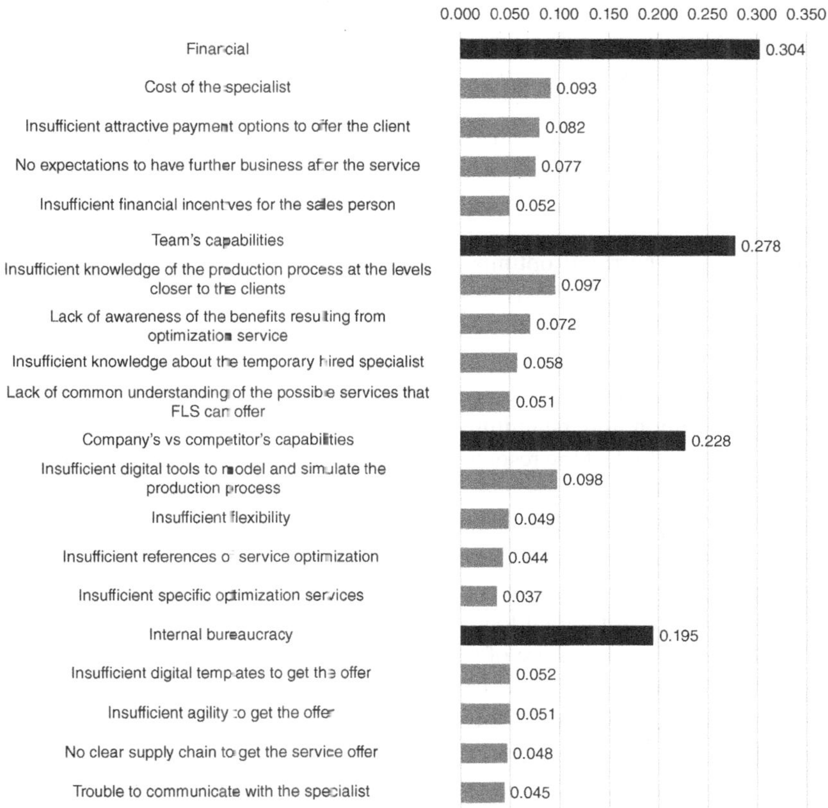

Figure 2.8 Scores of the main barrier and sub-barrier to sell a process optimization service.

foreign workforce. The sub-barrier "insufficient attractive payment options to offer the client" comes in second with a score of 0.082. According to one respondent:

> Payment based on performance is something clients are always pursuing.

Following the payment option sub-barrier, the second and third sub-barriers are "no expectations to have further business after the service" with a score of 0.077 and "insufficient financial incentives for the salesperson" with 0.052. Regarding financial incentives, the supporting qualitative data suggests that the sales personnel see optimization services as a starting-point to proceed with equipment sale. This is well illustrated in the comment bellow:

> [we] use it [process optimization service] as a teaser/door opener for process upgrades, equipment retrofit.

Furthermore, comments from the sales representatives indicate the need to make the benefits of optimization services more tangible, both internally and to the client.

> (…) we could have more marketing products to explain both ABC internally and our customers what is the real value proposition of the Process Optimization, show real cases to people, achievements, etc.

The second biggest barrier to optimization service sales is the "team's capabilities." Its most impactful sub-barrier is the "insufficient knowledge of the production process at the levels closer to the clients" with 0.097, followed by the "lack of awareness of the benefits resulting from the optimization service" with 0.072, "insufficient knowledge about the temporarily hired specialist" with 0.058 and lastly the "lack of common understanding of the possible services that ABC can offer" with 0.051. A close look at the responses reveals a consensus regarding the need to improve the staff's technical knowledge about the production process and how the services impact the clients' plant operations. Furthermore, the respondents believe that hiring third-parties to execute the services does not significantly impact sales.

The barrier "company's vs. competitor's capabilities" comes in third place. The sub-barrier "insufficient digital tools to model and simulate the production process" ranks in first with a score of 0.098, almost double the score of the second ranked "insufficient flexibility" with 0.049. Next come the "insufficient references of service optimization" with 0.044 and "insufficient specific optimization services" with 0.037. The respondents' comments indicate that the enhancement of digital tools will significantly impact the optimization services, but the lack of flexibility and the lack of references must be addressed soon so ABC can successfully compete with small local enterprises that have been permeating the optimization service markets.

The last ranked barrier was "internal bureaucracy." Yet, with a score of 0.195, it is responsible for 19.5% of the negative impact on optimization service sales considering the barriers investigated in this study. The nested sub-barriers presented balanced scores. "Insufficient digital templates to get the offer" comes in first place, with a score of 0.052, followed by "insufficient agility to get the offer" with 0.05, "no clear supply chain to get the service offer" with 0.048 and lastly "trouble to communicate with the specialist" with 0.045. It is possible to infer from the comments that although the internal bureaucracy does not offer major barriers to the selling of optimization services, it maximizes the difficulties in making the sale when combined with other factors. One comment illustrates this particularly well:

> The complex internal procedure and doubtfulness of customers hinder the priority of resources to sell process optimization services.

An analysis of the answers from the sales personnel and the technical specialists was performed. In total, eight sets of responses were available, four from each group. The data suggests significant differences between them, with the barrier rankings differing significantly as illustrated in Figure 2.9. The sales force top ranks the barrier "financial" with a score of 0.385, followed by "company's vs.

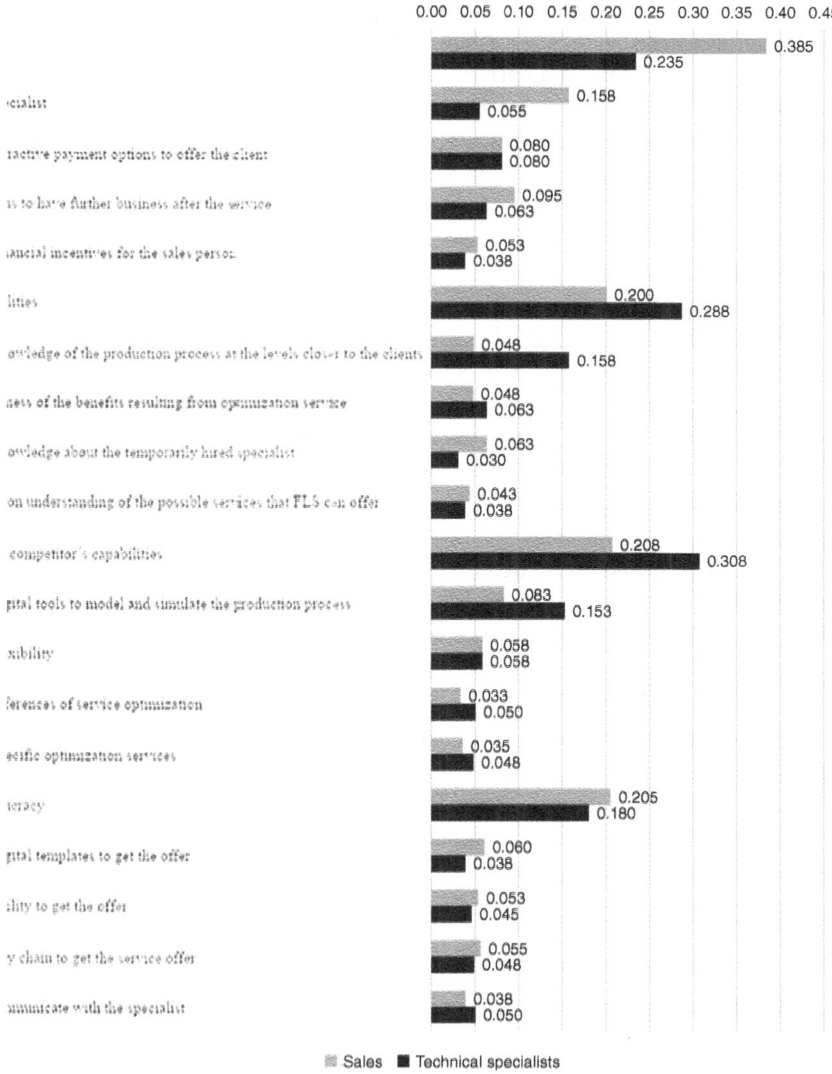

Figure 2.9 Scores of the main barrier and sub-barrier to sell a process optimization service given by sales and technical specialists.

competitor's capabilities" with 0.208, "internal bureaucracy" with 0.205 and lastly "team's capabilities" with 0.200. On the other hand, the technical specialists the top barrier is "company's vs. competitor's capabilities" with 0.308, followed by the team's capabilities' with 0.288, and only then "financial" with 0.235 and lastly "internal bureaucracy" with 0.180.

The biggest difference between the two groups lies in the barrier "financial." The sales representatives assigned it a score of 0.385, while the technical specialists assigned it a score of 0.235, a difference of 0.15 that comes mainly from the different evaluations of the sub-barrier "cost of the specialist." While the technical specialists assign a score of only 0.055 to this sub-barrier, the sales representatives see it as a much bigger barrier with a score of 0.158. This indicates that the specialists may not be aware of their cost impact in the sale of process optimization services.

The barrier with the second biggest discrepancy between the sales representatives and the technical specialists was the "company's vs competitor's capabilities." While the technical specialists assigned a score of 0.308 to it, the sales representatives assigned 0.208. This difference can be explained by the score assigned to the sub-barrier "insufficient digital tools to model and simulate the production process." While the technical specialists gave it a high importance with a score of 0.153, the sales representatives gave it a score of only 0.083. The logic used by the technical specialists refers to the improved assertiveness in the prediction of the service's impact on the clients' operations, which in turn facilitate the sales.

The score difference referring to the barrier "team's capabilities" was also significant between the two groups. The technical specialists assign a higher importance to this barrier than the sales representatives. This discrepancy is driven by the sub-criteria "insufficient knowledge of the production process at the levels closer to the clients" to which the technical specialists assigned a score of 0.158 while the sales representatives attributed a score of 0.048 to it.

From the comments inserted by the technical specialists, the improvement of technical capabilities of the sales team could positively impact the optimization services sales. What's more, their comments indicate the existence of a knowledge gap between the two groups, illustrated by quotes from the technical and sales team below, respectively:

> (…) they [sales team] do not have enough knowledge to clearly translate the client's problem to the specialists.
>
> (…) there is no clear source of this information [about process optimization service]. The brochures and the website are not clear to help the sales team.

Likewise, the sales representatives assign a higher weight to the barrier "internal bureaucracy" than the technical specialists, with scores of 0.205 and 0.180, respectively.

Regarding the sub-barriers, "insufficient digital templates to get the offer" presents the biggest score disparity, with the sales representatives assigning a score of 0.060 while the technical specialists assigned 0.038. Also, the sub-barrier "trouble to communicate with the specialist" is more important to the specialists (0.050) than it is to the sales (0.038).

2.4 Discussion

Cement plays a fundamental social and economic role worldwide (Habert et al. 2020) and is the most used man-made material in construction. However, the benefits of this versatile and cost-attractive material come at the price of environmental degradation, especially regarding to GHG emissions (5–8% of global CO_2 emissions) (Kajaste and Hurme 2016) and other environmental impacts through its value chain (e.g. mining, power consumption, and transportation).

To date, no attempts to substitute cement in the construction industry were successful. Furthermore, any new technologies aiming at replacing cement would not be scalable within the Paris Agreement timeframe (Claisse 2015; Davis et al. 2018). Thus, the urgent need to find solutions to mitigate the environmental impact of cement production.

One prominent solution to this problem lies within process optimization, which aims at increasing resource efficiency and curbing emissions to enhance the industry's environmental performance. Besides the potential contribution to environmental global goals, process optimization aims at enhancing operations by ensuring efficiency and productivity, and assuring quality, thus being an important asset for any cement producer in a highly competitive global market.

Any optimization service, however, deals with potentially conflicting decision-making criteria (e.g. budget vs. quality). To further develop its optimization services, a major cement industry player, namely ABC, wishes to investigate how its internal and external stakeholders weigh these conflicting decision-making criteria. To do so, this study surveyed internal and external stakeholders according to two hierarchical models developed using the HDM and informed by interviews with purposely selected stakeholders. The Hierarchical Model: Client's point of view contains the criteria and sub-criteria relevant in the decision making to invest in service optimization, while the Hierarchical Model: Staff's point of view contains barriers and sub-barriers relevant to internal stakeholders to sell an optimization service.

Three different surveys were conducted: a complete external survey, where the respondents were asked to make pairwise comparisons between the criteria and between the sub-criteria within each criteria; a simplified external survey, where the respondents were asked to make pairwise comparisons between the criteria only and with the purpose of increasing the number of responses obtained; and an internal survey, where respondents made pairwise comparisons between barriers, and between sub-barriers within each barrier. The data were fed into the Hierarchical Decision Model software version 2.0 developed by the Portland State University (HDM n.d.), which provided average weight allocations for the criteria and barriers considered in this work.

The results indicate that the external stakeholders are result-driven, having assigned an average weight of 0.40 to the criteria "resulting benefits of the service." Based on the weight assigned to each criteria, it is possible to infer that the clients' priority is to achieve a reliable and sustainable result. The high importance given to the three sub-criteria "potential impact of the recommendations," "reliability of the results," and "potential sustainability of the results" is connected to the lack of a clear outcome of the process optimization service. Although the client staff might understand the problem with the process, they do not know how and how much the possible solutions will improve the process. Therefore, their reluctance in investing in a process optimization service without knowing the potential benefits it will bring, and most importantly the prioritization of this criteria in the survey that indicates that the market is hungry for this feature.

The market opportunity for this feature is further corroborated by the supporting qualitative data, which indicate that most of the process optimization services available in the market do not have guaranteed results or a follow up with the specialists to ensure optimal results the recommendations are implemented. Following this line, a comment from a sales representative suggests the development of tools to better explain "internally and to our customers what is the real value proposition of the Process Optimization, show real cases to people, achievements, etc.," thus conferring tangibility to the benefits coming from such services and potentially facilitating the investment decision.

Today, ABC does not formally follow up on the results of optimization upon completion of the service. Thus, it is not possible to confirm the sustainability and reliability of implemented recommendations. Given that the clients' main concern when investing in optimization services is the assurance or evidence-based expectation of results, it would be important to create a follow up plan after the services. Besides being relevant to building a better relationship with the clients, it would also enable data gathering to build a solid knowledge base regarding selling points for each service in the portfolio. This would be done by using the data to improve the models used in process optimization, providing reliable guarantees of the service results and potentially increasing trust and sales.

With a reliable knowledge base allowing us to foresee the resulting benefits of the service, the client would also be able to calculate the ROI, making the financial benefits of the service tangible. This is aligned with the results referring to the criterion "budget for the process optimization service," as its highest ranked sub-criteria refers to "return on investment." The implementation of this solution may count on the development of a digital tool to support the sales staff in calculating the ROI based on the guarantees discussed in the previous paragraph.

The data also reveals that the "capabilities of the specialist team" play an important role in the hiring of optimization services, especially regarding the specialists' knowledge transfer abilities. Furthermore, there is indication from the supporting qualitative data that the "field engineer is not always trained enough to interact with the client." To address this issue, ABC can develop a training program to further develop the specialists' social and communication skills to ensure better interactions with clients and consequently strengthen the relationship with the clients. Any optimal solution would require further investigation of the skills currently missing and those that require polishing.

The external stakeholders attributed the lowest score to the criteria "supplier's reputation." This is a potential red flag for ABC because it has long considered its reputation to be one of the company's strengths in the market. This is especially relevant because new players entered (and continue to enter) the market, with strong focus on higher-ranked criteria such as the lowest price, especially coming from China and neighboring areas. Thus, it is imperative that ABC further develops its services to meet the criteria with higher ranking (such as the ones discussed above) to complement its good reputation and maintain its market-leading position.

Another relevant factor is the identification of different weights assigned by regional and global clients. Results from the simplified external survey reveal that both groups agree that the "resulting benefits of the service" is the most important criteria but they disagree on the ranking of all other criteria. Clients operating on a regional level give much more importance to the "capabilities of the specialist team," possibly because they count on less internal expertise and could benefit from third-party specialized services. On the other hand, global corporations count on a larger talent pool with specialists from different fields, which possibly reduces their need for seeking external specialized insights, leaving room for the budget criteria to come in second place in the ranking. This indicates a business opportunity for ABC to strengthen partnerships with regional players, who are more inclined to hire external specialist services as long as the specialists have proven capabilities. To better serve the needs of regional players, ABC can develop services with focus on providing external expertise, such as technical and sales onsite visits to assess prominent improvement opportunities, proactively developing business opportunities at these locations.

Another important aspect in the development and delivery of optimization services relates to the understanding of the client persona and their corresponding priorities and restraints. The analysis of responses according to seniority level indicated that higher ranked employees (supervisors and managers) are more concerned with the "capabilities of the specialist team" than they are with the "budget for the process optimization service." On the other hand, lower ranked professionals give more importance to budget than the capabilities. This reflects the strict budget restraints typically imposed on the operations teams in the cement industry in order to control production costs. The operations team, however, own most of the plant-specific technical knowledge, and therefore they are the ones who can better inform the specialist about issues. Thus, it is important to engage the professional level stakeholders to ensure accurate intel to inform the optimization strategy and the project scope, as well as the higher level employees to ensure budget availability and full understanding of the expected benefits and ROI.

The internal survey revealed a great concern regarding the "financial" barrier. In fact, most respondents added comments to this section of the survey, which did not occur in the other sections. Of all financial concerns (e.g. cost, payment options, future business opportunities, and financial incentives to the sales personnel), costs ranked in first place. This, however, disagrees with the results obtained in the external services, where "budget" came in third position while the "resulting benefits of the service" came in first place. This further supports the market opportunity for ABC to develop tangible measurements to predict and guarantee the impact of each service.

The highest-ranked sub-barrier within the "financial" barrier was the "cost of the specialist." According to data collected in the internal experts' interviews during the development of the HMs, the specialist hour pricelist used to calculate the total service price should be more flexible, making it possible to adjust prices according to the size, region, and other characteristics of the client, and thus making the offers more attractive to each client.

The sub-barrier "insufficient attractive payment options to offer the client" ranked in second place within the barrier "financial." The supporting qualitative data indicates that clients often ask about performance-based payments. Yet, the internal experts argue that service performance depends on plant-specific variables such as raw material chemical variations or (in)adequate equipment maintenance. Thus, performance-based payments require ABC to have a thorough understanding of the plant setup and operations. In this regard, digitalization can enable real-time plant monitoring and data collection. With complete and reliable information in hands, ABC may create accurate models specific to each operation/plant that will inform the setting of payment terms and conditions.

It is important to mention that the external respondents assigned a rather low score to the criterion "level of digitalization" (0.02). Putting this into perspective, this does not mean that digitalization will not benefit the clients. On the contrary, digitalization creates the necessary infrastructure to enable real-time accurate data collection, thus optimizing technical analysis that will translate into high-quality services and optimal plant operation.

The sub-barriers nested in the barrier "internal bureaucracy," namely "insufficient digital templates to get the offer," "insufficient agility to get the offer," "no clear supply chain to get the service offer," and "trouble to communicate with the specialist," presented very similar scores, indicating that the staff is generally concerned about the internal processes' efficiency and complexity. Once more, digitalization may be the key to conferring agility and simplicity while ensuring satisfactory communication flow and appropriate information logging. Viable improvements to digital tools include the development of commercial offer templates and a clear definition of process and decision owners with appropriate levels of approval and editing embedded in the tool.

The sub-barrier "insufficient financial incentives for the sales person," ranked last in the "financial" barrier. The same happened with the sub-barrier "lack of common understanding of the possible services that ABC can offer" in the barrier "team's capabilities." It is possible to infer from the respondent's comments that these scores are possibly related. Some sales representatives mentioned that the lack of financial incentives to sell optimization services is justified or even compensated by the possibility to use the services as a teaser to sell equipment and then be rewarded with commission. Yet, ABC offers monitoring and improvement services that do not require or lead to equipment renewal, indicating that the sales team has an inaccurate understanding about the services provided by the company. For instance, services aiming at enhancing clinquer quality may focus on chemical variables of raw materials without involving any investment in new equipment while reducing power consumption.

The barrier "team's capabilities" ranked second in the internal survey, with the sub-barrier "insufficient knowledge of the production process at the levels closer to the clients" ranking in first place. An analysis of the comments related to this sub-barrier score suggests the existence of a knowledge gap between the technical specialist team and the sales team, with the sales team claiming that they lack technical information resources and sources, while the technical specialist team denounces that this lack of knowledge undermines the ability of the sales team to communicate the client's problems to the technical team. An enhanced information management tool could improve communication between the sales personnel and the technical specialists. For instance, a database containing the client-provider interaction log would be helpful to eliminate communication noise. The database could store ABC's case studies and past experiences, allowing the teams to filter and search by topic, thus creating a knowledge resource available to the interested parties and facilitating internal communication. Additionally, ABC can bring the sales and the technical teams closer together by offering joint training sessions in which they can discuss their challenges, find solutions and build stronger communication channels based on empathy and collaboration.

Within the barrier "company's vs competitor's capabilities," the sub-barrier "insufficient digital tools to model and simulate the production process" ranked first, indicating that the investment in digital modeling tools may positively impact sales of optimization services. Furthermore, the technical specialists assigned more weight to this sub-barrier than the sales team, which is comprehensible as these types of digital tools are part of the technical team's routine, and therefore they are well aware of the limitations of the currently available tools.

2.5 Conclusions

This study reveals that ABC's clients prioritize the achievement of reliable and sustainable results with the process optimization services above the other criteria assessed in this research, namely capabilities, budget, quality and reputation. There is evidence of clients seeking for result assurance or evidence-based expectation of results, but no evidence of such practices in the market. To take advantage of this business opportunity, ABC can create after service follow up plans to collect data and create a knowledge base that will improve the models used in process optimization, thus allowing for evidence-based service results forecast and service guarantees, facilitating the calculation of ROI on each service, and potentially increasing trust and sales.

The results also suggest that two aspects of the team's capabilities could be improved: social and communication skills for interacting with clients, and the technical skills of the sales team. To address this issue, ABC can develop a specific training program targeted at the salespeople and the employees that interact with clients. Furthermore, the internal results suggest the existence of a communication distance and a knowledge gap between the technical specialist team and the sales team. One way to tackle this issue is through the creation of an information management tool that integrates and enhances the existing ones, which stores technical data and logs interactions with clients, and with broad access to the ABC staff. Additionally, ABC can strengthen the communication channels between the two teams by creating scheduled opportunities where the sales and technical teams can discuss their challenges and find solutions together. While the external stakeholders do not see budget as the most important aspect when hiring a process optimization service, the internal stakeholders think that the financial aspect is the most difficult barrier to overcome when selling such services, with the cost of specialists being the most relevant sub-barrier. To address this issue, ABC can redefine the specialist hour pricelist used to calculate the total service price, which will allow for more flexibility in pricing with adjustments according to the size, region, and other characteristics of the client.

The study also revealed different needs according to company size, with regional players valuing the capabilities of the specialist team more than the global players. This signals a business opportunity for ABC to strengthen partnerships with regional players by filling in the gap they experience in specialized technical knowledge. To do so, the company can develop services with a focus on providing external expertise to assess prominent improvement opportunities. The development of these measures require broad, reliable and up-to-date data from clients. This can be achieved through digitalization, which will create the necessary infrastructure to enable faster and better process optimization services. Thus, ABC should continue improving the digital tools combined with the education of the client about its importance. Lastly, the further development and implementation of the recommended measures may benefit from further qualitative investigation of the client's needs, spamming all markets in which ABC operates

2.A Interview Guide

Interviews: Guide questions to the experts to inform the creations of the Hierarchical Decision Models

1. What is your experience with process service optimization?
2. In the current situation of the cement industry, what is the importance of process optimization services?
3. What is the main challenge for process service optimization nowadays?
4. In your opinion, what are the main criteria of the client when investing in a process optimization service?
5. Which kind of outcomes the client is expecting of the process optimization services?
6. What do you see as internal barriers for FLSmidth to overcome in order to improve the sales of the process optimization services?
7. Do you have any personal experience, good or bad, regarding process optimization services?
8. Have you or someone on your team already implemented any strategy to boost the process optimization services? If yes, can you share the lessons learned?

References

ATech (2021). A TEC – innovative technologies for tailor-made solutions. https://www.atec-ltd.com/en (accessed July 22, 2021).

Benhelal, E., Zahedi, G., Shamsaei, E., and Bahadori, A. (2013). Global strategies and potentials to curb CO2 emissions in cement industry. *Journal of Cleaner Production* 51: 142–161.

Britannica Encyclopedia on-line: https://www.britannica.com/technology/cement-building-material/The-major-cements-composition-and-properties (accessed July 22, 2021). 2021.

Chow, L.K. and Ng, S.T. (2006). A fuzzy gap analysis model for evaluating the performance of engineering consultants. *Automation in Construction* 16 (4): 425–435.

Claisse, P.A. (2015). Cements and cement replacement materials. In: *Civil Engineering Materials*, 155–176. Butterworth-Heinemann.

Daim, T.U. and Dash, P. (2011). Roadmapping the convergence of technologies for services over broadband. *Benchmarking An International Journal* 18 (5): 668–693. https://doi.org/10.1108/14635771111166811.

Daim, T.U., Udbye, A., and Balasubramanian, A. (2012). Use of analytic hierarchy process (AHP) for selection of 3PL providers. *Journal of Manufacturing Technology Management* 24 (1): 28–51. https://doi.org/10.1108/17410381311287472.

Daim, T.U., Chan, L., and Estep, J. (2018). *Infrastructure and Technology Management*. Cham: Springer International Publishing.

Davis, S.J., Lewis, N.S., Shaner, M. et al. (2018). Net-zero emissions energy systems. *Science* 360 (6396).

Edwards, P. (2021). Towards net-zero: Low CO2 cement production. Global Cement Magazine (March 4), https://www.globalcement.com/magazine/articles/1203-towards-net-zero-low-co2-cement-production. (accessed on July 22 2021).

FLSmidth(2021a). Process optimization. https://www.britannica.com/technology/cement-building-material/The-major-cements-composition-and-properties FLSmidth Process optimization. https://www.flsmidth.com/en-gb/solutions/process-optimisation. (accessed on July 22 2021).

FLSmidth (2021b). History: More than 135 years of engineering and innovation legacy. Retrieved from: https://www.flsmidth.com/en-gb/company/about-us/history. (accessed on July 22, 2021).

Govindan, K., Mangla, S.K., and Luthra, S. (2017). Prioritising indicators in improving supply chain performance using fuzzy AHP: insights from the case example of four Indian manufacturing companies. *Production Planning & Control* 28 (6–8): 552–573. https://doi.org/10.1080/09537287.2017.1309716.

Habert, G., Miller, S.A., John, V.M. et al. (2020). Environmental impacts and decarbonization strategies in the cement and concrete industries. *Nature Reviews Earth and Environment* 1 (11): 559–573.

Hierarchical Decision Model (HDM). (n.d.). http://research1.etm.pdx.edu/HDM2/.

ICR Research (2020). Decarbonising the industry. Cemtech Webinar Conference in International Cement Review (September 24).

International Energy Agency (2020). Cement: Tracking report. https://www.iea.org/reports/cement. (accessed on July 21 2021).

International Energy Agency (2021). Driving energy efficiency in heavy industries: Global energy efficiency benchmarking in cement, iron & steel. https://www.iea.org/articles/driving-energy-efficiency-in-heavy-industries. (accessed on July 21 2021).

Kabir, G. and Sumi, R.S. (2014). Integrating fuzzy analytic hierarchy process with PROMETHEE method for total quality management consultant selection. *Production & Manufacturing Research* 2014 (1): 380–399.

Kajaste, R. and Hurme, M. (2016). Cement industry greenhouse gas emissions–management options and abatement cost. *Journal of Cleaner Production* 112: 4041–4052.

Thyssen Krupp (2021). Cement plants and life cycle services. https://www.thyssenkrupp-industrial-solutions.com/en/industries/cement. (accessed on July 21 2021).

Martin, H. and Daim, T.U. (2012). Technology roadmap development process (TRDP) for the service sector a conceptual framework. *Technology in Society* 34 (1): 94–105.

Razi, P.Z., Ramli, N.I., Ali, M.I., and Ramadhansyah, P.J. (2019). Selection of best consultant by using Analytical Hierarchy Process (AHP). *IOP Conf. Ser.: Mater. Sci. Eng.* 712: 12016.

Rodrigues, F.A. and Joekes, I. (2011). Cement industry: sustainability, challenges and perspectives. *Environmental Chemistry Letters* 9 (2): 151–166.

Smil, V. (2016). *Making the Modern World: Materials and Dematerialization*, 43. Lulu Press, Inc.

Soh, S. (2010). A decision model for evaluating third-party logistics providers using fuzzy analytic hierarchy process. *African Journal of Business Management* 4 (3): 339–349.

United Nations (n.d.). Global issues: Population. https://www.un.org/en/global-issues/population. (accessed on July 22 2021).

Van den Heede, P. and De Belie, N. (2012). Environmental impact and life cycle assessment (LCA) of traditional and 'green' concretes: literature review and theoretical calculations. *Cement and Concrete Composites* 34 (4): 431–442.

Vayvay, O., Ozcan, Y., and Cruz-Cunha, M.M. (2012). ERP consultant selection problem using AHP, fuzzy AHP and ANP: a case study in Turkey. *E3 Journal of Business Management and Economics* 3 (3): 106–117.

Weetch, B. (2021). GCP Applied Technologies: sustainability, savings, and strength go hand in hand. World Cement. Retrieved from

World Steel Association (2021). Global crude steel output decreases by 0.9% in 2020. https://www.worldsteel.org/media-centre/press-releases/2021/Global-crude-steel-output-decreases-by-0.9--in-2020.html. (accessed on July 21 2021).

3

Technology Assessment: Energy Storage Technologies

Aynur Yarga[1], Tugrul U. Daim[2], Saeed Alzahrani[3], and Serhat Burmaoglu[1]

[1] *Izmir Katip Celebi University, Izmir, Turkey*
[2] *Mark O. Hatfield Cybersecurity and Cyber Defense Policy Center, Department of Engineering and Technology Management, Portland State University, Portland, OR 97201, USA*
[3] *King Saud University, Riyadh, 11451 Saudi Arabia*

3.1 Introduction

In the contemporary landscape, the increasing demand for energy across households and industries has sparked growing apprehensions about the environmental ramifications linked to conventional power generation methods. With sectors like energy and transportation contributing significantly to greenhouse gas emissions, global efforts have been launched to mitigate pollution and curtail carbon dioxide production (Zhu et al. 2022; Rohit and Rangneker 2017; Yapıcıoğlu and Dincer 2023). This has triggered a shift toward renewable energy sources, driven by the imperative to lessen dependence on fossil fuels and mitigate their associated environmental impacts. However, the intermittent nature of renewables such as solar and wind energy introduces challenges due to their reliance on weather patterns and seasonal variations (Hadjipaschalis et al. 2009).

Addressing the gap between fluctuating energy production and consistent demand, the emergence of energy storage technologies offers a pivotal solution, enabling the seamless integration of renewables into existing electricity grids. Nevertheless, this integration necessitates comprehensive evaluations across a spectrum of dimensions, encompassing social, technical, economic, environmental, and political factors. Negotiating this intricate landscape of decision-making is facilitated by methodologies like the Hierarchical Decision Model (HDM), which allows for thorough analyses of multiple criteria when assessing the feasibility of emerging energy storage technologies (Kocaoglu 1983). Leveraging insights from field experts, these evaluations provide insights into the relative strengths and weaknesses of diverse technologies, offering essential guidance for policy-making and investment strategies.

Future-Oriented Technology Assessment: A Manager's Guide with Case Applications, First Edition.
Edited by Haydar Yalçın and Tugrul U. Daim.

Against this backdrop, the present study delves into the domain of energy storage technologies, highlighting their role in driving the transition toward sustainable energy systems. Through meticulous examination of the pros and cons of alternative energy storage solutions and their prioritization across diverse criteria, the study seeks to equip decision-makers with insights for making informed choices that contribute to a cleaner and more efficient energy landscape. This paper presents the outcomes of a rigorous evaluation process and sheds light on the intricate dynamics shaping the future integration of renewable energy with energy storage technologies.

3.2 Literature Review

3.2.1 Renewable Energy Sources

The demand for energy from households and industries is on the rise in the modern world. However, the use of current power generation methods is presenting a significant issue, namely environmental pollution. The transportation and energy sector, which causes the most environmental pollution in the United States, is responsible for 55% of greenhouse gas emissions. For this reason, while many international agreements foresee reducing pollution levels, global warming and carbon dioxide production, they decide to decommission old nuclear power plants and not to build new ones (Zhu et al. 2022). However, since the energies produced by the power plants through fossil fuels have to accelerate the production to meet the increasing energy production, it makes electricity more expensive when electricity is used intensively. Such situations bring along the need to replace traditional energy production methods with renewable energy sources (Hadjipaschalis et al. 2009). With the increasing importance of clean energy production in recent years, studies such as financial incentives, renewable energy standards and clean energy policies have gained momentum. It is planned that 95% of the increase in primary energy consumption in Europe between 2005 and 2030 will come from renewable energy sources. It is emphasized that renewable energy sources should become widespread in Europe in order to achieve an 80–95% reduction in greenhouse gas emissions after 2030 until 2050, which was decided by the EU member states (EU-27; Zhu et al. 2022). In addition, European countries have identified energy storage technologies as one of the most important technology areas within the scope of the European Union, as they have the greatest impact on preventing carbon emissions. Solutions to energy storage problems are characterized as a key element to achieve the EU's energy policy objectives (González et al. 2015). In this context, energy storage technologies are seen as the most important component of policies that can maximize the benefits of clean energy systems.

Renewable energy sources (wind and solar energy) are intermittent since there is no continuous production like fossil fuels and it is based on utilizing the power of nature directly, the energy produced shows large fluctuations in power output in monthly and even annual cycles depending on the weather and seasons (Olabi et al. 2021). To cope with fluctuations in a wind turbine's output (discharge time), energy must be stored when overproduced and released when production levels are less than required demand (Hadjipaschalis et al. 2009). There is therefore a need for energy storage technologies that can store and release large amounts of energy for a long period of time when climatic conditions are not suitable for energy production (Hall and Bain 2008). Moreover, regulations regarding storage of the electricity obtained from renewable energy sources are needed to ensure the proper integration of renewable energy technologies into electricity grids (González et al. 2015).

3.2.2 Assessment of the Energy Storage Technologies

In the landscape of rapidly evolving technologies, the practice of technology assessment emerges as a critical undertaking, particularly in domains such as energy storage and power management. These domains hold escalating significance due to the pressing requirement for renewable energy sources as a countermeasure against the imminent threats of global warming and climate change. Across the globe, numerous nations have acknowledged the paramount importance of harnessing electrical energy from sustainable sources, aligning their efforts with the commitment to sustainable energy solutions (Luo et al. 2015).

Within the realm of technology assessment, a series of diverse studies has cast light on pivotal facets of technological evaluation. In one such study, Van Blommestein and Daim (2013) conducted research on the adoption of energy-efficient devices in South Africa, unraveling the intricate web of factors that sway their acceptance within the framework of sustainable energy practices. This exploration extended further through the work of Daim et al. (2014 and 2012), who meticulously assessed energy storage solutions complementing the integration of wind power. This inquiry accentuated the indispensable role of storage technologies in the proficient management of intermittent renewable energy sources. These studies collectively underscore the multidimensional nature of technology assessment, encapsulating not only technical feasibility but also societal dynamics and environmental considerations.

Kim et al. (2017) contribute a comprehensive perspective by elucidating technology assessment as a systematic approach for gauging the feasibility, advantages, and potential risks of specific technologies. This methodological framework empowers decision-makers with the insights needed to make well-informed choices regarding the adoption, implementation, and fusion of technology. Additionally, Daim et al. (2018) introduce an innovative methodology that amalgamates bibliometrics, social network analysis, and HDM. This research sheds light on the strategic roadmap for robotics technologies within the power industry, showcasing how a holistic assessment can guide the formulation of development and execution strategies.

In a distinct context, Cho et al. (2019) delve into the realm of solar photovoltaic (PV) technology adoption, unearthing the correlations between socioeconomic variables and geographical factors to enrich the understanding of adoption trends. Their findings underscore the significance of incorporating local factors into technology assessment, thereby informing policy decisions and bolstering sustainable energy practices. Collectively, these studies emphasize the pivotal role of technology assessment in steering decision-making, strategic planning, and the assimilation of innovative technologies across a spectrum of domains.

Moreover, against the backdrop of our technology-driven era, the domains of energy storage and power management take on amplified importance. This elevation is largely attributed to the dire need for renewable energy sources, acting as a bulwark against the impending perils of climate change. Notably, electrical energy, although abundant, ranked second in consumption after oil in 2012 (Aneke and Wang 2016). This predicament necessitates energy storage systems due to the intermittency of electricity from renewable sources and the call for adaptable power and energy density to cater to evolving demands (Hadjipaschalis et al. 2009). It is within these contexts that energy storage technologies emerge as transformative assets with the potential to reshape our energy landscape.

These technologies span diverse categories, including mechanical, chemical, electrochemical (supercapacitor energy storage, battery energy storage), superconducting magnetic, and thermal energy storage (Aneke and Wang 2016; Li et al. 2023). Their superior attributes, encompassing high efficiency, capacity, durability, and environmental friendliness, position them as invaluable alternatives to traditional methods (Kozak and Kozak 2012). Notably, energy storage technologies assume a pivotal role in modern energy supply chains, allowing the utilization of renewable energy precisely when needed, rather than solely when generated (Hadjipaschalis et al. 2009). Their contribution to grid stability, renewable energy penetration, and energy system efficiency bears immense significance, enabling the effective integration of renewable sources like wind, solar, and tidal into the energy matrix.

In a world grappling with the challenges of intermittent power generation, energy storage systems offer a reliable solution. These technologies hold the potential to balance load and generation fluctuations, mitigate the need for immediate backup power plants, and enable efficient energy management (Olabi et al. 2021; Li et al. 2023). A central aspect of their role lies in network load leveling, wherein they bridge the gap between supply and demand, aligning electricity generation with consumption patterns (Aneke and Wang 2016). Furthermore, energy storage systems stand as versatile backup energy sources during power plant interruptions, bolstering grid stability, and reinforcing energy security (Hadjipaschalis et al. 2009).

Luo et al. (2015) elaborate on the myriad benefits of energy storage technologies, ranging from meeting peak electricity demands and managing time-varying energy to enhancing renewable energy utilization and augmenting power quality and reliability. These technologies are pivotal in enhancing energy management by minimizing waste and increasing efficiency, leading to reduced carbon emissions and conserving finite fossil fuel resources (Mahlia et al. 2014). In microgrid systems, they play a vital role in improving power quality. In essence, energy storage technologies signify a transformative force that holds the potential to redefine energy systems.

In conclusion, the synergy between technology assessment and the imperative for renewable energy sources is manifest. Studies elucidating various facets of technological evaluation provide a roadmap for informed decision-making and strategic planning. Energy storage technologies emerge as vital components of modern energy systems, combating the challenges of intermittent energy sources and propelling us toward a sustainable energy future.

3.3 Methodology

3.3.1 Hierarchical Decision Model

HDM, introduced by Kocaoglu (1983) in the early 1980s, is one of the multi-criteria decision-making methods that facilitates decision-making in the context of more than one criterion in complex situations. It helps to choose the best among the alternatives by evaluating the criteria determined in order to solve the problem that has more than one alternative (Hwang and Masud 1979). HDM is similar to the Analytical Hierarchy Process (AHP) presented by Saaty (1997). In both methods, decision items at each level are linked to other decision items above and below. The difference between the two models is; While AHP uses the eigenvector calculation method, HDM uses a different calculation approach by using constant sum (Constant Sum Method) calculations instead of eigenvectors (Kocaoglu 1983). For n elements whose constant sum method is examined, $n(n-1)/2$ pairwise comparisons are made. Then, the decision-makers are asked to evaluate the two items using a fixed total calculation over 100 points, rather than making comparisons between these items using a scale in the range of 0–9 such as AHP (Kocaoglu 1983; Abbas and Kocaoglu 2016). The reason for evaluating the criteria over 100 points in the HDM model is to digitize any criterion in the

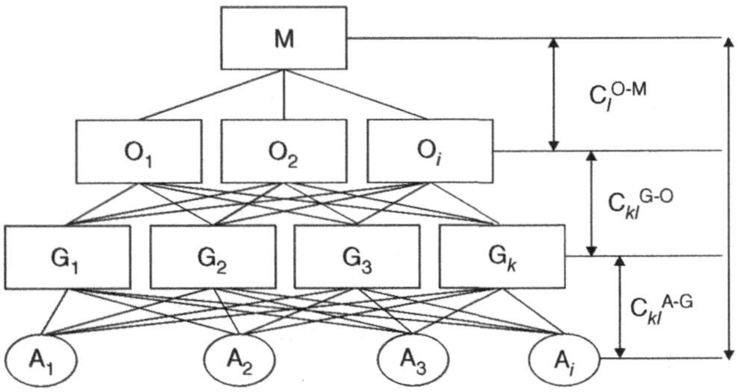

Figure 3.1 Hierarchical decision model.

HDM model through the desirability curves. Thus, for example, the economic values of the alternatives in the model can be determined much more reasonably than the 0–9 comparison scale used in AHP (Sheikh 2013).

As can be seen in Figure 3.1, Kocaoglu (1983) argues that decision criteria should be presented in a multilevel hierarchy.

In order to define the HDM in more detail, the following steps are defined (Estep 2017; Khanam and Daim 2021; Barham 2019).

3.3.2 HDM Steps

3.3.2.1 Creating Model

Looking at the hierarchical order in Figure 3.1, first of all, in order to achieve the main goal, certain criteria (O_j) and sub-criteria directly or closely related to this criterion (G_k) are determined according to the efficiency criteria with the help of experts. Alternatives to be evaluated or alternatives to be compared (Ai) are evaluated or prioritized in line with the determined criteria and sub-criteria (Khanam and Daim 2021; Cleland and Kocaoglu 1981; Kocaoglu 1983; Saaty 1977; Estep 2017).

When the HDM formulas mentioned in Figure 3.1 are examined:

It defines the criteria in the range of $O_\iota : i = 1, 2..., l$

C_l^{O-M} : L defines the relative contribution of the th criterion to the goal.

$G_k : k = 1, 2\ldots., K$ defines the sub-criteria that is in the range and determined under the criteria.

Defines the relative contribution of the C_{kl}^{G-O} : Kth sub-criterion to the Lth criterion.

Defines alternatives in the range $A_i : i = 1, 2..., l.$

It determines the overall contribution of the C_i^{A-M} : i alternative to the goal.

C_{ik}^{A-G} : $i$defines the relative contribution of the ith alternative to the kth sub-criterion.

C_{il}^{A-O} : i defines the relative contribution of the ith alternative to the ith criterion.

The formula obtained according to this approach defined in the model is as in Eq. (3.1):

$$C_i^{A-M} = \sum_{l=1}^{L} \sum_{k=1}^{K} C_l^{O-M}.C_{kl}^{G-O}.C_{ik}^{A-G} \tag{3.1}$$

Thus, thanks to the HDM, it helps to make the best decision by considering the effects and importance of each element. Comparing the alternatives in terms of the determined criteria helps to identify the alternative that makes the best contribution to achieving the goal (Lingga 2016; Barham 2019).

3.3.2.2 Pairwise Comparison

After the hierarchical model has been created as shown in Figure 3.1, decision-makers need to put forward their judgments based on the comparison they have made between the two criteria (Abbas and Kocaoglu 2016). Thus, when comparing the criteria, sub-criteria, and alternatives determined in line with the purpose, 100 points are distributed between the two criteria by the field experts. In other words, as can be seen in Figure 3.8, in this step, within the scope of the value curves approach, experts must determine and evaluate the common levels/metrics, i.e. the desirability matrix, by giving a quantitative value (between 0 and 100) between each two criteria (Barham 2019).

The relative value/contribution/weight of the criteria is determined by making use of the personal judgments of the experts regarding the criteria determined in the HDM process. The personal judgments of the experts are expressed by pairwise comparisons, which are converted into relative weights in the ratio scale of the criteria. Thus, the weights obtained as a result of the 100 points distributed by the experts by pairwise comparison are calculated by the ratio scale, and their relative weights are determined in the ratio matrix table (Abbas and Kocaoglu 2016). Therefore, the ratio of each column in the ratio matrix of the items obtained is divided by the ratio of the next column, the relative weight matrix table (relative weight matrix) is obtained, and the mean and standard deviation of the data in each column are calculated. To calculate the relative weights of the criteria, the last value in the table is assumed to be 1, and the other values are calculated. The values obtained in the final relative weight matrix are normalized to obtain the final weight of each element. In short, it is revealed which criterion is more meaningful and prioritized than the others (Khanam and Daim 2021; Barham 2019).

The calculation of relative values is based on the formula in eq. 2:

$$VA_m = \sum_{k=1}^{K} \sum_{L=1}^{L} V(C_k).V(S_{LK}).VA_{mLk} \tag{3.2}$$

C_k: represents the criteria in the range of 1 to K.
S_{lk}: represents the sub-criteria of the K criterion in the range of 1 to L.
VA_m: Represents the alternatives evaluated in the model.

Pairwise comparison evaluation can be done through HDM software developed by Portland State University. Pairwise comparison depends on the total number of elements (n). The total number of pairwise comparisons to be made by experts is calculated as $n(n-1)/2$.

3.3.2.3 Inconsistency and Disagreement

HDM relies heavily on expert judgments, where inputs from the experts in the quantification phase are validated against inconsistency and disagreement. Inconsistency occurs when contradictory results appear in the judgments of an expert. Inconsistency may arise from the lack of skill, experience, time, and preference of experts (Abotah 2014; Khanam and Daim 2021; Abbas and Kocaoglu 2016). The value is above 0.10, that indicates that expert judgments are unreliable since the threshold is 0.10 (Saaty 1980). In cases of inconsistency, experts should be asked to redo the pairwise comparisons again and explain why the inconsistency occurred (Lingga 2016; Khanam and Daim 2021). On the other hand, disagreements between the experts' judgments occur due to factors such as ideology, competition, different perspectives, weak feedback, insufficient data, and insufficient participation (Gibson et al. 2019; Iskin and Daim 2014; Barham 2019). This disagreement should be calculated, treated, and can be accepted only if it is below the threshold of 0.10. When the disagreement value is high, it is necessary to separate the experts who do not have common views, that is, to ensure that two-thirds of them are in harmony. With the help of the Delphi method, there are other alternatives to eliminating the disagreeing opinion or to accept the disagreement and present the conflict as a report (Lingga 2016; Barham 2019).

3.3.3 Technology Development Envelop (TDE)

This research employs the concept of technology development envelope (TDE) (Daim et al. 2011; Gerdsri 2005; Gerdsri and Kocaoglu 2003; Gerdsri 2007). TDE enable organizations to identify emerging technologies relevant to their domain, evaluate their values with respect to the organization's objective under assessment criteria. The TDE approach integrates both technology and business strategies. TDE enables high-level decision-makers in companies and government officials responsible for policy-making to include emerging and disruptive energy storage technologies when formulating technology strategies. It tracks emerging technologies over a specific period of time. The technologies are evaluated based on their value to organizations for each time period. For each technology a path of its value is presented over a period of time. The highest value for any technology for each time period represents the TDE. The expert judgments are collected in order to provide these values for each technology at each time period.

3.4 Model Development

This section will present the details of the model development (Figure 3.2).

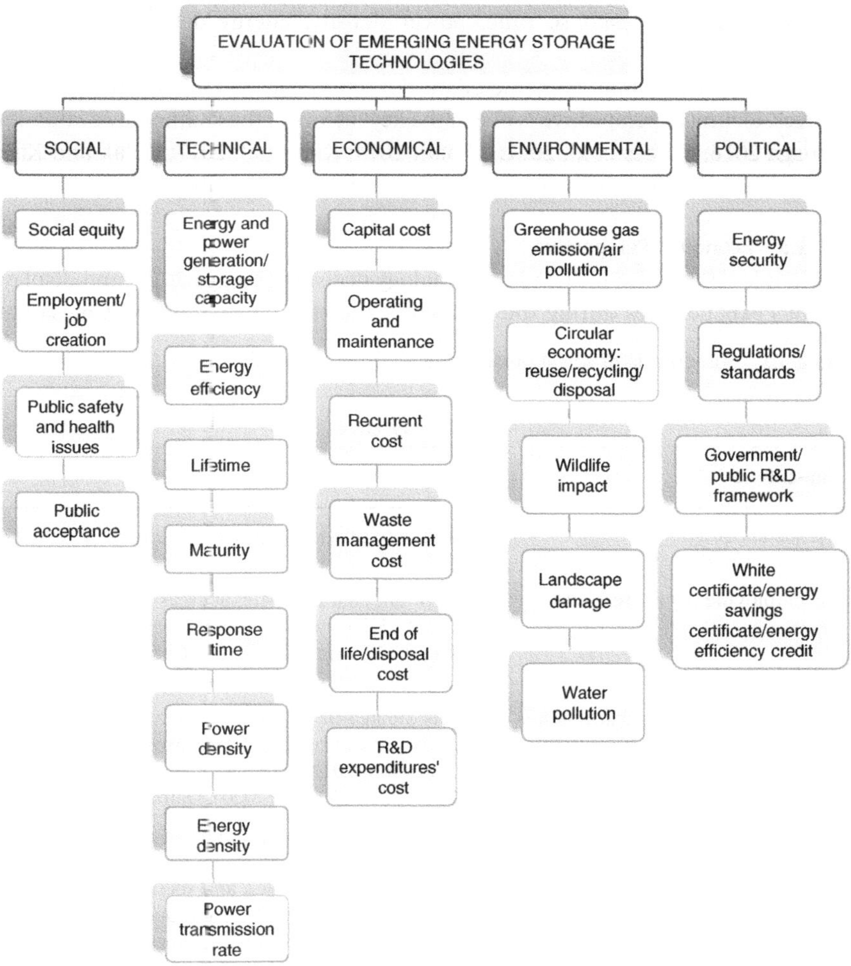

Figure 3.2 Criteria/perspectives.

3.4.1 Criteria/Perspectives

3.4.1.1 Social Perspective

The human factor that is social structure significantly impacts energy usage (Lingga 2016). The social perspective assesses the social effects of the development of the storage technology (Daim et al. 2014), such as satisfying consumers, personal gain, and making users prefer a particular technology (Sheikh 2013; Abotah 2014; Kim et al. 2017; Colak and Kaya 2020) (Table 3.1).

3.4.1.2 Technical Perspective

Technical perspective has various sub-criteria, such as efficiency, maturity, lifetime, capacity, response time, power density, energy density, and power transmission.

To identify the common characteristics or differences among technology alternatives, technical perspective is one of the key factors of the technology evaluation (Lingga 2016; Garces et al. 2021; Abotah 2014; Kim et al. 2017; Colak and Kaya 2020; Zhao et al. 2019; Zhao et al. 2020; Daim et al. 2014) (Table 3.2).

3.4.1.3 Economic Perspective

Economic perspective influences technology decision (Lingga 2016) and displays various expenditures during the lifespan of the storage technology (Chiesa and

Table 3.1 Sub-criteria of social perspective.

	Social perspective
Sub-criteria	**Definition**
S1: social equity	The ability to provide affordable energy to everyone (DOE)
S2: employment/job creation	Employment characteristics such as workforce availability, job creation and poverty alleviation Sheikh (2013), Kim et al. (2017), Colak and Kaya (2020), Daim et al. (2014)
S3: public safety and health issues	Protects the health, safety, and welfare of human beings by preventing long-term hazardous health effects Sheikh (2013), Lingga (2016), Kim et al. (2017), Colak and Kaya (2020), Daim et al. (2014)
S4: public acceptance	Public acceptance, perception or awareness has an impact on adoption of emerging technologies. Social benefits, lifestyle, esthetics, and social acceptance of technology can both improve and decrease opposition or support level of technology Sheikh (2013), Kocaoglu et al. (2016), Lingga (2016), Barham (2019), Kim et al. (2017), Colak and Kaya (2020), Daim et al. (2014)

Table 3.2 Sub-criteria of technical perspective.

Technical perspective	
Sub-criteria	**Definition**
T1: energy and power generation/storage capacity	"The quantity of available energy in the storage system after charging" Kim et al. (2017). Required energy and power generation capacity can be changed based on the actual and future needs Lingga (2016), Colak and Kaya (2020), Zhao et al. (2019), Daim et al. (2014).
T2: energy efficiency	"Efficiency is an indicator of the amount of useful energy from a renewable energy source or the output productivity in the production of the source" Sheikh (2013, p. 101). It's measured in the "ratio between released energy and stored energy" Kim et al. (2017). For example, while compressed air energy storage system has 85% effective efficiency, flywheel has between 85% and 95% efficiency Colak and Kaya (2020), Daim et al. (2014).
T3: lifetime	Lifetime represents the duration of the useful life of energy storage technologies. Some factors such as location, design and application have impacts on lifetime of technology Lingga (2016). For example, the lifetime for flywheel energy storage is 20 years while compressed air energy storage can last for over 30 years Daim et al. (2014)
T4: maturity	If the technology has been in use for a while and many of the associated defects and problems have been corrected it can be considered as mature. Mature represents the readiness of a technology for implementation. Maturity includes: density level of patents, modularity, scalability, flexibility, and obsolescence resistance; while at the same time maturity represents: the stage of technology, trends and persistence ability Sheikh (2013), Lingga (2016), Kim et al. (2017). It "measures how long energy storage technology has been in the market and field tested" Kim et al. (2017), Daim et al. (2014).
T5: response time	"The interval time a storage unit takes to react to a given input" Kim et al. (2017)
T6: power density	"The rated output power divided by the volume of the storage device" Kim et al. (2017)
T7: energy density	"The stored energy divided by the volume of the storage device" Kim et al. (2017); Colak and Kaya (2020)
T8: power transmission rate	"The ratio of the quantity of electricity removed from a storage device to its rated capacity" Kim et al. (2017)

Franzo 2015). Thus, it is one of the main components in evaluating alternatives (Daim et al. 2014). Economic impact perspective is divided into six sub-criteria: capital cost, operating and maintenance costs, end-of-life cost, recurrent cost, waste management cost, and R&D expenditures cost, which are some of the main financial characteristics of the emerging energy-storage technologies (Garces et al. 2021; Abotah 2014; Kim et al. 2017; Colak and Kaya 2020; Zhao et al. 2019; Daim et al. 2014) (Table 3.3).

Table 3.3 Sub-criteria of economic perspective.

Economic perspective	
Sub-criteria	**Definition**
E1: capital cost	Capital cost includes: cost of essential equipment and facility, cost of land and complementary technologies to be used in the storage unit. Kocaoglu et al. (2016), Kim et al. (2017). It includes all fiscal costs to use in the set-up process of the storage systems Daim et al. (2014). For example, capital cost of flywheels range from \$600–\$2000/kW Mongird et al. (2019), Zhao et al. (2019).
E2: operating and maintenance cost	O&M cost (fixed and variable) contain all the spending to operate and maintain the system for the first year. In addition, O&M cost includes: raw material costs, labor cost, replacement costs, product and service costs Daim et al. (2014), Kocaoglu et al. (2016), Colak and Kaya (2020), Zhao et al. (2019), Daim et al. (2014).
E3: recurrent cost	"The cost per unit energy divided by the cycle life" Kim et al. (2017)
E4: waste management cost	Some process outputs of energy technologies can spread harmful toxic to the environment that is why they should be contained and disposed of safely. Waste management cost items are associated with collecting, transporting, processing, recycling, storing or disposing of waste; after energy production process Kocaoglu et al. (2016).
E5: end of life/disposal cost	Disposal or end of life cost covers all expenditures to get rid of end-of-life equipment Daim et al. (2014), Kim et al. (2017), Daim et al. (2014).
E6: R&D expenditures' cost	Emerging energy storage technologies have intense R&D expenditures. Thus, in order to develop emerging energy storage technologies, organizations need to deal with financial restraints Kocaoglu et al. (2016).

Table 3.4 Sub-criteria of environmental perspective.

Economic perspective	
Sub-criteria	**Definition**
EN1: greenhouse gas emission/air pollution	Greenhouse gases such as carbon dioxide (CO2) emissions, methane (CH4), nitrous oxide (N2O), sulfur hexafluoride (SF6), hydrofluorocarbons (HFC) and perfluorocarbons (PFC) Kocaoglu et al. (2016) cause global warming that is a serious worldwide environmental problem Hanan et al. (2017), Kim et al. (2017), Colak and Kaya (2020), Zhao et al. (2019).
EN2: circular economy: reuse/recycling/disposal	End of life of a product needs to be considered in terms of recycling, chemical, or gas based; product disposal and biodegradability Sheikh (2013)
EN3: wildlife impact	The harmful effects on wildlife and nature Kim et al. (2017), Colak and Kaya (2020)
EN4: landscape damage	Energy related land degradation could cause negative effects on ecosystem such as inappropriate agricultural practices, vegetation removal, hydrology alteration and acid deposition. To sustain soil quality and nutrient balance these variables need to be considered Kocaoglu et al. (2016), Colak and Kaya (2020).
EN5: water pollution	Energy related technologies can cause pollutants in water sources that affects water quality by draining of the waste water into natural water sources Kocaoglu et al. (2016), Kim et al. (2017), Colak and Kaya (2020).

3.4.1.4 Environmental Perspective

Environmental criteria are related to the environmental ecosystem that considers environmental harmfulness, damages, and pollution (Lingga 2016; Garces et al. 2021; Abotah 2014; Kim et al. 2017; Colak and Kaya 2020; Zhao et al. 2019). (Table 3.4)

3.4.1.5 Political Perspective

Political effects on developing storage technology includes regulations, strategies, and policies (Daim et al. 2014; Sheikh 2013; Lingga 2016; Abotah 2014; Kim et al. 2017; Colak and Kaya 2020) (Table 3.5).

Table 3.5 Sub-criteria of political perspective.

Economic Perspective	
Sub-criteria	**Definition**
P1: energy security	Availability of adequate-stable-secure energy supply and stabile energy prices. Governments need to control and keep stable both energy prices and energy supplies Sheikh (2013). Energy sources have to be resilient/invulnerable and reliable. While resilience handle shocks and changes on energy system, reliability copes with the accessibility to required energy safely Kocaoglu et al. (2016), Sheikh (2013), Kim et al. (2017).
P2: regulations/standards	When all state-owned and private companies invest in any energy storage technology they need to consider regulation and standards that obligate some certain expectations Kocaoglu et al. (2016), Sheikh (2013), Lingga (2016).
P3: government/public R&D framework	Government or national laboratory funded research and development, private sector supported technology transfer activities, and strategy based technology roadmaps can increase the positive impacts of the technology deployment and development Sheikh (2013); Lingga (2016).
P4: white certificate/energy savings certificate (ESC)/energy efficiency credit (EEC)	The aim of white certificate is to achieve maximum utilization of energy saving and energy efficiency (KYOS). White certificate obligates energy distributors and suppliers to follow energy efficiency goals over a period of time Kocaoglu et al. (2016).

3.4.2 Alternative Technologies

There are certain systems for the storage of renewable energy sources. Energy storage systems are classified based on their mechanical, electrical, chemical, or thermal properties (Li et al. 2023). These different systems are separated from each other according to the transformation stages of the energy obtained before and after storage (Tetteh et al. 2021). These technologies include rechargeable batteries, lithium-ion batteries, electrochemical supercapacitors, superconducting magnetic energy storage, and kinetic energy storage (Zhang et al. 2021). For example, electricity can be stored directly as electrical energy in supercapacitors, which is one of the energy storage types. The electricity stored as mechanical energy is indirectly converted into potential energy in a pumped storage hydroelectric power plant (Tetteh et al. 2021). These technologies provide a clean energy

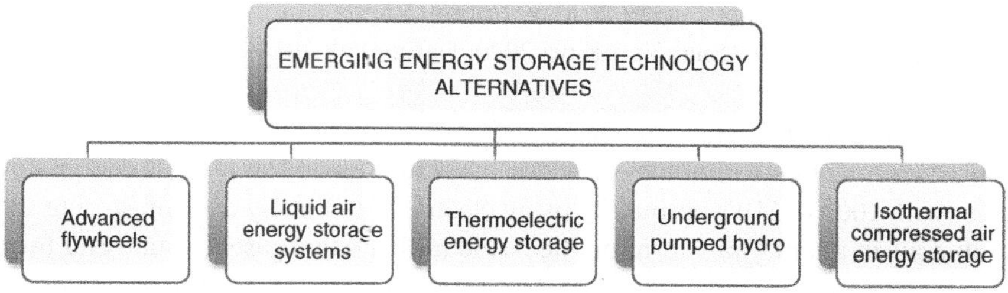

Figure 3.3 Alternative technologies.

environment for the future (Hall and Bain 2008). While some are in heavy use today, others are still under intensive research and development (Hadjipaschalis et al. 2009). Emerging energy storage technologies are still being investigated (Figure 3.3).

In general, when these technologies are classified according to storage systems, flywheels, pumped hydroelectric energy storage, isothermal compressed air energy storage (I-CAES), and liquid air energy storage technologies belong to mechanical energy storage types (Olabi et al. 2021; Aneke and Wang 2016; Luo et al. 2015). Mechanical energy storage types are electromechanical systems that can easily store electrical energy and convert it into potential and kinetic energy forms. The biggest disadvantage of mechanical storage systems is that they need to be installed in areas with certain geographical features, such as water sources or underground caves (Zhang et al. 2021; Tetteh et al. 2021).

Thermoelectric energy storage (TEES), another emerging energy storage technology discussed in current research, can be shown as an example of a thermal energy system. The TEES system, on the other hand, can directly store thermal energy with a large capacity. These systems can use energy for heating and cooling by considering seasonal conditions (Tetteh et al. 2021).

3.4.2.1 Flywheels

To provide a new solution for serious energy problems, energy developments, more effectively existing energy use, the advantages of advanced flywheels that is an electromechanical system are high energy conversion efficiency (85–95%), long and high energy storage density, more power in a short time, pollution-free (environmentally friendly), absence of noise, short charging time, long lifetime, width in the application, easy maintenance, and continuous working. It can be used for electric energy, solar energy, and wind energy (Liu and Jiang 2019; Mahlia et al. 2014).

Energy can be stored in advanced flywheels energy storage mechanisms in high-speed mode for a long time without any considerable loss. For example, only a 5% loss in speed after 24 hours uninterrupted operation (Bherde et al. 2018).

While Power range goes up to 20 MW, Energy Range can increase up to 5 MWh (Mongird et al. 2019; Colak and Kaya 2020; Daim et al. 2014).

3.4.2.2 Liquid Air Energy Storage Technologies

This energy storage technology provides long-duration and large-scale energy storage (with 100 s of MWs output) by using electricity to liquefy cool air, storing this liquid air in a tank, transforming the liquid air back to a gaseous state and turning that gas into a turbine to generate electricity, respectively. It can reduce the storage space of air and the storage pressure volatility of renewable energy power generation (An et al. 2020). It can also help to reduce dependency on geographical conditions because this system has subsystems such as a compression subsystem, cold storage subsystem and expansion power generation system. For example, the cold storage subsystem that is a component of the liquid air energy storage system stores and recovers the cold energy of the low-temperature liquid air to cool the compressed air (An et al. 2020).

3.4.2.3 Thermoelectric Energy Storage

TEES enables the conversion of heat stored in thermal power plants to electricity by reducing storage system costs (Vinnemeier et al. 2016). TEES is normally classified as hot storage but also can be classified as cold storage depending on the environmental temperature level. For that reason, it is widely free from geographical and geological constraints (Vinnemeier et al. 2016; Ayachi et al. 2016). "During periods of excess electricity generation, a vapor compression heat pump consumes electricity and transfers heat between a low-temperature heat source and a higher temperature heat sink" (Ayachi et al. 2016). This temperature difference can be protected for several hours in order to generate electricity during the peak hours (Ayachi et al. 2016). These two different applications have different technologies and system costs (Morandin et al. 2011). For that reason, multimegawatt TEES as a large-scale energy storage technology is an alternative to existing technologies such as pumped storage hydroelectricity (PSH) and compressed air energy storage (CAES) systems (Mahlia et al. 2014; Ayachi et al. 2016; Daim et al. 2014).

3.4.2.4 Underground Pumped Hydro

While the underground pumped hydro storage reduces the grid electricity consumption costs, it also enhances the use of underground water reservoirs (Maruf et al. 2022). This storage system is a great alternative for remote districts and farms or flat topological areas that have sufficient underground water resources (Mahlia et al. 2014; Akinyele and Rayudu 2014; Shirinda et al. 2020; Stutz et al. 2020).

"The system operates by utilizing excess energy to store gravitational potential energy by inflating a subsurface reservoir with water from a nearby pond" (Stutz et al. 2020: pp. 2). It provides advantages against conventional pumped hydro energy storage such as reducing the land use and water evaporation so that it can also reduce the construction costs for reservoirs (Guo et al. 2020; Colak and Kaya 2020; Daim et al. 2014).

3.4.2.5 Isothermal Compressed Air Energy Storage (I-CAES)

Isothermal compressed air energy storage system (I-CAES) was developed by Enairys Powertech and LightSail energy (Dib et al. 2021). I-CAES is a competitive technology, an integrated version of CAES technologies into renewable sources that prevents renewable energy intermittence. The I-CAES system tries to keep the temperature constant during discharging and charging to have an isothermal process (Zhao et al. 2020). To have high-energy efficiency performance, Isothermal CAES considers a quasi-isothermal compressor/expander (Akinyele and Rayudu 2014; Dib et al. 2021). This system as a conventional gas turbine technology uses electricity in order to compress air and store it in underground caverns (Akinyele and Rayudu 2014; Mongird et al. 2019; Colak and Kaya 2020).

3.5 Results Analysis and Discussion

3.5.1 Model Validation

In the HDM stage, firstly, the verification of the HDM created by the experts was carried out. In the second step, the criteria and sub-criteria were compared by the experts determined as a result of the social network analysis, and the importance levels of the criteria and sub-criteria for each technology were determined. With the help of the Delphi technique, 12 field experts, with the help of social network analysis, were asked to verify the criteria, sub-criteria, and alternative technologies in terms of content validation. Qualtrics survey link, which is a web-based survey tool, was sent to the experts for the validation of the model. Which criteria should or should not be included was determined by the experts. A Qualtrics questionnaire link was sent to the experts for the evaluation of energy storage technologies, where all criteria, sub-criteria, and definitions of alternative technologies were sent, and they were also asked whether there was a model created and if there were other elements that should be removed or added.

In order not to change the criterion in the model, the majority criterion of 67% and above was taken into account (Lingga 2016). In terms of all other criteria, since at least two-thirds (67%) of the experts deemed the model appropriate, the model was accepted and no changes were made. In sum, the determined criteria and sub-criteria were found suitable for newly developing energy storage technologies. In the next step, the criteria and sub-criteria determined in the model are evaluated by the expert.

3.5.2 Model Quantification

In this step, in order to evaluate the developing energy storage technologies in multiple perspectives, a pairwise comparison of the criteria, sub-criteria, and alternative technologies was made by the subject matter experts. Participants were invited to participate in the research by communicating via email with field experts discovered as a result of social network analysis or social platforms such as LinkedIn.

Although a total of 31 people participated in the expert group created for the pairwise comparison of the criteria, 15 experts completed their evaluations. Since the Delphi technique was used in this study, it was defined as a sufficient number for 15 experts to evaluate and verify the criteria in the model. The criteria according to which the experts evaluated are given in Table 3.1. Then, expert evaluations over 100 points were uploaded to HDM software by the researcher. Relative prioritization of each element was made by 15 experts. At the same time, inconsistencies and disagreements between experts above 0.10 were also determined in the analysis results below. Inconsistencies with a value greater than 0.10 were excluded from the study.

3.5.2.1 Results of the Perspectives and Criteria

As can be seen in Table 3.6, the weights of the criteria were determined after each expert evaluated five different criteria by making pairwise comparisons. Since all these values are normalized weighted values, their sum is equal to 1. The economic perspective, which has the highest average among the criteria (0.27), was evaluated by experts as the most important criterion for energy storage technologies. The prominence of the economic criterion in the evaluation of energy storage technologies is scientifically grounded in its pivotal role, influencing resource allocation, long-term sustainability, market competitiveness, regulatory compliance, and risk mitigation.

Then, in order of importance, environmental (0.22), technical (0.21), political (0.18), and social (0.12) criteria follow, respectively. The relatively lower weight assigned to the social criterion in the evaluation of energy storage technologies

Table 3.6 Weights of criteria.

Criteria	Weights
Social	0.12
Technical	0.21
Economic	0.27
Environmental	0.22
Political	0.18

can be attributed to the intrinsic focus of this particular field, where economic, technical, and environmental considerations tend to hold greater sway due to the pronounced emphasis on efficiency, cost-effectiveness, and sustainability.

The disagreement rate for the criteria was 0.082, and it was determined that there was no disagreement between expert opinions. As a result of the answers given by each expert, it was determined that there was no inconsistency because the inconsistency rate for each expert is below 10%, which is the acceptable limit.

In the context of social considerations, the provided table assigns weights to various criteria and sub-criteria, offering a structured framework for decision-making. The global weights indicate the overall importance of each criterion within the broader social category. Notably, Public Safety and Health Issues emerged as the most crucial aspect, carrying the highest global weight of 0.037. Following closely is Employment with a global weight of 0.031, emphasizing its significance in the social context. Public Acceptance holds a weight of 0.029, underscoring the importance of societal approval in decision-making processes. Lastly, Social Equity is assigned a global weight of 0.023, reflecting its role as a contributing factor within the social criteria. This hierarchy of global weights provides valuable insight into the prioritization of social considerations, aiding in informed and weighted decision-making processes.

Table 3.7. and Figure 3.5. details technical criteria, emphasizing the paramount importance of Energy Efficiency and Energy and Power Generation/Storage Capacity, each holding the highest global weights of 0.036 and 0.034, respectively. Moderately influential factors include Lifetime, Response Time, and Power Density, each contributing with global weights of 0.029. Criteria such as Maturity, Energy Density, and Power Transmission Rate carry lower global weights of 0.023, indicating their relatively lesser impact in the overall technical evaluation. This hierarchy provides a concise guide for prioritizing technical considerations based on their global weights, aiding in efficient decision-making processes.

In economic considerations, Capital Cost is identified as the most influential factor with the highest global weight of 0.070. Following closely are Operating and Maintenance Cost and R&D Expenditures Cost with global weights of 0.049 and 0.051, respectively. Other factors such as Recurrent Cost, Waste Management Cost, and End of Life/Disposal Cost contribute to the overall economic evaluation with relatively lower global weights, providing a comprehensive perspective on economic aspects for decision-making.

Within the environmental criterion, Water Pollution is identified as the most significant factor, holding the highest global weight of 0.057. Greenhouse Gas Emission follows closely with a global weight of 0.048, emphasizing the importance of these factors in the environmental assessment. Subsequent considerations include Circular Economy: Reuse/Recycling/Disposal, Wildlife Impact, and Landscape Damage, each contributing to the overall environmental

Table 3.7 Global weights of criteria and sub-criteria.

Criteria	Weights	Sub-criteria	Local weights	Global weights
Social	0.12	**Social equity**	0.19	0.023
		Employment	0.26	0.031
		Public safety and health issues	0.31	0.037
		Public acceptance	0.24	0.029
Technical	0.21	**Energy and power generation/storage capacity**	0.16	0.034
		Energy efficiency	0.17	0.036
		lifetime	0.14	0.029
		maturity	0.11	0.023
		response time	0.14	0.029
		Power density	0.11	0.023
		Energy density	0.11	0.023
		Power transmission rate	0.08	0.017
Economic	0.27	**Capital cost**	0.26	0.070
		Operating and maintenance cost	0.18	0.049
		Recurrent cost	0.13	0.035
		West management cost	0.12	0.032
		End of life/disposal cost	0.11	0.030
		R&D expenditures cost	0.19	0.051
Environmental	0.22	**Greenhouse gas emission**	0.22	0.048
		Circular economy: reuse/recycling/disposal	0.19	0.042
		Wildlife impact	0.17	0.037
		Landscape damage	0.16	0.035
		Water pollution	0.26	0.057
Political	0.18	**Energy security**	0.38	0.068
		Regulations/standards	0.23	0.041
		Government/public R&D framework	0.18	0.032
		White certificate/energy saving Certificate/energy efficiency credit	0.21	0.038

evaluation with their respective global weights, providing a comprehensive overview for decision-making concerning environmental aspects.

In the political criterion, Energy Security is identified as the most influential factor, holding the highest global weight of 0.068. Regulations/Standards and White Certificate/Energy Saving Certificate (ESC)/Energy Efficiency Credit (EEC) follow with global weights of 0.041 and 0.038, respectively. Government/Public R&D Framework contributes to the overall political evaluation with a global weight of 0.032. This hierarchy provides a clear perspective on the political considerations and their respective weights, aiding in decision-making processes related to political aspects.

In sum, the table highlights three pivotal sub-criteria with the highest global weights: Capital Cost (0.070), Water Pollution (0.057), and Energy Security (0.068). Within economic considerations, Capital Cost takes precedence, signifying its substantial impact on the overall evaluation. Simultaneously, environmental concerns are underscored by the significance of Water Pollution, reflecting the emphasis on environmental sustainability. Energy Security stands out as the most critical sub-criterion across the board, holding the highest global weight, emphasizing its paramount importance in decision-making processes. This triad of Capital Cost, Water Pollution, and Energy Security serves as a focal point for strategic considerations, providing valuable insights for comprehensive and informed decision-making.

In the comparative analysis of alternative energy storage technologies, as presented in Table 3.8., distinct strengths and weaknesses become evident. Flywheel technologies exhibit the highest weight (0.034) across all social criteria, establishing them as the leading choice in terms of social equality, employment opportunities, public safety, and public acceptability. Following closely are TEES (0.022), I-CAES (0.022), and liquid air energy storage (0.021) with their weights closely aligned. Notably, the sub-criterion of public safety and health issues, with the highest global weight (0.037) among the social criteria, holds utmost importance, particularly for flywheel (0.009) and I-CAES (0.007) technologies.

Interestingly, the underground pumped hydro storage technology emerges as a candidate for substantial development, as it displays the lowest average score among the alternatives (0.018), signaling a need for focused improvements. In the case of liquid air energy storage (0.021) and I-CAES (0.022), public safety and health issues and employment/job creation criteria have the most substantial impact. These findings underscore the preeminence of flywheel technologies in the social dimension and highlight areas necessitating further focus and investment, particularly for underground pumped hydro storage.

In this comprehensive analysis, the assigned weights to social criteria have revealed that flywheel technologies are the preferred choice among alternative energy storage options, emphasizing their strengths in promoting social equality,

Table 3.8 Pairwise comparison of criteria and sub-criteria based on alternative technologies.

Criteria	Weights	Sub-criteria	Local weights	Global weights	Alternative technologies				
					Flywheels	Liquid air energy storage	Thermoelectric energy storage	Underground pumped hydro energy storage	Isothermal compressed air energy storage
					Average of Weight				
Social	**0.12**	**Social equity**	0.19	0.0228	0.006 156	0.004 104	0.005 244	0.003 648	0.003 648
		Employment	0.26	0.0312	0.008 424	0.005 928	0.005 616	0.004 992	0.006 552
		Public safety and health issues	0.31	0.0372	0.009 672	0.007 068	0.007 068	0.00 558	0.007 812
		Public acceptance	0.24	0.0288	0.010 368	0.004 896	0.004 896	0.004 608	0.00 432
		Weights of the alternative technologies based on social criteria			**0.03 462**	**0.02 1996**	**0.022 824**	**0.018 828**	**0.022 332**
Technical	**0.21**	**Energy and power generation/storage capacity**	0.16	0.0336	0.00 504	0.007 056	0.006 384	0.0084	0.00 672
		Energy efficiency	0.17	0.0357	0.01 071	0.005 712	0.008 568	0.004 998	0.005 712
		lifetime	0.14	0.0294	0.00 4704	0.006 174	0.005 586	0.00 735	0.00 588
		maturity	0.11	0.0231	0.00 462	0.00 4389	0.004 389	0.005 775	0.003 927
		Response time	0.14	0.0294	0.00 882	0.00 4998	0.007 938	0.003 822	0.004 116
		Power density	0.11	0.0231	0.00 6468	0.003 696	0.00 462	0.00 462	0.003 465
		Energy density	0.11	0.0231	0.00 5082	0.003 927	0.005 313	0.005 082	0.003 696
		Power transmission Rate	0.08	0.0168	0.003 024	0.002 856	0.00 504	0.003 192	0.002 856

		Weights of the alternative technologies based on technical criteria			**0.04 8468**	**0.038 808**	**0.047 838**	**0.043 239**	**0.036 372**
Economic	**0.27**	**Capital cost**	0.26	0.0702	0.013 338	0.015 444	0.01 053	0.014 742	0.016 146
		Operating and maintenance cost	0.18	0.0486	0.006 804	0.009 234	0.011 178	0.010 692	0.010 206
		Recurrent cost	0.13	0.0351	0.005 967	0.005 967	0.009 828	0.007 371	0.006 318
		West management cost	0.12	0.0324	0.005 184	0.007 452	0.005 184	0.00 648	0.007 776
		End of life/disposal cost	0.11	0.0297	0.006 831	0.00 594	0.004 752	0.00 594	0.006 534
		R&D expenditures cost	0.19	0.0513	0.009 747	0.01 026	0.009 234	0.01 026	0.011 799
		Weights of the alternative technologies based on economic criteria			**0.047 871**	**0.054 297**	**0.050 706**	**0.055 485**	**0.058 779**
Environmental	**0.22**	**Greenhouse gas emission**	0.22	0.0484	0.00 968	0.010 648	0.008 228	0.009 196	0.010 648
		Circular economy: reuse/recycling/disposal	0.19	0.0418	0.008 778	0.008 778	0.007 942	0.007 524	0.00 836
		Wildlife impact	0.17	0.0374	0.008 602	0.00 6732	0.007 106	0.006 732	0.008 228
		Landscape damage	0.16	0.0352	0.008 448	0.005 632	0.005 984	0.0088	0.006 336
		Water pollution	0.26	0.0572	0.013 156	0.01 144	0.012 012	0.008 008	0.013 156
		Weights of the alternative technologies based on environmental criteria			**0.048 664**	**0.04 323**	**0.041 272**	**0.04 026**	**0.046 728**
Politic	**0.18**	**Energy security**	0.38	0.0684	0.014 364	0.010 944	0.01 368	0.017 784	0.010 944
		Regulations/standards	0.23	0.0414	0.009 936	0.007 038	0.009 522	0.006 624	0.00 828

(continued)

Table 3.8 (Continued)

Criteria	Weights	Sub-criteria	Local weights	Global weights	Alternative technologies: Flywheels	Liquid air energy storage	Thermoelectric energy storage	Underground pumped hydro energy storage	Isothermal compressed air energy storage
					Average of Weight				
		Government/public R&D framework	0.18	0.0324	0.010 692	0.005 832	0.005 832	0.00 486	0.005 508
		White certificate/energy saving certificate/energy efficiency credit	0.21	0.0378	0.009 072	0.006 426	0.007 938	0.006 048	0.008 316
		Weights of the alternative technologies based on politic criteria			**0.044 064**	**0.03 024**	**0.036 972**	**0.035 316**	**0.033 048**
					0.2237	**0.1886**	**0.1996**	**0.1931**	**0.1973**
		The ranking			**1**	**5**	**2**	**4**	**3**

employment opportunities, public safety, and general acceptability. These weighted criteria offer valuable insights for decision-makers seeking the most suitable energy storage technology for diverse applications.

In the evaluation of technical criteria, as detailed in Table 3.8, the analysis reveals distinctive strengths and weaknesses among the alternative energy storage technologies. When the average scores for each technical criterion are computed, flywheels (0.048) emerges as the leading technology, followed by TEES (0.047), underground pumped hydro (0.043), liquid air energy storage (0.038), and I-CAES (0.036).

A closer examination of these technologies reveals their specific strengths and areas for improvement. Flywheel technology demonstrates excellence in energy efficiency (0.010), and response time (0.008). Isothermal compressed air technologies excel in storage capacity (0.006), lifetime and energy efficiency (0.005), but they exhibit the lowest scores in other areas.

Liquid air energy storage technologies need enhancements in various aspects, particularly power transmission rate (0.002) and power density (0.003), whereas TEES should focus on improving maturity and power density (0.004).

These findings offer valuable insights for tailoring improvements to each alternative storage technology, allowing for a more comprehensive understanding of their technical strengths and areas requiring development.

The technical evaluation highlights the strengths and weaknesses of each alternative energy storage technology. Underground pumped hydro technology excels in energy and power capacity, lifespan, and maturity. Flywheels demonstrate remarkable energy efficiency, response time, and power density. However, flywheels have limitations in power transmission rate, while liquid air energy storage struggles with power transmission rate and power density. I-CAES needs improvement in power transmission rate. These insights can guide focused improvements in each technology for a more comprehensive understanding of their technical attributes.

When the energy storage technologies are evaluated from an economic perspective, flywheel technologies have the lowest cost (0.47) as a result of the general average, while the storage technologies with the highest costs are I-CAES (0.058), underground pump hydro (0.055) and liquid air energy storage technologies (0.054). When the economically weak and strong aspects of technologies are evaluated; end of life/disposal cost (0.004), west management cost (0.005) and R&D expenditure cost (0.009) are low for TEES technologies. However, operating and maintenance cost (0.011) and capital costs (0.010) also show the highest cost among other technologies. R&D expenditure cost (0.009), end of life/disposal cost and operating and maintenance cost (0.006) indicate low cost elements in flywheel technologies. On the other hand, the cost of capital (0.013) is costly for I-CAES technologies.

The economic analysis reveals varying costs associated with different energy storage technologies. Flywheel technologies are identified as the most cost-effective option, while TEES, liquid air energy storage, underground pump hydro, and I-CAES technologies are considered more costly. Examining cost breakdowns, it becomes evident that each technology has its economic strengths and weaknesses. These insights can guide investment decisions and highlight areas for cost optimization.

In terms of environmental criteria, flywheel technologies (0.048), I-CAES (0.046), liquid air energy storage technologies (0.043), TEES (0.041), and underground pump hydro (0.040) have the order of importance. When the strengths and weaknesses of alternative storage technologies are evaluated; while flywheel technologies have a negative impact compared to other technologies in terms of landscape damage and wildlife impact (0.008), they have a positive impact in terms of water position (0.01).

While liquid air energy storage technologies and I-CAES (0.010) has a significant impact on greenhouse gas emission/air pollution, TEES (0.008) has a higher negative impact level. Storage technologies that have little negative impact on wildlife impact and land-landscape damage are liquid air energy storage technologies (0.006–0.005). Environmental considerations are crucial, given the growing focus on sustainability. Flywheel technologies, while strong in some aspects, are found to have a negative impact on landscape damage and wildlife. In contrast, they are more aligned with water pollution principles.

As a result of evaluating the technologies from a political perspective, flywheel technologies (0.044), TEES (0.036), I-CAES (0.033), underground pump hydro (0.035), and liquid air energy storage technologies (0.030) are located in order of importance. In terms of energy security, the underground pump hydro (0.017) has high security, while liquid air energy storage technologies and I-CAES (0.10) have low energy security features. While flywheel (0.009) technologies are the leading storage technologies, Underground Pump Hydro (0.006) technologies need regulations and standards. Within the framework of the Government/Public R&D criteria, the highest supports are transferred to flywheel (0.010) technologies, while the lowest R&D supports are made to underground pump hydro (0.004) technologies. While Flywheel Technologies (0.009) does not have a problem with white certificate/ESC/EEC, it needs regulations on underground pump hydro (0.006) certification. The political evaluation ranks flywheel technologies as the most politically supported option, with underground pump hydro requiring more regulations and standards. Energy security is notably strong for underground pump hydro and weak for liquid air energy storage and I-CAES. The allocation of Government/Public R&D support also varies, with flywheel technologies receiving the highest support and underground pump hydro the lowest. These insights can inform policy makers and industry stakeholders regarding the political landscape for each technology.

In conclusion, the pairwise comparison phase of this study has provided a comprehensive understanding of the strengths and weaknesses of developing energy storage technologies across various criteria. These findings offer valuable guidance for stakeholders in making informed decisions, shaping policy frameworks, and directing future research efforts to advance the state of energy storage technologies in an ever-evolving energy landscape.

In addition to these definitions, the year-based development of each technology is defined on the basis of the next step.

3.5.3 TDE Diagram

Emerging technologies present organizations with significant challenges in understanding the evolution of these technologies over time and how these advancements impact an organization's objectives. These challenges stem from the scarcity of data in emerging fields and the complex process of integrating qualitative and quantitative factors into the decision-making process (Gerdsri 2007). For each time period, each technology with the help of experts' inputs was evaluated based on its relevant value to the organization at a specific period of time and prioritize the evaluated technologies for the organizations as shown in Figure 3.4. The technologies were evaluated based on the 27 sub-criteria validated and quantified by the subject matter experts. The TDE diagram shows the five energy storage technologies over five time periods between the years of 2021 and 2030. This presents a significant value to organizations in order to understand the value of each technology at each time period.

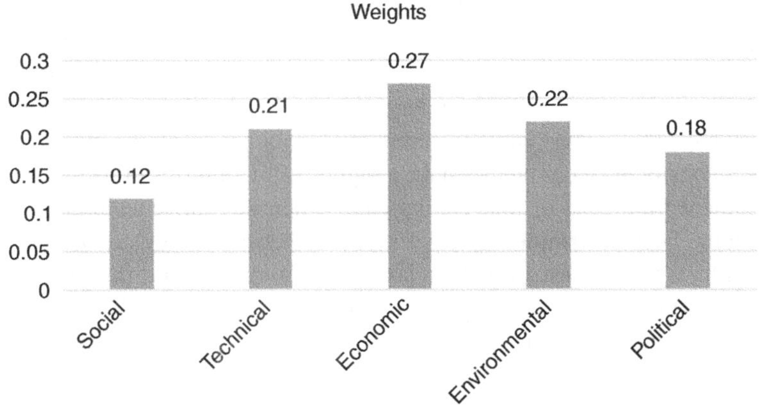

Figure 3.4 Weights of criteria.

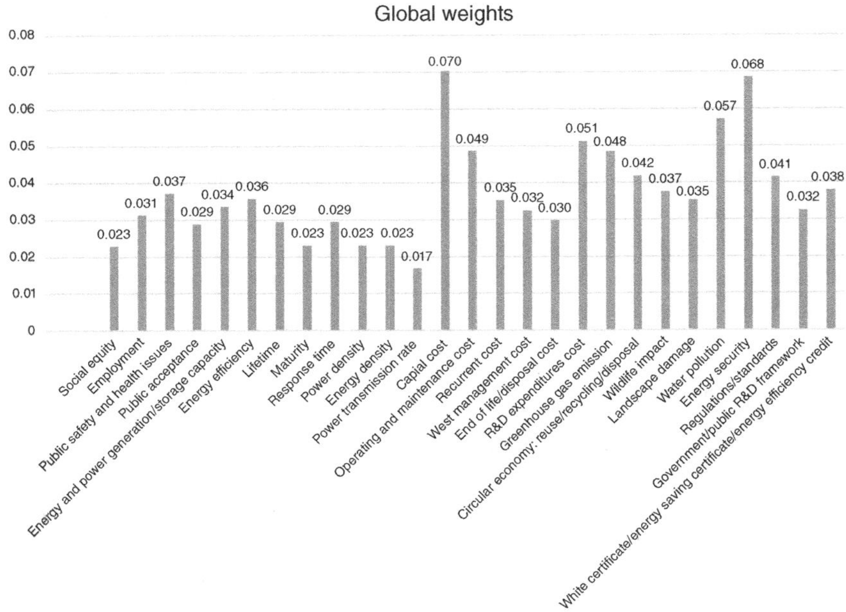

Figure 3.5 Global weights of subcriteria.

The estimated year-based development course for each alternative technology for which technology intelligence is presented in the TDE (step is defined in Figure 3.6).

As can be seen in Figure 3.6., the technological development course of all alternative emerging energy storage technologies between 2021 and 2030 has been determined by experts. Based on a total of seven expert opinions, flywheel, TEES, I-CAES, and underground pump hydro technologies show increasing technological development. However, liquid air energy storage technologies are characterized by experts that there will be a decrease in technology levels starting from 2027 to 2028. Between 2021 and 2022, underground pumped hydro energy storage technologies have the highest value in terms of technology development, while liquid air energy storage technologies have a higher value in terms of technological development from 2023 to 2026 compared to other alternative technologies. As of 2027, it has been determined by experts that isothermal energy storage technologies will have the highest value by a large margin. As a result, the TDE for all alternative technologies was defined by experts on a yearly basis. Thus, it can be defined which of the alternative technologies will develop or have a stagnant course. Thus, decision-makers can make their investments by considering which technologies are more valuable in technical terms.

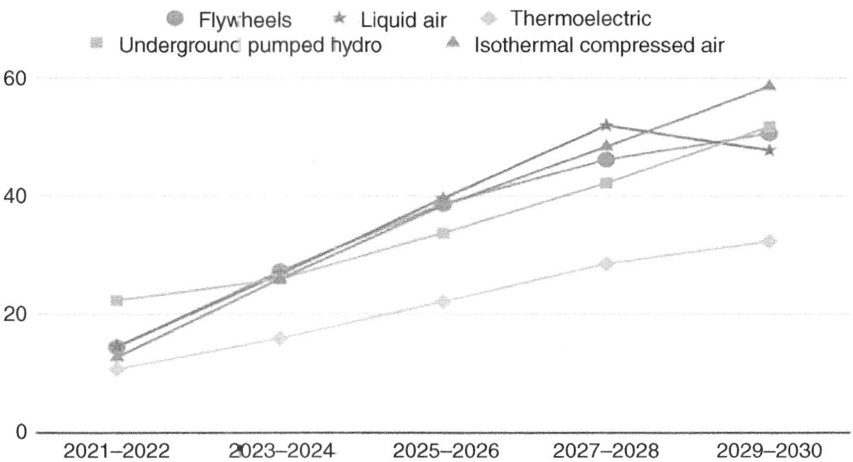

Figure 3.6 TDE diagram.

3.6 Conclusion

As a result, it has been determined as a result of expert opinions that among the social, technical, economic, environmental, and political criteria defined in order to evaluate the developing energy storage technologies in detail, the economic criterion is the most important factor in evaluating energy storage technologies. Then, it was defined as a result of expert answers that these technologies should be evaluated according to environmental, technical, political, and social factors, respectively. The main reason why the economic criterion has the highest degree of importance is that these technologies have investments that require very high costs. Because energy storage technologies include many cost elements rather than a single cost element: capital cost, operation and maintenance cost, repetitive cost, life cycle/disposal cost, and R&D expenditure costs. On the other hand, the criterion with the lowest degree of importance is defined as the social criterion. It has been understood as a result of expert opinions that social inequality, employment, public safety, health problems, and general acceptability are the last factors to be considered in the evaluation of energy storage technologies.

As a result of the pairwise comparison of technologies, flywheel technologies have the highest degree of importance among other alternative technologies. Then, respectively, liquid air energy storage technologies and underground pumped hydro technologies have the same importance, while the technologies with the lowest importance are TEES and I-CAES technologies.

When alternative energy storage technologies are evaluated in more detail on the basis of the determined criteria, the weaknesses and strengths of the technologies are determined. When the weaknesses and strengths are evaluated from a social perspective, flywheel technologies are among the most mature energy storage technologies in terms of social equality, employment creation, public safety and health problems, and general acceptability, while underground pumped hydro energy storage technology is the alternative technology that needs development most in these areas.

Since the technical perspective is among the most important criteria for energy storage technologies, these technologies are discussed separately from a technical perspective. As a result of expert evaluations, while TEES technologies are technically adequate, I-CAES technologies require more technical infrastructure. When the technical weaknesses and strengths of all technologies are considered separately, the flywheel technologies have high performance in terms of power density, but these technologies have not reached a sufficient level in terms of maturity. While liquid air energy storage technologies are of high importance in terms of life span and energy – power generation/storage capacity, the sub-criteria of energy efficiency and power density are of low importance as a result of the technical evaluation. While the strength of TEES technologies is fast response time, the lack of power density shows the weakness of these technologies. Another energy storage technology, underground pumped hydro technology, is technically strong in its energy and power generation/storage capacity, lifespan, and maturity characteristics. In terms of power density, it is necessary to develop these technologies in order to increase the power density by defining the underground pumped hydro as a result of expert evaluations that it does not have sufficient maturity. I-CAES technologies, which are among the alternative technologies, have reached a certain level of maturity and have a high power transmission rate, but it has been determined as a result of the evaluations of the experts that energy and power generation/storage capacity and life span are among their weaknesses. If the weaknesses of each technically defined technology are taken into consideration and the necessary investment evaluations are made, companies will be able to provide a competitive advantage.

The economic criterion has been determined by field experts as the most important criterion for energy storage technologies. At this point, the cost of flywheel technologies has been determined to be lower compared to other alternative storage technologies, while the cost of I-CAES technologies has been defined as higher. On the basis of economic sub-criteria, Flywheel technologies have a high lifetime/disposal cost, while low usage and maintenance costs. Recurring costs for I-CAES technologies, which were determined to have the highest cost as a result of expert evaluation, are at a reasonable level, while waste management costs are high. Liquid air energy storage technologies, on the other hand, show the opposite feature compared to I-CAES technologies, and it can be interpreted that while the waste management cost is high, the repetitive cost is low. The cost items for which TEES and underground pump hydro technologies are costly, respectively, are repetitive costs and operating and maintenance costs. Since energy storage technologies are very costly, it is necessary to invest in all these technologies, taking into account the low and high cost elements presented above as intelligence.

When all alternative technologies are evaluated from an environmental perspective, it has been determined that the energy storage technology that causes the most environmental pollution is the flywheel technologies, while the technology that causes the least environmental damage is the TEES technology. Flywheel and underground pumped hydro technology cause the most land and landscape damage, while liquid air energy storage, TEES and I-CAES technologies all cause the most water pollution. Considering the lowest environmental damage caused by alternative technologies, flywheel and TEES cause low environmental damage at the point of greenhouse gas emission/air pollution. In terms of land and landscape damage, there are also TEES and I-CAES technologies. In terms of water pollution, underground pumped hydro technology is the technology with the lowest environmental damage.

The last of the evaluation criteria is the political perspective. As a result of the evaluation of alternative technologies from a political perspective, flywheel technologies are the most equipped technology compared to other alternative technologies in terms of energy security, regulations/standards, state public R&D framework and white certificate, ESC and EEC required by energy storage technologies. Within the framework of these political criteria, the alternative technology with the lowest importance is liquid air energy storage technology. This technology needs regulation on standards from a political point of view.

This research makes substantial theoretical contributions by introducing a comprehensive evaluation framework through the application of the HDM. By including diverse dimensions such as social, technical, economic, environmental, and political factors, the study enhances the theoretical foundations for assessing emerging energy storage technologies. The reliance on expert opinions adds depth to the theoretical framework, providing nuanced insights into the strengths and weaknesses of various technologies. Pairwise comparison techniques within the HDM framework contribute to systematic priority setting, particularly emphasizing the economic criterion as pivotal in the evaluation process. The technology-specific analysis distinguishes this study, offering a more tailored theoretical understanding that acknowledges the diversity within the energy storage landscape.

From a practical perspective, this research offers valuable guidance for policymakers and investors involved in renewable energy and energy storage. The prioritization of economic considerations and detailed cost analyzes of different technologies provide practical insights for formulating effective policies and investment strategies. The technology-specific development pathways identified in the study offer practical guidance for researchers, engineers, and industry professionals seeking to address specific weaknesses and enhance the overall maturity of energy storage technologies. The assessment of environmental impacts contributes practically by highlighting areas of concern and suggesting mitigation strategies, informing industry practices and regulatory frameworks. The evaluation from a political perspective provides practical insights for policymakers, aiding in the development of regulatory frameworks, standards, and incentives to create a conducive political environment for energy storage technology deployment.

In conclusion, this research successfully bridges the gap between theory and practice, enriching academic discourse while offering actionable insights for stakeholders in the renewable energy landscape. Theoretical contributions encompass a comprehensive evaluation framework, technological forecasting theories, and an innovative approach to technology evaluation. On the practical front, stakeholders are armed with actionable intelligence to guide policies, direct R&D efforts, and make optimal technology investments, ultimately steering toward sustainable energy futures.

The study acknowledges a limitation in the process of engaging experts, which may have implications for the diversity and representativeness of opinions incorporated. The challenge in finding experts could potentially introduce biases and impact the generalizability of findings. Future research efforts may benefit from addressing this limitation by employing strategies to enhance expert participation and ensure a broader spectrum of perspectives in the evaluation of emerging energy storage technologies.

Acknowledgments

This study was supported by the Scientific and Technological Research Council of Turkey (TUBITAK). We are sincerely grateful for TUBITAK 2214-A support.

References

Abbas, M.S. and Kocaoglu, D.F. (2016). Consistency thresholds for hierarchical decision model. In: *2016 Portland International Conference on Management of Engineering and Technology (PICMET)*, 566–575. IEEE.

Abotah, R. (2014). *Evaluation of Energy Policy Instruments for the Adoption of Renewable Energy: Case of Wind Energy in the Pacific Northwest US.*

Akinyele, D.O. and Rayudu, R.K. (2014). Review of energy storage technologies for sustainable power networks. *Sustainable Energy Technologies and Assessments* 8: 74–91.

An, B., Chen, J., Deng, Z. et al. (2020). Design and testing of a high performance liquid phase cold storage system for liquid air energy storage. *Energy Conversion and Management* 226: 113520.

Aneke, M. and Wang, M. (2016). Energy storage technologies and real life applications–A state of the art review. *Applied Energy* 179: 350–377.

Ayachi, F., Tauveron, N., Tartière, T. et al. (2016). Thermo-electric energy storage involving CO2 transcritical cycles and ground heat storage. *Applied Thermal Engineering* 108: 1418–1428.

Barham, H.A. (2019). *Development of a Readiness Assessment Model for Evaluating Big Data Projects: Case Study of Smart City in*. Portland, OR, USA: Portland State University.

Bherde, R., Palandurkar, M., and Tarnekar, S. (2018). Ultra-low frequency low power oscillator using ultra-capacitor and advanced flywheel. In: *2018 International Conference on Smart Electric Drives and Power System (ICSEDPS)*, 18–21. IEEE.

Chiesa, M. and Franzò, S. (2015). The economic evaluation of energy efficiency in industry: an innovative methodology. In: *ACEEE Summer Study Energy Efficiency in Industry*, 1–11. ACEEE.

Cho, Y., Shaygan, A., and Daim, T. (2019). Energy technology adoption: Case of solar photovoltaic in the Pacific Northwest USA. *Sustainable Energy Technologies and Assessments* 34 (August): 187–199.

Cleland, D.I. and Kocaoglu, D.F. (1981). *Engineering Management*. McGraw-Hill Companies.

Çolak, M. and Kaya, İ. (2020). Multi-criteria evaluation of energy storage technologies based on hesitant fuzzy information: a case study for Turkey. *Journal of Energy Storage* 28: 101211.

Daim, T.A.I.M. et al. (2011). Technology development envelope approach for the adoption of future powertrain technologies: a case study on ford otosan roadmapping model. *Journal of Transportation Systems Engineering and Information Technology* 11 (2): 58–69.

Daim, T.U., Li, X., Kim, J., and Simms, S. (2012). Evaluation of energy storage technologies for integration with renewable electricity: quantifying expert opinions. *Environmental Innovation and Societal Transitions* 3: 29–49.

Daim, T.U., Lim, D., Gomez, F.A. et al. (2014). Storage technologies for wind power in the Columbia river gorge. *International Journal of Sustainable Energy* 33 (1): 1–15.

Daim, T.U., Yoon, B.S., Lindenberg, J. et al. (2018). Strategic roadmapping of robotics technologies for the power industry: a multi criteria technology assessment. *Technological Forecasting and Social Change* 131 (June): 49–66.

Dib, G., Haberschill, P., Rullière, R., and Revellin, R. (2021). Thermodynamic investigation of quasi-isothermal air compression/expansion for energy storage. *Energy Conversion and Management* 235: 114027.

Estep, J. (2017). Development of a technology transfer score for evaluating research proposals: Case study of demand response technologies in the Pacific Northwest. Doctoral dissertation. Portland State University.

Garces, E., Daim, T.U., and Dabić, M. (2021). Evaluating R&D projects in regulated utilities: the case of power transmission utilities. *IEEE Transactions on Engineering Management*.

Gerdsri, N. (2005, DF,). An analytical approach on building a technology development envelope (TDE) for roadmapping of emerging technologies. In: *Technology Management: A Unifying Discipline for Melting Boundaries* (ed. T.R. Anderson, T.U. Daim, and Kocaoglu), 123–135. Piscataway NJ: IEEE.

Gerdsri, N. (2007). An analytical approach to building a technology development envelope (TDE) for roadmapping of emerging technologies. *International Journal of Innovation and Technology Management* 4: 121–135.

Gerdsri, N. and Kocaoglu, D.F. (2003). An analytical approach on building a technology development envelope (TDE) for roadmapping of emerging technologies: A case study of emerging electronic cooling technologies for computer servers. In: *Technology Management for Reshaping the World* (ed. D.F. Kocaoglu and T.R. Anderson), 380–387. Piscataway NJ: IEEE.

Gibson, E., Daim, T., and Dabic, M. (2019). Evaluating university industry collaborative research centers. *Technological Forecasting and Social Change* 146 (September): 181–202.

González, E.L., Llerena, F.I., Pérez, M.S. et al. (2015). Energy evaluation of a solar hydrogen storage facility: comparison with other electrical energy storage technologies. *International Journal of Hydrogen Energy* 40 (15): 5518–5525.

Guo, Z., Ge, S., Yao, X. et al. (2020). Life cycle sustainability assessment of pumped hydro energy storage. *International Journal of Energy Research* 44 (1): 192–204.

Hadjipaschalis, I., Poullikkas, A., and Efthimiou, V. (2009). Overview of current and future energy storage technologies for electric power applications. *Renewable and Sustainable Energy Reviews* 13 (6–7): 1513–1522.

Hall, P.J. and Bain, E.J. (2008). Energy-storage technologies and electricity generation. *Energy Policy* 36 (12): 4352–4355.

Hannan, M.A., Hoque, M.M., Mohamed, A., and Ayob, A. (2017). Review of energy storage systems for electric vehicle applications: Issues and challenges. *Renewable and Sustainable Energy Reviews* 69: 771–789.

Hwang, C.L. and Masud, A.S.M. (1979). Methods for multiple objective decision making. In: *In Multiple Objective Decision Making—Methods and Applications*, 21–283. Berlin, Heidelberg: Springer.

Iskin, I. and Daim, T. (2016). An assessment model for energy efficiency program planning in electric utilities: case of Northwest U.S. *Sustainable Energy Technologies and Assessment* 15: 42–59.

Khanam, M. and Daim, T. (2021). A market diffusion potential (MDP) assessment model for residential energy efficient (EE) technologies in the US. *Renewable and Sustainable Energy Reviews* 144: 110968.

Kim, J., Suharto, Y., and Daim, T.U. (2017). Evaluation of electrical energy storage (EES) technologies for renewable energy: a case from the US Pacific northwest. *Journal of Energy Storage* 11: 25–54.

Kocaoglu, D.F. (1983). A participative approach to program evaluation. *IEEE Transactions on Engineering Management* 3: 112–118.

Kocaoglu, D.F., Daim, T.U., Iskin, I., and Alizadeh, Y. (2016). Technology assessment: criteria for evaluating a sustainable energy portfolio. In: *Hierarchical Decision Modeling*, 3–34. Cham: Springer.

Kozak, M. and Kozak, Ş. (2012). Enerji depolama yöntemleri. *Uluslararası Teknolojik Bilimler Dergisi* 4 (2): 17–29.

Li, L., Wang, B., Jiao, K. et al. (2023). Comparative techno-economic analysis of large-scale renewable energy storage technologies. *Energy and AI* 100282.

Lingga, M.M. (2016). Developing a hierarchical decision model to evaluate nuclear power plant alternative siting technologies.

Liu, S. and Jiang, J. (2019, July). Technical evolution of advanced flywheel energy storage system. In: *2019 Chinese Control Conference (CCC)*, 6588–6593. IEEE.

Luo, X., Wang, J., Dooner, M., and Clarke, J. (2015). Overview of current development in electrical energy storage technologies and the application potential in power system operation. *Applied Energy* 137: 511–536.

Mahlia, T.M.I., Saktisahdan, T.J., Jannifar, A. et al. (2014). A review of available methods and development on energy storage; technology update. *Renewable and Sustainable Energy Reviews* 33: 532–545.

Maruf, M.N.I., Morales-España, G., Sijm, J. et al. (2022). Classification, potential role, and modeling of power-to-heat and thermal energy storage in energy systems: a review. *Sustainable Energy Technologies and Assessments* 53 (Part B): 102553.

Mongird, K., Viswanathan, V.V., Balducci, P.J. et al. (2019). *Energy Storage Technology and Cost Characterization Report* (No. PNNL-28866). Pacific Northwest National Lab: (PNNL), Richland, WA (United States).

Morandin, M., Henchoz, S., and Mercangöz, M. (2011). Thermo-electrical energy storage: a new type of large scale energy storage based on thermodynamic cycles. *World Engineers' Convention*, Geneva.

Olabi, A.G., Onumaegbu, C., Wilberforce, T. et al. (2021). Critical review of energy storage systems. *Energy* 214: 118987.

Rohit, A.K. and Rangnekar, S. (2017). An overview of energy storage and its importance in Indian renewable energy sector: part II–energy storage applications, benefits and market potential. *Journal of Energy Storage* 13: 447–456.

Saaty, T.L. (1977). A scaling method for priorities in hierarchical structures. *Journal of Mathematical Psychology* 15 (3): 234–281.

Saaty, T.L. (1980). *The Analytic Hierarchy Process*. McGraw Hill International.

Saaty, T.L. (1997). That is not the analytic hierarchy process: what the AHP is and what it is not. *Journal of Multi-Criteria Decision Analysis* 6 (6): 324–335.

Sheikh, N.J. (2013). Assessment of solar photovoltaic technologies using multiple perspectives and hierarchical decision modeling.

Shirinda, K., Kusakana, K., and Koko, S.P. (2020). Optimal power dispatch of a grid-connected photovoltaic with groundwater pumped-hydro storage system supplying a farmhouse. In: *2020 International Conference on Smart Grid and Clean Energy Technologies (ICSGCE)*, 5–9. IEEE.

Stutz, H.H., Norlyk, P., Sørensen, K. et al. (2020). Finite element modeling of an energy-geomembrane underground pumped hydroelectric energy storage system. In: *E3S Web of Conferences*, vol. 205, 07001. EDP Sciences.

Tetteh, S., Yazdani, M.R., and Santasalo-Aarnio, A. (2021). Cost-effective electro-thermal energy storage to balance small-scale renewable energy systems. *Journal of Energy Storage* 41: 102829.

Van Blommestein, K.C. and Daim, T.U. (2013). Residential energy efficient device adoption in South Africa. *Sustainable Energy Technologies and Assessments* 1: 13–27.

Vinnemeier, P., Wirsum, M., Malpiece, D., and Bove, R. (2016). Integration of heat pumps into thermal plants for creation of large-scale electricity storage capacities. *Applied Energy* 184: 506–522.

Yapicioglu, A. and Dincer, I. (2023). A newly developed renewable energy driven multigeneration system with hot silica sand storage for power, hydrogen, freshwater and cooling production. *Sustainable Energy Technologies and Assessments* 55: 102938.

Zhang, Z., Ding, T., Zhou, Q. et al. (2021). A review of technologies and applications on versatile energy storage systems. *Renewable and Sustainable Energy Reviews* 148: 111263.

Zhao, H., Guo, S., and Zhao, H. (2019). Comprehensive assessment for battery energy storage systems based on fuzzy-MCDM considering risk preferences. *Energy* 168: 450–461.

Zhao, P., Lai, Y., Xu, W. et al. (2020). Performance investigation of a novel near-isothermal compressed air energy storage system with stable power output. *International Journal of Energy Research* 44 (14): 11135–11151.

Zhu, Z., Jiang, T., Ali, M. et al. (2022). Rechargeable batteries for grid scale energy storage. *Chemical Reviews* 122 (22): 16610–16751.

4

Technology Forecasting: A Secure Solar Power Generation Forecasting Framework for Recurrent Neural Networks

Murat Kuzlu[1], Salih Sarp[2], Ferhat O. Catak[3], Umit Cali[4], and Yanxiao Zhao[2]

[1] *Batten College of Engineering & Technology, Old Dominion University, VA 23507, USA*
[2] *Electrical & Computer Engineering Department, Virginia Commonwealth University, VA 23284, USA*
[3] *Electrical Engineering & Computer Science, University of Stavanger, Stavanger, Rogaland, Norway*
[4] *School of Physics, Engineering and Technology, University of York, Heslington, York, YO10 5DD, UK*

4.1 Introduction

Digital decarbonization, using digital technologies to minimize carbon emissions, is becoming more crucial in the battle against climate change. Digital decarbonization uses machine learning (ML) and other AI methods to anticipate renewable energy supply and production. This enhances renewable energy consumption and integrates it into the energy system through renewable energy forecasting. By precisely forecasting wind and solar power production, utilities may better manage energy supply and demand, which decreases the need for nonrenewable backup power. Renewable energy forecasting uses ML algorithms, data analysis, and sophisticated sensors and monitoring systems. These technologies allow utilities and energy firms to gather and analyze weather, energy consumption, and renewable energy data to better estimate future energy demands.

Furthermore, stability, safety, and reliability of the grid are three crucial factors that must be satisfied to be able to provide accessible and risk-free electrical power to all people at all times (Hauser et al. 2005). However, the task is more challenging and complex than expected. Distributed energy resources (DER), i.e. renewable energy and energy storage technologies, have a profound impact on the way electricity grids operate (US DoE FEMP 2002). The solar photovoltaic (PV) capacity in the United States reached 113.5 GWdc in the third quarter of 2021 after residential solar installations contributed an additional 5.4 GWdc to the total. Solar energy is now the most promising alternative energy resource available in comparison to other forms of fuel. However, solar energy is dependent on weather and thus cannot be reliably sent. This makes solar power unstable (Guizzi et al. 2015).

Future-Oriented Technology Assessment: A Manager's Guide with Case Applications, First Edition.
Edited by Haydar Yalçın and Tugrul U. Daim.

Therefore, accurate forecasting of solar power generation is essential to provide reliable information about power generation to ensure the safe functioning of the power system. There is significant scientific research focusing on AI-based solar power forecasting models and other applications closely related to this field. In the smart grid setting, AI-based technologies have seen widespread use for the purpose of alleviating concerns about solar integration through the provision of accurate forecasting of power generation (Kuzlu et al. 2020; Mellit 2008; Wan et al. 2015).

Time series solar power datasets are often used by AI-based models and can be defined as a collection of solar power data points gathered in a certain order over a given amount of time (usually 1 minute, 1 hour, 1 day, etc.) (Sorkun et al. 2017). As a means of making accurate forecasts, time series analysis requires an in-depth familiarity with many elements of series' fundamental properties.

Using support vector regression (SVR), polynomial regression, and Lasso, Alfadda et al. (2017) came up with several models to forecast solar PV power 1 hour in advance. From their results, the SVR forecasting model is more accurate than the other two. Recent improvements to deep learning (DL) models based on long- and short-term memory (LSTM) show promise for forecasting time-series solar generation.

In Zhang et al. (2018), the authors suggest solutions to hybrid algorithms that use DL and artificial neural network (ANN) algorithms such as deep belief networks (DBN), auto encoder, and LSTM to improve the accuracy of the model. The results show that LSTM performs better at making predictions than ANN and other reference models, such as physical models. With adversarial training, they look at both white-box attacks and black-box attacks. Based on the results, adversarial attacks can do black-box attacks and pose a serious threat to regression problems with similar data structures. In Chitalia et al. (2020), researchers offer a reliable framework for predicting short-term electrical demand in commercial structures. The study looks at the nine RNN-based model hybrids currently in use. The models created, as shown by the findings, improve the forecast of hours of load by 30%. There is a lack of recent research on the vulnerabilities of models to cyberattacks, particularly adversarial attacks.

This chapter delves into RNN-based models used to forecast solar PV power production and assesses how adversarial attacks affect model performances. Using a safe solar forecasting framework, the study reveals that AI-based algorithms can be tricked by adversarial assaults into making inaccurate forecasts. Contrarily, the research shows that adversarial training may successfully protect RNN-based models of solar power production from such cyberattacks. This study makes three main contributions: (i) the development of a secure solar generation framework hosting adversarial attack and training models, (ii) the creation of RNN-based forecasting models (i.e. simple RNN, LSTM, BiLSTM, GRU, and LSTM/– BiLSTM with attention), and (iii) an evaluation of the performance of the developed RNN-based models under adversarial attacks with and without adversarial training.

The remainder of the chapter is structured as follows. The proposed secure solar power forecasting framework is described in Section 4.2, and the DL approaches employed in this research are described in Section 4.3. Section 4.4 covers adversarial attack and adversarial training techniques, followed by the explanation of the dataset and performance measures in Section 4.5. Section 4.6 contains case studies using RNN-based solar forecasting algorithms with adversarial attack and training. Section 4.7 summarizes the findings, and Section 4.8 wraps up the chapter.

4.2 Proposed Secure Solar Power Generation Forecasting Framework

The proposed secure solar power forecasting framework is shown in Figure 4.1. The framework includes the following steps:

1) *Application Selection:* The framework supports forecasting and cybersecurity applications. The forecasting application covers only solar power forecasting, while the cybersecurity one covers adversarial attack and training methods.

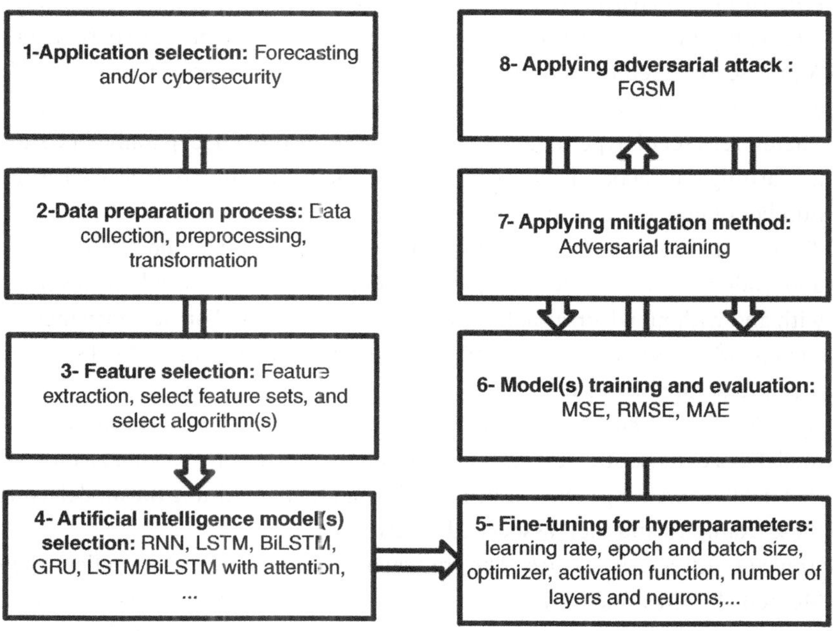

Figure 4.1 The proposed secure solar power generation forecasting framework.

2) *Data Preparation:* This process begins by obtaining data either from publicly accessible sources or user uploads. The collected data is then refined through pre-processing to ensure accurate analysis and decision-making. This includes addressing missing values, identifying outliers, and normalizing the data.
3) *Feature Selection:* This step includes selecting the feature sets to be used for a solar power forecasting model. The main objective of this step is to remove redundant information, reduce the input data dimension, and combat overfitting problems by focusing only on the relevant features. In this study, the input vector, i.e. X, consists of 13 features, includes weather data (X_w), historical solar power generation data (X_s), and time index as an hour (X_t), while (y) is the output vector referring to the forecasting solar power generation. The selection of features involves two internal steps: (i) using Pearson's correlation to determine the most relevant features among all input features and (ii) running the selected model(s) with different rolling window sizes ($m = 1$–24), that is, the number of consecutive observations per rolling window, to determine the optimum rolling window size that offers the best forecasting performance.
4) *Artificial Intelligence Model(s) Selection:* Six different RNN-based solar power generation forecasting models are embedded in the framework, i.e. simple-RNN, LSTM, BiLSTM, GRU, and LSTM/BiLSTM with attention.
5) *Fine-Tuning for Hyperparameters:* The performance of a solar power generation forecasting model is significantly dependent on the model hyperparameters. A grid search method determines the best hyperparameters on each possible combination. However, it requires too much time and a computing process. Therefore, a set of hyperparameters is used to tune the model parameters, such as learning rate, epoch/batch size, optimizer, activation function, as well as the number of layers and neurons.
6) *Training and Evaluating the Model(s):* The solar forecasting models are trained and evaluated to measure their performance against adversarial attacks, both with and without adversarial training. Evaluation is based on accuracy metrics such as Mean Squared Error (MSE), Root Mean Squared Error (RMSE), and Mean Absolute Error (MAE).
7) *Applying Mitigation Method:* The framework includes an adversarial training model as a defense strategy. Adversarial training is a widely recognized method that involves creating adversarial examples and then retraining the model using these examples and their associated labels.
8) *Applying Adversarial Attack:* The framework features the Fast Gradient Sign Method (FGSM), one of the earliest and most widely used adversarial attack models, which aims to deceive models by providing craftily manipulated input with a slight difference.

The proposed framework allows us to predict solar power generation using RNN-based forecasting models and only includes the FGSM-based attack and adversarial training-based mitigation method in its current version. In the future, we plan to expand our framework to support a wider range of datasets, AI models, adversarial attacks, and mitigation techniques for various applications and tools, including classification, anomaly detection, and explainable AI (XAI).

4.3 Deep Learning Techniques

This study aims to utilize various types of RNNs, including simple RNN, LSTM, BiLSTM, GRU, and LSTM/BiLSTM, with attention to predicting solar power generation and analyze the effects of adversarial attacks on these models for solar power generation forecasting.

4.3.1 Recurrent Neural Networks (RNN)

RNNs, originally proposed by David Rumelhart in 1986 (Hf et al. 2020), are designed to process sequential data and learn patterns to make predictions. Sequential data can take various forms, such as text, audio, and video. In an RNN, the layers of the network are directly associated with positions in the sequence, typically represented by time-stamps. These layers are then repeated over time. Compared to a feed-forward network, which only considers the current input, an RNN takes into account both the current input and what has been learned from previous inputs. Recurrent neural networks and feed-forward neural networks are shown in Figure 4.2. Figure 4.3 also illustrates the structure of an RNN cell.

The governing equations of an RNN model are shown in Eq. (4.1) with present x_t and past states h_{t-1}.

$$
\begin{aligned}
h_t &= \sigma_h\left(i_t\right) \\
h_t &= \sigma_h\left(U_h x_t + V_h h_{t-1} + b_h\right) \\
y_t &= \sigma_y\left(o_t\right) \\
y_t &= \sigma_y\left(W_y h_t + b_h\right)
\end{aligned}
\tag{4.1}
$$

where

x_t: input vector
i_t: input gate vector
h_t: hidden layer vector
o_t: output vector
b_h: bias vector
U, W, V: parameter matrices

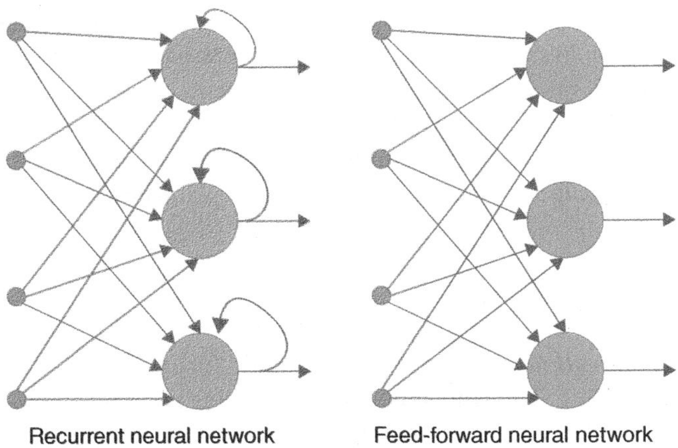

Figure 4.2 Recurrent neural networks and Feed-forward neural networks.

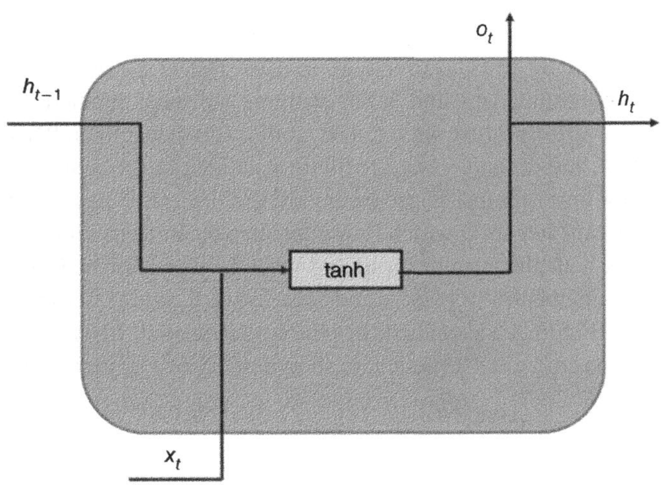

Figure 4.3 RNN cell structure.

σ_h and σ_y: activation functions
y_t: output vector

When training an RNN network, backpropagation is utilized over time. At each step during training, the gradient is calculated and used to update the network's weights. If the effect of the previous layer on the current layer is expected to be small, the gradient value will be small, and vice versa. The gradient of the previous layer may also be smaller, causing the gradients to shrink exponentially as backpropagation progresses. As a result, the network may not learn the impact of previous inputs, leading to short-term memory problems or vanishing gradients. This problem was resolved with the introduction of the concept of LSTM by Sepp Hochreiter and Juergen Schmidhuber (Jiang et al. 2018). Another issue that may arise during training is exploding gradients, which occurs when the derivatives or slopes become larger and larger as it goes backward with every layer during backpropagation – the exact opposite of vanishing gradients.

4.3.2 Long Short-term Memory (LSTM)

LSTM is a variant of RNN that is effective in predicting inventory values based on historical data. Introduced in 1997 by Hochreiter and Schmidhuber (Gers et al. 2002), LSTM is designed to address the issue of long-term dependency in RNN, where gradients propagated over multiple time steps tend to vanish or explode during training, making it challenging for the model to store long time steps in its memory (Goodfellow et al. 2016; Yu et al. 2019). To achieve more accurate trend predictions, the LSTM model depends on longer time steps. LSTMs are particularly suitable for modeling sequence data, such as time series, as they maintain an internal state that keeps track of data that have already been observed.

LSTMs use gates that can open and close to store, write, and read information. The memory cell, c_t, is a key feature of LSTMs, which functions as a container for state information. Unlike RNNs, LSTMs have multiple gates, including the forget, input, and output gates, which determine the signals that are passed on to the next node. The forget gate, f_t, leverages sigmoid activation, σ, to determine which information should be prioritized and which can be ignored. The input gate, i_t, determines when new data should be read into the memory cell. When the input gate, i_t, is activated, new information is accumulated in the cell. The output gate, o_t, determines the value of the subsequent hidden state, h_t, which includes information from previous inputs. A significant advantage of using memory cells, c_t, and gates to manage information flow is that the gradient is trapped in the cell, preventing it from vanishing too quickly, which is a crucial challenge for vanilla RNN models. Figure 4.4 illustrates the structure of an LSTM cell. The governing equations of an LSTM cell are shown in (Eq. 4.2) (The notation "$\odot$" represents the Hadamard product, which refers to the element-wise multiplication of matrices).

$$
\begin{aligned}
i_t &= \sigma\left(x_t U^i + h_{t-1} W^i + b_i\right)\\
f_t &= \sigma\left(x_t U^f + h_{t-1} W^f + b_f\right)\\
o_t &= \sigma\left(x_t U^o + h_{t-1} W^o + b_o\right)\\
\hat{C}_t &= tahn\left(x_t U^g + h_{t-1} W^g + b_c\right)\\
C_t &= f_t \odot C_{t-1} + i_t \odot \hat{C}_t)\\
h_t &= o_t \odot tanh\left(C_t U^o\right)
\end{aligned}
\tag{4.2}
$$

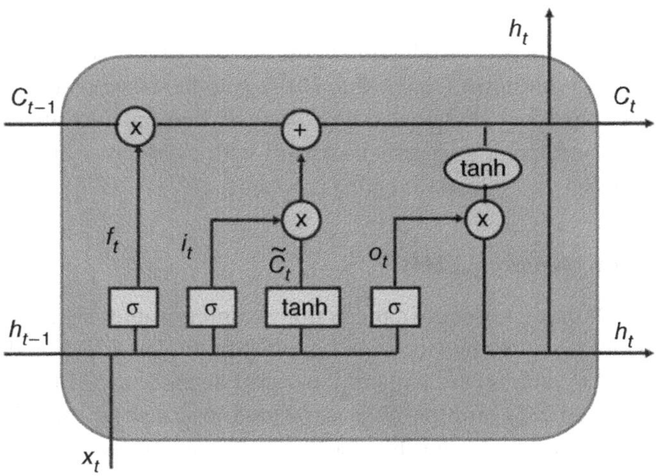

Figure 4.4 LSTM cell structure.

Let i_t be the input gate vector, c_t the cell state vector, f_t the forget gate vector, o_t the output gate vector, h_t the hidden state vector, x_t represent the input vector, and b_f, b_i, b_o, b_c the bias vectors. Additionally, let U, W denote parameter matrices, and σ_h, σ_y denote the activation functions.

4.3.3 Bidirectional LSTM

The bidirectional LSTM (BiLSTM) is an enhancement to the conventional LSTM approach that aims to improve model performance. It allows a neural network to store sequence information in both directions, i.e. backward and forward, by running the input in both directions. When all timesteps of the input sequence are available, two LSTMs are trained on the input sequence in bidirectional LSTMs. The first LSTM processes the input sequence as it is, while the second processes a reversed copy of the input sequence. This enables the network to gather additional context and learn more comprehensively and efficiently, resulting in faster and more thorough learning. Previous research has demonstrated the superior performance of BiLSTM networks in comparison to traditional LSTM models, especially for classification and regression tasks (Li et al., 2020). The BiLSTM architecture, which comprises a forward and a backward layer, is depicted in Figure 4.5. The two hidden layers of LSTM are connected to the output layer, allowing for improved learning of long-term dependencies and overall model performance enhancement. An unfolded BiLSTM layer consists of both a forward and a backward LSTM layer.

The BiLSTM layer output sequence is shown in the Figure 4.5, where h represents the forward LSTM layer output sequence obtained in the usual way as

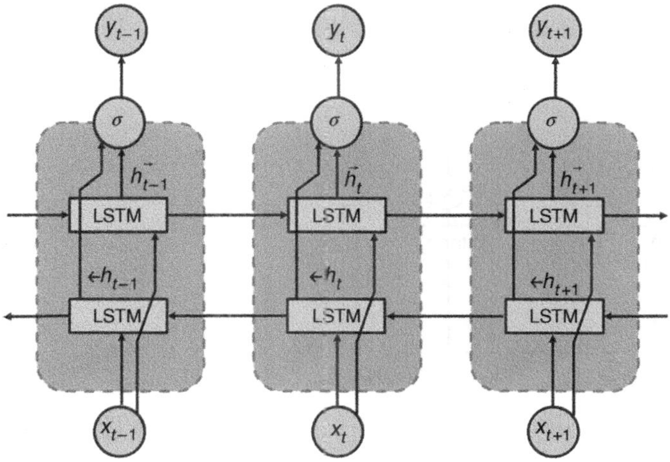

Figure 4.5 The bidirectional LSTM (BiLSTM) architecture with three consecutive steps.

unidirectional LSTM. On the other hand, *h* represents the backward LSTM layer output sequence, which is calculated by reversing the input sequence from time *t-1* to *t-k*. The output of both layers is fed into a sigmoid activation function σ to combine them into an output vector y_t. The final output of the BiLSTM layer can be represented by a vector Y_t, which consists of elements from y_{t-1}, the first element of the vector, to y_{t-k}, the last element of the vector.

4.3.4 Gated Recurrent Units (GRU)

The gated recurrent unit (GRU) is a variant of the LSTM model that was proposed by Kyunghyun Cho et al. in 2014 (Cho et al. 2014). It aims to simplify the LSTM architecture by eliminating the need for an explicit cell state. The GRU uses a single reset gate, whereas the LSTM uses separate forget and output gates. Due to its simplified architecture, the GRU is faster in terms of computation. It consists of two gates, the update gate and the reset gate, which determine whether the cell state should be updated with the candidate state and whether the previous cell state is important, respectively. The final cell state is passed directly from the activation to the next cell. A GRU uses two simple nonlinearities, sigmoid (ranging from 0 to 1) and tanh (ranging from −1 to 1). Although the GRU and LSTM have similar goals, the GRU is simpler in structure. The workflow of a GRU is the same as that of an LSTM, but the internal operations of the GRU unit are different. Figure 4.6 illustrates the structure of a GRU cell. Eq. (4.3) presents the key equations for a GRU cell:

$$z_t = \sigma\left(x_t U^z + h_{t-1} W^z + b_z\right)$$

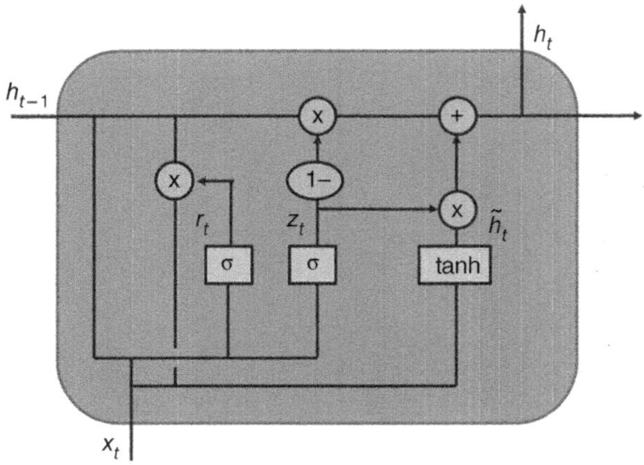

Figure 4.6 GRU cell structure.

$$r_t = \sigma\left(x_t U^r + h_{t-1} W^r + b_r\right)$$
$$\hat{h}_t = tanh\left(x_t U^h + \left(r_t \odot h_{t-1}\right) W^h\right)$$
$$h_t = \left(1 - z_t\right) \odot h_{t-1} + z_t * \hat{h}_t \tag{4.3}$$

Let x_t be input vector, r_t reset gate vector, $\hat{h}_t$ candidate activation vector, h_t output vector, z_t update gate vector, b_z, b_r the bias vectors. Additionally, let U, W, V denote the parameter matrices.

4.3.5 LSTM and BiLSTM with Attention

Attention models are a set of techniques that allow for selective processing of complex inputs, especially for long source inputs. These models have been widely used in AI-based systems to improve performance, particularly in computer vision and natural language processing applications (You et al. 2016). The attention mechanism assigns weights to different features in order to highlight their importance for model prediction. This approach allows the model to focus on the most relevant time-series features, while suppressing interference from non-important features, effectively addressing the problem of judging the influence of different time-series features. LSTM-BiLSTM models have been shown to offer superior performance on long time-series datasets, and any RNN-based models, including LSTM and BiLSTM, can be combined with an attention block. The architecture of the attention block, as shown in Figure 4.7, generates a context vector that is a weighted combination of the input states.

The output of the attention block is shown in Eq. (4.4):

$$c_i = \sum_{j=0}^{T_x} \alpha_{ij} h_j \tag{4.4}$$

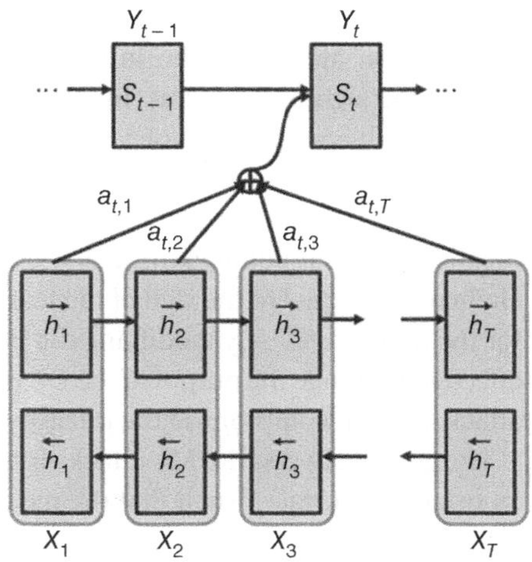

Figure 4.7 The architecture of the attention block.

Let c_i denote the output of the attention block, h_j the global features and the attention block assigns weights to the features, and $\alpha_i j$ the resulting weights represented by the variable.

4.4 Adversarial Attack and Mitigation Methods

This section presents fundamentals of the adversarial attack and training methods, followed by a modified FGSM attack for solar power generation forecasting.

This section discusses the proposed framework to train robust neural networks against adversarial ML attacks. First, we discuss the attack and training methods. Then, we give details about how to train the robust neural network.

4.4.1 Adversarial Attack

Adversarial ML is used to manipulate neural network models by entering precisely modified data. An approach to this technique is called the FGSM, which uses the gradients of the neural network to generate an adversarial example that can trick the model. To generate an adversarial instance, FGSM takes an input instance $\mathbf{x}$ and uses the gradients ∇_x of the loss value ℓ for $\mathbf{x}$ to create a new instance $\mathbf{x}^{adv}$ that maximizes the loss value of the classifier hypothesis h. The new instance produced is known as the adversarial instance, and the FGSM process can be summarized with Eq. (4.5).

$$\mathbf{x}^{adv} = \mathbf{x} + \varepsilon \cdot sign\left(\nabla_x \ell(\theta, \mathbf{x}, y)\right) \tag{4.5}$$

In this equation, ε is a small constant that determines the magnitude of the modification applied to the input instance, and *sign* is the sign function that returns the sign of the input. By utilizing the FGSM technique, adversaries can generate adversarial examples that can mislead the neural network models, potentially leading to incorrect predictions.

An attacker can manipulate the output of a DL model by introducing slight modifications to the input data through a small noise vector $\eta \in \mathbb{R}^n$. The noise vector's elements are equal to the sign of the features of the gradient of the cost function ℓ for the input $\mathbf{x} \in \mathbb{R}^n$. By modifying the input data in this manner, the attacker can effectively change the output of the DL model. This type of attack is an adversarial attack and can compromise the integrity and reliability of DL models.

The goal of adversarial ML attacks is to create an adversarial example x^{adv} from an original example x such that the model classifies it incorrectly (Tuna OF et al. 2022a, 2022b). The adversarial example can be generated by adding a small perturbation r to the original example x, i.e. $x^{adv} = x + r$. The small perturbation r is calculated to minimize the difference between the original example x and adversarial example x^{adv} while maximizing the classification error of the adversarial example x^{adv}.

Several adversarial ML attacks have been proposed in the literature. For example, the FGSM (Goodfellow et al. 2014) is a one-step gradient-based method. It generates an adversarial example, shown in Eq. (4.6):

$$x^{adv} = x + \varepsilon \text{ sign} \left(\nabla_x J(\theta; x, y)\right) \tag{4.6}$$

where ε controls the magnitude of the perturbation and $\nabla_x J(\theta; x, y)$ is the gradient of the loss function $J(\theta; x, y)$ with respect to the input x with label y and model parameters θ.

Figure 4.8 shows the details of the FGSM attack.

4.4.2 Adversarial Training

Adversarial training is a widely recommended defense technique that involves generating adversarial instances using the gradient of the victim classifier and then retraining the model with the adversarial instances and their respective labels. This technique is effective in defending models from adversarial attacks, as demonstrated in several studies (Madry et al. 2017; Benz et al., 2021; Tuna OF et al. 2022a, 2022b). It can learn to become more robust to such attacks by exposing the model to adversarial instances during training. Adversarial training is considered one of the most effective defense mechanisms against adversarial attacks, but it can also be computationally expensive and require a large amount of data. However, recent studies have proposed more efficient and effective adversarial training methods that can improve the defense mechanism's

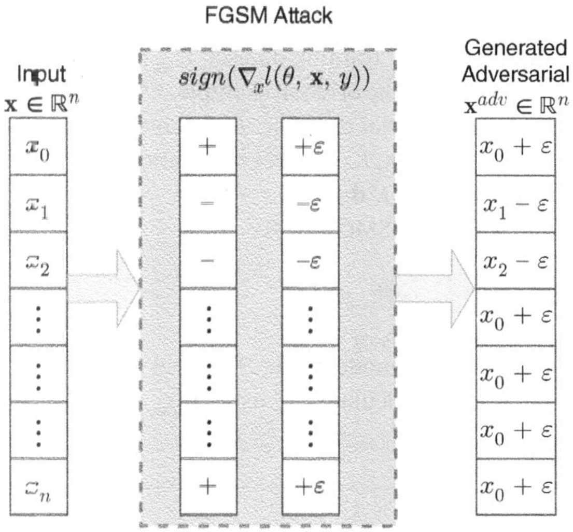

Figure 4.8 FGSM attack steps. The input vector $\mathbf{x} \in \mathbb{R}^n$ is poisoned with an FGSM attack to generate adversarial, x^{adv} (i.e. loss maximization direction).

efficiency while reducing the computational costs (Benz et al. 2021; Tuna OF et al., 2022a, 2022b).

Consider a typical classification problem with training instances $X \in \mathbb{R}^{m \times n}$ of dimension d and a label space Y. Assume that the classifier h_θ has been trained to minimize a loss function ℓ given by Eq. (4.7).

$$\min_{\theta} \frac{1}{m} \sum_{i=1}^{m} \ell\left(h_\theta\left(\mathbf{x}_i, y_i\right)\right) \tag{4.7}$$

An adversarial instance x^{adv} is an input instance x that satisfies the following two conditions: (i) the classifier $h_\theta(x^{adv})$ produces a different output than the correct label y, and (ii) the distance between the original input instance x and the adversarial instance x^{adv} is smaller than a predefined threshold ε. This is formulated as Eq. (4.8).

$$h_\theta\left(x^{adv}\right) \neq y \wedge d\left(x, x^{adv}\right) < \varepsilon \tag{4.8}$$

To generate an adversarial instance, attackers often solve an optimization problem that aims to maximize the loss between the predicted output $h_\theta(x^{adv})$ and the correct label y, subject to a distance constraint. This can be expressed as a constrained optimization problem, as shown in Eqs. (4.9) and (4.10):

$$\arg\max_{x} \ell\left(h_\theta\left(x^{adv}\right), y\right) \tag{4.9}$$

$$s.t. \; d\left(x, x^{adv}\right) < \varepsilon \tag{4.10}$$

Algorithm 1 Adversarial Training

Require: D_{train}: training dataset, D_{test}: testing dataset, x: an original example, y: the label of the original example, x^{adv}: an adversarial example, y^{adv}: the label of the adversarial example, L_{adv}: adversarial training loss function, L_{test}: testing loss function, K: number of iterations.

```
1: for k = 1 to K do
2:   x^adv ← FGSM(x)
3:   y^adv ← y
4:   D_adv ← (x^adv, y^adv)
5:   D_train ← D_train ∪ D_adv
6:   Train the model using D_train and L_adv
7:   L_test ← Calculate L_test using D_test
8:   D_train ← D_train \ D_adv
9: end for
```

To defend against adversarial attacks, a min-max objective function is used instead of the conventional training procedure given by Eq. (4.7). This function aims to minimize the expected value of the maximum loss at each training step, as expressed by Eq. (4.11):

$$\min_{\theta} \mathbb{E}_{(x,y)} \left(\max_{d(x,x^{adv})<\epsilon} \ell \left(h \left(x^{adv} \right), y \right) \right) \tag{4.11}$$

This approach forces the model to consider the worst-case scenario and improve its robustness against adversarial attacks.

Algorithm 4.1 shows the pseudocode of adversarial training.

4.5 Dataset Description, Feature Selection, and Performance Metrics

4.5.1 Dataset Description

In this study, a publicly available benchmark dataset was obtained from the 2014 GEFCOM, i.e. Global Energy Forecasting Competition. The dataset includes hourly solar power generation data along with numerical weather data from April 2012 to July 2014 and is publicly accessible to promote the reproducibility of the study (Hong et al. 2016). The dataset includes weather variables provided by the European Centre for Medium-Range Weather Forecasts (ECMWF), which are listed in Table 4.1. In this study, 98% of the dataset is used for training and 2% for testing.

Table 4.1 Feature parameters.

Parameter	Description	Unit
TI	Hourly time index	h
TCLW	Total column liquid water	$kg\,m^{-2}$
TCIW	Total column ice water	$kg\,m^{-2}$
SP	Surface pressure	Pa
HUM	Relative humidity	%
TCC	Total cloud cover	0–1
U	10-m U wind component	$m\,s^{-1}$
V	10-m V wind component	$m\,s^{-1}$
TP	Total precipitation	m
TEMP	2-m temperature	K
SSRD	Surface solar radiation	$J\,m^{-2}$
STRD	Surface thermal radiation	$J\,m^{-2}$
TSR	Top net solar radiation	$J\,m^{-2}$
POW	Solar power generation	kW

4.5.2 Feature Selection

Feature selection is an important process for AI-based models to improve performance in terms of accuracy and computing time. This process includes two internal steps for solar power generation models. i.e. selection of the most relevant weather features and rolling window size. The Pearson correlation (ρ) method is used to determine the difference between solar power generation (S) and weather parameters (W), including the time index as an hour. The Pearson correlation coefficients between loads and weather parameters for solar power generation are given in Figure 4.9a. According to the results, HOUR, HUM, TEMP, SSRD, and TSR parameters are highly correlated, i.e. (ρ) > 0.3, with solar power generation data (kW).

However, the other parameters are low correlated with solar data, (ρ) < 0.3. Furthermore, this study finds that the size of the rolling window for historical solar power generation (i.e. the values of $i = (t-1, t-2, t-3,...)$) is highly correlated with future forecasts. This means that any combination of historical solar power generation values can be used as features for forecasting. The study also examines various rolling window sizes, ranging from 1 to 24 hours, for historical solar power generation and selected weather parameters in order to identify the best feature sets for forecasting. The results show that the best feature sets for solar power

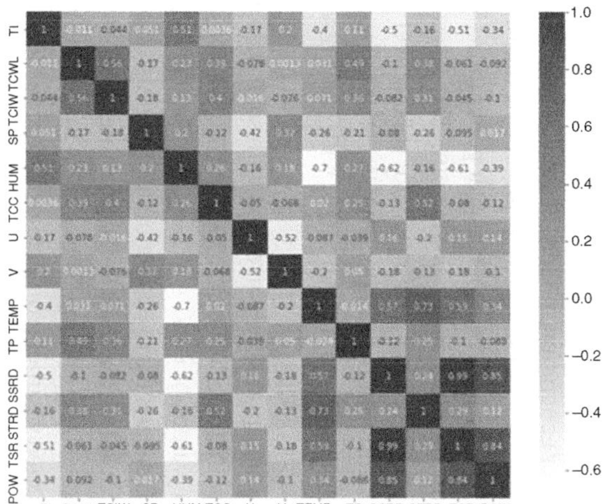

(a) Pearson correlation

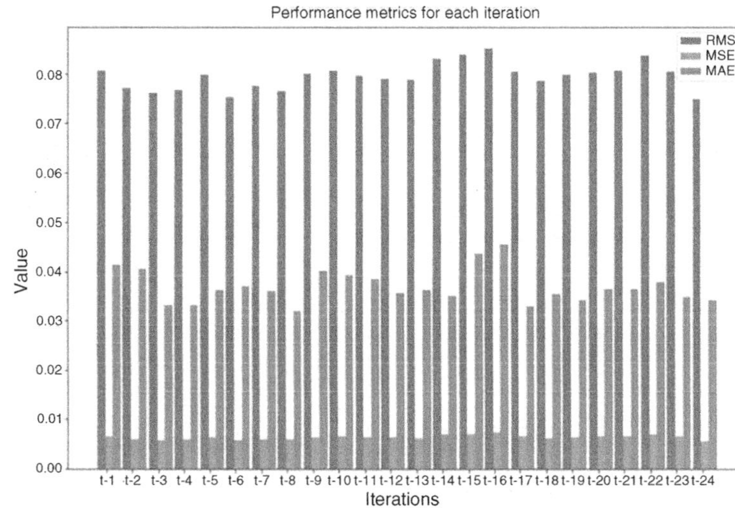

(b) Rolling windows size

Figure 4.9 Pearson correlation coefficients matrix and the performance metrics for each window size. (a) Pearson correlation. (b) Rolling windows size.

generation forecasting can be achieved by exploring different rolling window sizes for historical data.

Figure 4.9b shows MSE, RMSE, and MAE values of the solar power generation forecasting for 24 feature sets used as the input features. LSTM model is selected to conduct the case studies for all input features. As shown in Figures 4.9b, (t)-4.6, and t-24 yield the best comparable results, i.e. MSE value of around 0.02. However, t-6 is selected for the rolling window size, i.e. 6, due to needing less computing process.

The extracted features through the feature selection process are taken as the input of LSTM layers to obtain the solar power forecasting of the next hour. Through several parameters fine-tuning experiments, when the input historical data are chosen for the past 9 hours, the trained model has the highest prediction accuracy, so the input is the HOUR, HUM, TEMP, SSRD, and TSR parameters per hour of the past 9 hours.

4.5.3 Performance Metrics

The evaluation and comparison of RNN-based models is performed using several performance metrics, such as RMSE/MSE/MAE. The equations for these metrics are presented in Eqs. (4.12), (4.13), and (4.14).

$$RMSE = \sqrt{\frac{\sum_{t=0}^{n} \left(y_t - \hat{y}_t\right)^2}{n}} \tag{4.12}$$

$$MSE = \frac{\sum_{t=0}^{n} \left(y_t - \hat{y}_t\right)^2}{n} \tag{4.13}$$

$$MAE = \frac{1}{n}\sum_{t=0}^{n} \left|y_t - \hat{y}_t\right| \tag{4.14}$$

The performance metrics RMSE/MSE/MAE are calculated using the actual and forecasted values. Calculation parameters are given as follows:

y_t: the actual t^{th} instance
$\hat{y}_t$: the forecasted t^{th} instance
n: the total number of instances.

4.6 Experiments

The purpose of this chapter is to examine the performance of RNN-based models to forecast solar photovoltaic generation in the presence of adversarial attacks, both

Table 4.2 The best hyperparameters found through a grid search process for the models.

Metric	RNN	LSTM	BiLSTM	GRU	LSTM (w/att.)	BiLSTM (w/att.)
# of Neurons-L1	202	410	724	330	330	234
# of Neurons-L2	42	80	–	202	170	10
# of Neurons-L3	74	120	–	266	106	266
Activation	relu	selu	relu	relu	selu	softplus
Last layer act	elu	sigmoid	selu	selu	softsign	tahn
Optimizer	Nadam	Nadam	Nadam	Adam	Adam	Nadam

with and without adversarial training. The models are first trained, and then their performance is evaluated using a selected test dataset. The discrepancy between predicted and actual values is measured using RMSE/MSE/MAE. Then, the same dataset is manipulated to simulate an adversarial attack and the RNN-based models are evaluated to determine their vulnerability to the attack, both with and without adversarial training. This study aims at providing a secure solar power generation forecasting framework embedded adversarial attack and training methods for RNNs, i.e. simple RNN, LSTM, BiLSTM, GRU, and LSTM/BiLSTM with attention. A publicly available time-series solar PV power generation dataset is used for all test cases. The first step is to find the best hyperparameter settings before running the experiments. These parameters are shown in Table 4.2, which provides information about the architecture of different models used for forecasting solar photovoltaic power generation. This includes details such as the number of layers and neurons and the activation functions used in each model. To determine the optimal hyperparameters for each model, a grid search was conducted.

4.6.1 RNN-based Solar Power Generation Forecasting

Solar power generation forecasting is essential in the smart grid environment because of increasing power injection into the power grid dramatically and its high impact on grid stability and reliability. The graphical prediction representation of each model is depicted in Figure 4.10. The figure combines hour-ahead forecasting results of six different solar PV power generation models with the actual values for a week, i.e. 24*7 data points. However, it is difficult to interpret the results due to them being similar to each other. Therefore, the performance metrics of each model will be investigated for a detailed analysis.

The performance of each model is given in Table 4.3. In the table, the lower scores are considered better. The performance of each model is similar to each

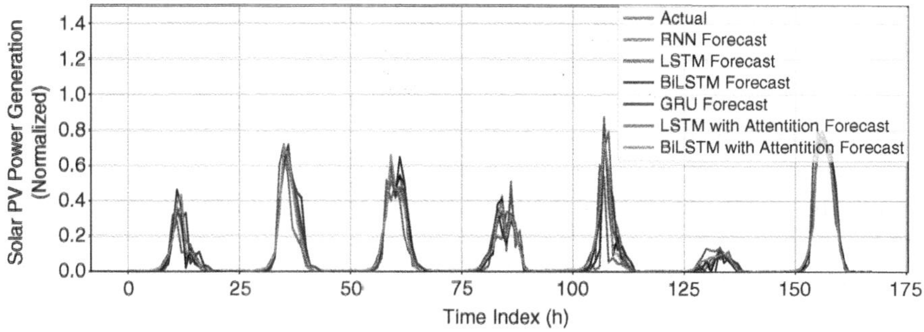

Figure 4.10 RNN-based model forecast results with actual values.

Table 4.3 RNN-based models' performance with RMSE, MSE, and MAE metrics.

Metric	RNN	LSTM	BiLSTM	GRU	LSTM (w/Att.)	BiLSTM (w/Att.)
RMSE	0.1010	0.0721	0.0862	0.0925	0.0694	0.0769
MSE	0.0102	0.0052	0.0074	0.0086	0.0048	0.0059
MAE	0.0411	0.0297	0.0367	0.0416	0.0324	0.0326

other, i.e. RMSE changes between 0.0694 and 0.1010. RNN has the lowest performance, while LSTM with attention has the highest performance in terms of the RMSE with a regular dataset. It is an expected result because the BiLSTM model enables additional training by traversing the input data twice, i.e. left-to-right and right-to-left. It outperforms regular unidirectional LSTM-based models with additional training capability. In addition, the attention layer focuses on specific aspects of the source input. GRU model trains faster due to using less training parameters and less memory, i.e. the computing time is lower, but the performance is also low in terms of RMSE, as expected. The other interesting finding is that the RNN-based model offers better performance than LSTM and GRU, i.e. around 1.0–1.3%. Overall, all performance results (RMSE, MSE, and MAE) are pretty close to each other, i.e. 0.0694–0.1010 for RMSE, 0.0048–0.0102 for MSE, and 0.297–0.0416 for MAE. Several reasons can be given to explain these performance results, such as the selected dataset and model hyperparameters. First, the selected dataset is a time series of hourly average solar power generation from April 2012 to July 2014. Second, the feature set consists of TI, HOUR, HUM, TEMP, SSRD, and TSR parameters per hour of the past 6 hours. Third, the grid search is applied to find the best hyperparameters for each model.

4.6.2 Adversarial Machine Learning Attack Against RNN-based Models

Adversarial attacks on forecasting have gained increased attention, with various attack methods being developed. In this study, the FGSM is utilized to generate adversarial examples. The performance of each model was previously evaluated using the RMSE, MAE, and MSE metrics, with each model being analyzed individually.

Figure 4.11 shows the RMSE/MSE/MAE results with FGSM-generated deceptive inputs from $\varepsilon = 0.00$ to $\varepsilon = 1.0$ for each model, i.e. (a) RNN, (b) LSTM, (c) BiLSTM, (d) GRU, (e) LSTM with Attention, and (f) BiLSTM with Attention. The RMSE, MSE, and MAE results are shown by the red line, the black line, and the blue line, respectively. The figure shows that RMSE is more sensitive to adversarial attacks than MSE and MAE. The figure also shows that the acceptable limit for RMSE is around 0.15. Among the models, the LSTM model is the most

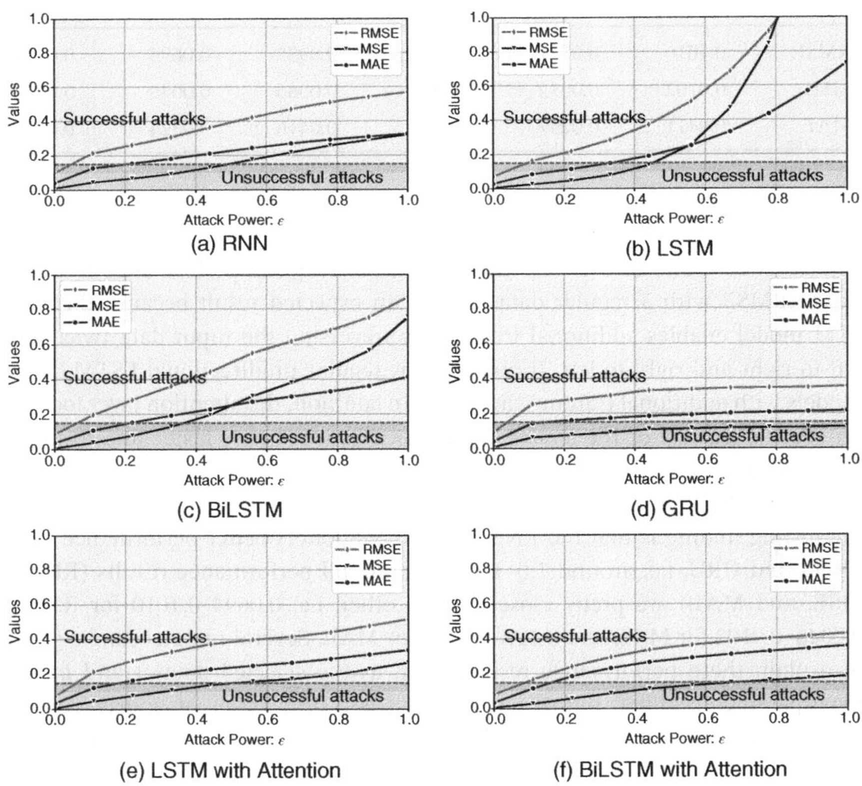

Figure 4.11 Prediction performance changes with various attack power values (ε) (a) RNN, (b) LSTM, (c) BiLSTM, (d) GRU, (e) LSTM with attention, (f) BiLSTM with attention.

Table 4.4 RMSE scores of RNN-based models for a specific ε value.

Metric	RNN	LSTM	BiLSTM	GRU	LSTM (w/Att.)	BiLSTM (w/Att.)
0.00	0.1010	0.0721	0.0862	0.0925	0.0745	0.0809
0.11	0.2151	0.1637	0.1963	0.2535	0.2133	0.1668
0.22	0.2621	0.2183	0.2811	0.2795	0.2779	0.2408
0.33	0.3074	0.2838	0.3604	0.3016	0.3286	0.2941
0.44	0.3631	0.3710	0.4429	0.3303	0.3735	0.3362
0.56	0.4207	0.5050	0.5504	0.3356	0.3986	0.3639
0.67	0.4671	0.6830	0.6172	0.3434	0.4156	0.3835
0.78	0.5084	0.9190	0.6792	0.3452	0.4412	0.4003
0.89	0.5395	1.2164	0.7509	0.3477	0.4780	0.4154
1.00	0.5670	1.5819	0.8672	0.3537	0.5154	0.4312
Average	0.3751	0.6014	0.4832	0.2983	0.3517	0.3113

vulnerable to adversarial attacks, with an average RMSE of 0.6014, while the bidirectional LSTM with attention model is the most resilient, with an average RMSE of 0.3113. The solar photovoltaic power forecasting model can withstand FGSM attacks when the value of ε is around 0.01. These results demonstrate the significant vulnerability of forecasting models to FGSM attacks.

The results in Table 4.4 demonstrate the relationship between the value of ε and the RMSE performance of each model. It can be seen that as ε increases, the RMSE performance of the models decreases, indicating that the adversarial attacks become stronger and more effective. The results highlight the importance of considering the effects of adversarial attacks in the development and evaluation of forecasting models. The value of ε ranges from 0 to 1.0, with a higher value indicating a stronger attack on the forecasting. When ε is 0, it means that the data are not altered.

4.6.3 RNN-based Solar Power Generation Forecasting with Adversarial Machine Learning Training

Adversarial ML training is widely used to increase robustness against adversarial attacks, especially for forecasts. In this study, the regular adversarial ML training method is applied to reduce its vulnerability against adversarial attacks. The performance of each model after adversarial training is evaluated in terms of RMSE, MAE, and MSE, as in the previous section.

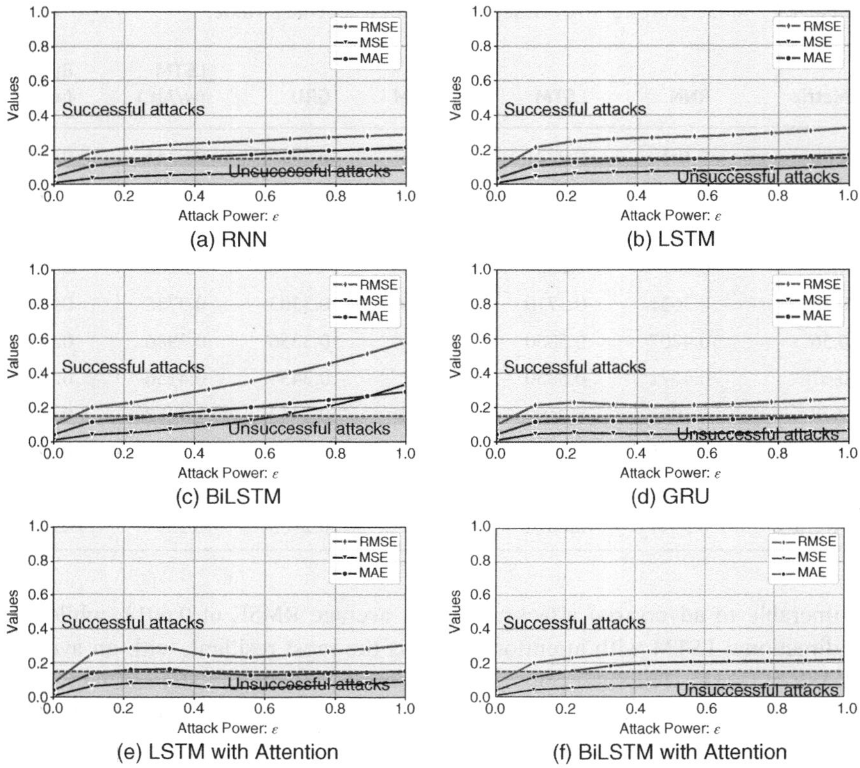

Figure 4.12 Prediction performance with various attack power values (ε) after adversarial training, (a) RNN, (b) LSTM, (c) BiLSTM, (d) GRU, (e) LSTM with attention, (f) BiLSTM with attention.

Figure 4.12 and Table 4.5 show the models' performance, i.e. RMSE/MSE/MAE values, against adversarial attacks from $\varepsilon = 0.00$ to $\varepsilon = 1.0$ after adversarial training. The RMSE, MSE, and MAE results are shown by the red line, the black line, and the blue line, respectively, as in the previous one. The figure shows that the model is still sensitive to adversarial attacks. However, the models' robustness is better against the adversarial attack. According to the figure, BiLSTM among models is the most vulnerable to adversarial attacks; i.e. the average RMSE is 0.3397, while GRU is the most resistible after the adversarial training, i.e. the average RMSE is 0.2123. GRU model can resist any adversarial attack with $\varepsilon = 0.44$. Results also indicate that adversarial training has a significant contribution to the model robustness solar PV power generation forecasting model against the adversarial attack.

Table 4.5 RMSE scores of RNN-based models for a specific ε value after adversarial training.

Metric	RNN	LSTM	BiLSTM	GRU	LSTM (w/Att.)	BiLSTM (w/Att.)
0.00	0.0959	0.0819	0.0942	0.0971	0.0824	0.0806
0.11	0.1838	0.2148	0.2004	0.2140	0.2531	0.2024
0.22	0.2125	0.2511	0.2271	0.2297	0.2887	0.2305
0.33	0.2283	0.2622	0.2644	0.2183	0.2872	0.2523
0.44	0.2412	0.2695	0.3063	0.2109	0.2417	0.2662
0.56	0.2514	0.2752	0.3532	0.2119	0.2339	0.2681
0.67	0.2623	0.2831	0.4034	0.2189	0.2480	0.2657
0.78	0.2726	0.2941	0.4546	0.2313	0.2674	0.2631
0.89	0.2808	0.3078	0.5136	0.2414	0.2855	0.2633
1.00	0.2880	0.3238	0.5799	0.2500	0.3056	0.2681
Average	0.2317	0.2564	0.3397	0.2123	0.2493	0.2360

4.7 Results and Discussion

This study applies adversarial training to the RNN-based models to reduce their sensitivity to the FGSM-based adversarial attack. The results show that adversarial training significantly reduces the impact of adversarial attacks on the RNN-based models, i.e. the average RMSE is reduced from 0.6014 to 0.3113. These results indicate that adversarial training is an effective method for enhancing the robustness of RNN-based models for solar PV power generation forecasting. However, all models are susceptible to FGSM-type adversarial attacks, e.g. the RMSE value is 0.8672 under a strong adversarial attack, i.e. $\varepsilon = 1$.

The results in Figure 4.12 show that adversarial training can mitigate the impact of adversarial attacks and improve the robustness of the RNN-based models. The results indicate that adversarial training can significantly reduce models' performance and make the models more resistant to FGSM attacks.

Key observations from the overall results can be given below:

- RNN-based models are vulnerable to adversarial attacks, and their performance metrics can be significantly impacted.
- Adversarial training can effectively defend models against adversarial attacks.
- LSTM model is more susceptible to adversarial attacks than other models, such as the bidirectional LSTM with attention model.

- The RMSE metric is more sensitive to adversarial attacks compared with MSE/MAE.
- The strength of adversarial attacks can be quantified using the ε factor, with a higher value indicating a stronger attack.
- It is important to consider the effects of adversarial attacks in the development and evaluation of forecast models of solar PV power generation.

4.8 Summary

The study proposes a secure solar power generation forecasting framework that incorporates adversarial attack and training methods. The proposed framework is based on widely used RNN models, such as simple-RNN, LSTM, BiLSTM, GRU, and LSTM/BiLSTM, with attention to forecasting solar PV power generation. The performance of these models is evaluated using metrics (i.e. RMSE/MSE/MAE). The impact of adversarial attacks on the model performance is also analyzed using the FGSM. The attack parameter, ε, is systematically increased from 0 to 1.0 to represent increasing levels of adversarial attacks ($\varepsilon = 0$ represents the case without any cyberattack, while $\varepsilon = 1.0$ represents the heaviest cyberattack). The study also investigates how adversarial training can reduce the sensitivity of the models to these attacks. The results demonstrate the impact of adversarial attacks on the performance of the solar PV power generation forecasting models and the effectiveness of adversarial training in reducing the impact of these attacks.

References

Alfadda, A., Adhikari, R., Kuzlu, M., and Rahman, S. (2017). Hour-ahead solar PV power forecasting using SVR-based approach. In: *2017 IEEE Power & Energy Society Innovative Smart Grid Technologies Conference (ISGT)*, 1–5. IEEE.

Benz, P., Zhang, C., Karjauv, A., and Kweon, I.S. (2021). Robustness may be at odds with fairness: an empirical study on class-wise accuracy. In: *NeurIPS 2020 Workshop on Pre-registration in Machine Learning*. Vol. 148 of Proceedings of Machine Learning Research (ed. L. Bertinetto, J.F. Henriques, S. Albanie, et al.), 325–342. PMLR https://proceedings.mlr.press/v148/benz21a.html.

Chitalia, G., Pipattanasomporn, M., Garg, V., and Rahman, S. (2020). Robust short-term electrical load forecasting framework for commercial buildings using deep recurrent neural networks. *Applied Energy* 278: 115410.

Cho, K., Van Merriënboer, B., Gulcehre, C. et al. (2014). Learning phrase representations using RNN encoder-decoder for statistical machine translation. arXiv arXiv:1406.1078.

Gers, F.A., Schraudolph N.N., and Schmidhuber, J. (2002). Learning precise timing with LSTM recurrent networks. *Journal of Machine Learning Research* 3 (Aug): 115–143.

Goodfellow, I.J., Shlens, J., and Szegedy, C. (2014). Explaining and harnessing adversarial examples. arXiv e-prints. arXiv:1412.6572. https://arxiv.org/abs/1412.6572. [stat.ML].

Goodfellow, I., Bengio, Y., and Courville, A. (2016). Sequence modeling: recurrent and recursive nets. *Deep Learning* 367–415.

Guizzi, G., Iacovella, L., and Manno, M. (2015). Intermittent non-dispatchable renewable generation and reserve requirements: historical analysis and preliminary evaluations on the Italian electric grid. *Energy Procedia* 81: 339–344.

Hauser, C.H., Bakken, D.E., and Bose, A. (2005). A failure to communicate: next generation communication requirements, technologies, and architecture for the electric power grid. *IEEE Power and Energy Magazine* 3 (2): 47–55.

Hf, D., Hw, B., Ry, W., and Ma, H (2020). An INS/GNSS integrated navigation in GNSS denied environment using recurrent neural network. *Defence Technology* 16 (2): 334–340.

Hong, T., Pinson, P., Fan, S. et al. (2016). Probabilistic energy forecasting: global energy forecasting competition 2014 and beyond. *International Journal of Forecasting* 32 (3): 896–913.

Jiang, Q., Tang, C., Chen, C. et al. (2018). Stock price forecast based on LSTM neural network. In: *International Conference on Management Science and Engineering Management*, 393–408. Springer.

Kuzlu, M., Cali, U., Sharma, V., and Güler, Ö. (2020). Gaining insight into solar photovoltaic power generation forecasting utilizing explainable artificial intelligence tools. *IEEE Access* 8: 187814–187823.

Li, Y.H., Harfiya, L.N., Purwandari, K., and Lin, Y.D. (2020). Real-time cuffless continuous blood pressure estimation using deep learning model. *Sensors* 20 (19): 5606.

Madry, A., Makelov, A., Schmidt, L. et al. (2017). Towards deep learning models resistant to adversarial attacks. *arXiv e-prints* arXiv:1706.06083. https://arxiv.org/abs/1706.06083.

Mellit, A. (2008). Artificial intelligence technique for modelling and forecasting of solar radiation data: a review. *International Journal of Artificial Intelligence and Soft Computing* 1 (1): 52–76.

Sorkun, M.C., Paoli, C., and Incel, Ö.D. (2017). Time series forecasting on solar irradiation using deep learning. In: *2017 10th International Conference on Electrical and Electronics Engineering (ELECO)*, 151–155. IEEE.

Tuna, O.F., Catak, F.O., and Eskil, M.T. (2022a). Exploiting epistemic uncertainty of the deep learning models to generate adversarial samples. *Multimedia Tools and Applications* 1–22.

Tuna, O.F., Catak, F.O., and Eskil, M.T. (2022b). Uncertainty as a Swiss army knife: new adversarial attack and defense ideas based on epistemic uncertainty. *Complex & Intelligent Systems* 1–19.

US DoE FEMP (2002). Using distributed energy resources-a how-to guide for Federal Energy Managers. *Cogeneration and Competitive Power Journal* 17 (4): 37–68.

Wan, C., Zhao, J., Song, Y. et al. (2015). Photovoltaic and solar power forecasting for smart grid energy management. *CSEE Journal of Power and Energy Systems* 1 (4): 38–46.

You, Q., Jin, H., Wang, Z. et al. (2016). Image captioning with semantic attention. In: *Proceedings of the IEEE Conference on Computer Vision and Pattern Recognition*, 4651–4659. IEEE.

Yu, Y., Si, X., Hu, C., and Zhang, J. (2019). A review of recurrent neural networks: LSTM cells and network architectures. *Neural Computation* 31 (7): 1235–1270.

Zhang, J., Chi, Y., and Xiao, L. (2018). Solar power generation forecast based on LSTM. In: *2018 IEEE 9th International Conference on Software Engineering and Service Science (ICSESS)*, 869–872. IEEE.

5

Technology Intelligence: Transformative Trends and Technological Synergies for the Smart Grid

Haydar Yalçın

Department of Business Administration, Ege University, Izmir, 35800, Turkey

5.1 Introduction

Smart grids can be described as a type of electrical grid. This technology has great potential in the field of sustainable energy and its management. It is among the most important scientific topics today and has several important characteristics described in the literature. Communication is one of these characteristics. Smart grids provide two-way communication between the electricity grid and its customers, allowing customers to monitor their electricity consumption in real time (Wang et al. 2011). When considered in terms of energy efficiency, it can be said that smart grids reduce the damage to the environment by reducing energy consumption (Moura et al. 2013). Smart grids, which enable the use of energy resources more effectively within the framework of renewable energy, enable more efficient allocation of energy obtained from renewable energy sources such as solar and wind. In terms of reliability, smart grids serve as a tool designed to prevent power outages (Hossain et al. 2016). In this way, they provide the infrastructure for a more reliable electricity grid. Smart grids have an important characteristic in terms of flexibility, as they can quickly respond to changes in electricity demand. The electricity grid becomes more flexible, efficient, reliable, and sustainable (Carmichael et al. 2021). The rising importance of smart grids has brought about a major transition and transformation in the fields of technology and engineering management (Mah et al. 2013). In the energy sector, there is an increased need for engineers and technologists to collaborate to design, implement, and maintain this innovative infrastructure (Land 2012). Technology management involves developing strategies to manage, optimize, and continuously update advanced features of smart grids, while engineering management deals with overseeing and improving the physical design and operation

Future-Oriented Technology Assessment: A Manager's Guide with Case Applications, First Edition.
Edited by Haydar Yalçın and Tugrul U. Daim.

of energy infrastructure (Tuballa and Abundo 2016). This new era reveals that the collaboration of technology and engineering management to implement smart grids can lead to more effective, efficient, and sustainable solutions in the energy sector (Borowski 2021; Lacy et al. 2010).

This dataset contains important descriptive statistics on smart grid research. First, the wide period between 1991 and 2024 shows that studies in the field of smart grid technology have been comprehensively examined: 20 908 documents obtained from 1872 different sources reflect a wide range of the subject. This shows that the academic community has contributed diverse perspectives and expertise to the subject. The average age of the documents in the data set was only 4.49 years. Smart grid research is increasing by 2.12% per year, and each document was cited an average of 30.01 times, indicating that these works have significant impact and recognition within the academic community. Additionally, the presence of 486 268 references highlights the deep and broad knowledge base in this field. Additional statistics on document content, such as keywords and author-assigned words, help us better understand the scope and topics of research. These words deepen the analysis by enriching the content of academic studies. The studies appear to contain many authors; 40 493 authors reflect the broad and diverse nature of the research community. Documents written by 891 authors alone show that individual contributions and expertise are important. Considering that each document was created with an average of 3.97 co-authors and that 33.54% of the documents included international collaboration, it is understood that smart grid research involves collaboration and a global perspective. To get detailed information about the cluster analysis conducted in the study and the tools used, you can review the book *Digital Transformations: New Tools and Methods for Mining Technological Intelligence* written by Daim and Yalçın (2022).

5.2 Cluster Analysis

Cluster analysis is among the important analytical tools for structuring and exploring scientific research. In this context, the LLR (log-likelihood ratio) algorithm, which stands out in the fields of text mining and information visualization, allows this method to be used effectively. The LLR algorithm is a statistical analysis method used in the context of text mining. This algorithm aims to determine the relationships of these terms with each other by measuring the prominence level of certain words or terms in texts. This feature is used in cluster analysis processes to determine the similarities of texts and important terms in their groups. Additionally, this analytical information can be used in visualization tools, helping to understand relationships between important subject areas,

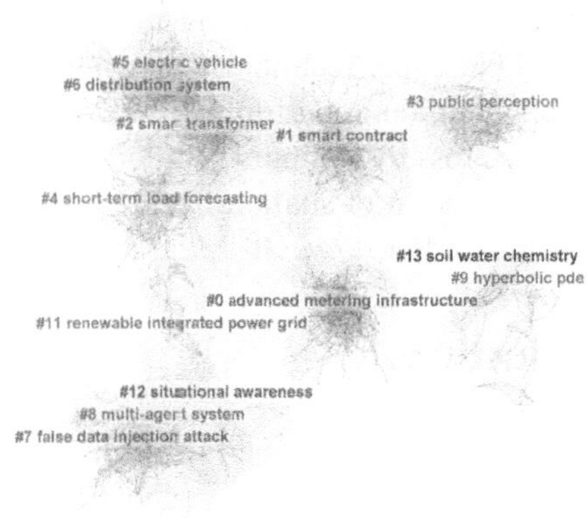

Figure 5.1 Major clusters.

keywords, and study groups in scientific fields. This method was used in our study to determine how the prominent research focuses were shaped (Figure 5.1).

The provided summary offers insights into the major thematic clusters within a network, each characterized by its size, silhouette score, cluster label (determined by LLR), and the average year associated with the research within that cluster (Figure 5.1). These clusters seem to pertain to distinct areas of study or research topics within a specific domain, each with its unique focus and temporal context. Notable clusters include "advanced metering infrastructure" (2015), "smart contract" (2015), "smart transformer" (2014), "public perception" (2015), "short-term load forecasting" (2017), and "electric vehicle" (2013). Additionally, "distribution system" (2015), "false data injection attack" (2016), and "situational awareness" (2013) represent other significant research themes, each offering valuable contributions to the field. Larger clusters, such as "advanced metering infrastructure" and "electric vehicle," suggest that these topics have received substantial attention and are significant in the field. Technology managers should pay special attention to these areas due to their high research volume. Higher silhouette scores, like those in clusters "electric vehicle" and "false data injection attack," imply that the research within these clusters is more internally consistent and focused. Technology managers might find these clusters useful for specialized and focused technology development efforts. Clusters with more recent years, such as "smart contract" and "false data injection attack," highlight emerging technology trends. On the other hand, clusters like "electric vehicle (1999)" indicate historical research trends. Technology managers should consider these time frames when planning for technology adoption or development. "Electric vehicle" in multiple clusters with varying years and citation counts suggest the long-term importance of a technology. Technology managers should consider both current and historically influential topics for strategic planning. The presence of multiple clusters with varying topics and metrics indicates a diversified technological landscape. Technology managers should appreciate this diversity and explore collaboration opportunities between different areas, potentially leading to innovative cross-disciplinary solutions. These clusters have been evaluated based on a combination of citation-based and network-based metrics, emphasizing their significance in the overall research landscape (Table 5.1).

5.3 Authors Productivity

List of researchers reveals a commendable performance in the realm of smart grid technology. Their h-indices, a measure of their scholarly impact, showcase their significant influence within the field. Researchers like H. Vincent Poor, Jianhui Wang, and Nadeem Javaid, each possessing h-indices of 33 or higher, have contributed substantially to the body of knowledge in this domain, as evidenced by their impressive citation counts. For instance, H. Vincent Poor's h-index of 35 underscores his extensive influence, with 4713 citations attributed to his 59 articles. These researchers have not only produced a substantial body of work but have also garnered the attention and recognition of their peers, underscoring their vital role in advancing smart grid technology. Their contributions are a testament to the

Table 5.1 Major clusters.

Cluster ID	Size	Silhouette	Label (LLR)	Average year
0	160	0.7	Advanced metering infrastructure (77 092.06, 1.0E-4)	2015
1	146	0.656	Smart contract (9752.96, 1.0E-4)	2015
2	142	0.715	Smart transformer (16 868.89, 1.0E-4)	2014
3	141	0.674	Public perception (9730.5, 1.0E-4)	2015
4	138	0.653	Short-term load forecasting (66 246.61, 1.0E-4)	2017
5	138	0.77	Electric vehicle (75 513.12, 1.0E-4)	2013
6	118	0.672	Distribution system (78 463.12, 1.0E-4)	2015
7	94	0.805	False data injection attack (98 265.3, 1.0E-4)	2016
8	63	0.783	Electric vehicle (13 535, 1.0E-4)	2014
9	41	0.984	Electric vehicle (6176, 1.0E-4)	1999
11	30	0.981	Electric vehicle (7042, 1.0E-4)	2007
12	17	0.956	Situational awareness (775.13, 1.0E-4)	2013
13	14	0.982	Soil water chemistry (882.89, 1.0E-4)	2005

Table 5.2 Author productivity.

h-index	Unit	Citation sum within h-core	All citations	All articles
35	POOR, H. VINCENT	4713	5017	59
33	WANG, JIANHUI	4746	4890	45
33	JAVAID, NADEEM	2330	3142	97
32	CATALAO, JOAO P. S	2813	3189	64
32	SIANO, PIERLUIGI	3379	3950	84
31	ZHANG, YAN	4156	4331	46
30	KUMAR, NEERAJ	2594	2849	58
26	VALE, ZITA	1718	2030	62
25	LU, RONGXING	2765	2899	37
24	YUEN, CHAU	2445	2557	33
23	DONG, ZHAO YANG	2934	3064	37
23	GUERRERO, JOSEP M	2609	2842	49
21	LEHTONEN, MATTI	1259	1431	40
21	WANG, SHENGWEI	1533	1591	30
20	RODRIGUES, JOEL J. P. C	1563	1787	40
20	SHEN, XUEMIN (SHERMAN)	2352	2388	24
20	GIANNAKIS, GEORGIOS B	1576	1611	23
19	CHOW, MO-YUEN	3586	3630	23
19	MOHAMMED, OSAMA A	1623	1728	29
19	WANG, LINGFENG	1763	1779	22
19	HAN, ZHU	2810	2937	34
19	GUIZANI, MOHSEN	1025	1111	32
19	MOHAMED, YASSER ABDEL-RADY I	1603	1606	20
18	MONTI, ANTONELLO	989	1135	40
18	SRINIVASAN, DIPTI	1770	1798	22

importance of their research in shaping the landscape of this innovative and vital field (Table 5.2).

5.4 Co-word Analysis

Keyword analysis in publications related to smart grids provides valuable information for understanding important concepts, trends, and focuses in this field. As a first step, the keywords in the data set are determined by extracting them from the "Keywords Plus (ID)" sections. These keywords include terms highlighted by authors that focus on specific topics. Frequency analysis evaluates the importance of these terms in representing general topics by determining which terms are used more frequently. In this step, researchers are provided with the opportunity to understand the weight of certain key concepts in the field (Sarsenbayeva et al. 2023). By visualizing the relationships between terms, keyword network analysis determines which concepts are used together and the strong relationships between these concepts. Centrality metrics are used to evaluate the importance of nodes in the network; this is helpful in understanding the impact and connections of certain terms within the field (Aksay and Yalçın 2023; Daim and Yalçın 2022; Doğrul et al. 2023; Sarsenbayeva et al. 2023; Tseng et al. 2023; Tu et al. 2023). Attribution to authors and institutions evaluates which authors or institutions use which keywords more frequently. This is used to describe authors or institutions that have developed expertise on topics. Finally, keyword relevance maps visualize relevance levels between terms. These maps are used to identify terms that have strong relationships with each other and topics where these terms are discussed together. All these steps provide researchers with a comprehensive perspective in understanding key concepts, trends, and research focuses in the literature on smart grids (Figure 5.2).

A set of keywords related to smart grid technology research are given with Social Network Analysis indicators in terms of various centrality and restrictedness values. These keywords represent critical focus areas in the smart grid field. "SYSTEMS" in particular ranks highest across a variety of metrics, highlighting the central role of system-related research in smart grids. "MODEL" and "MANAGEMENT" are ranked highly, highlighting the importance of modeling and effective management in this field. "OPTIMIZATION" and "DESIGN" highlight the importance of optimizing and designing smart grid systems for optimal performance. "POWER" and "PRODUCTION" reflect key aspects of energy production and distribution. "ALGORITHM" and "STRATEGY" refer to the importance of algorithm development and strategic planning. "ENERGY" and "TECHNOLOGY" underline the central role of energy-related technologies. "STORAGE" and "DEMAND RESPONSE" highlight the areas of energy storage and demand response management. "NETWORKS" and "IMPACT" emphasize networking and impact assessment. "INTEGRATION" and "FRAMEWORK" indicate the focus on integrating various components and creating structured frameworks within smart grids. These keywords are important as they collectively depict a comprehensive research landscape in the field of smart grid technology, ranging from basic technological concepts to operational and management strategies. The high aggregate constraints associated with keywords in smart grid technology research provide valuable insights into the future of the field. Concepts that have

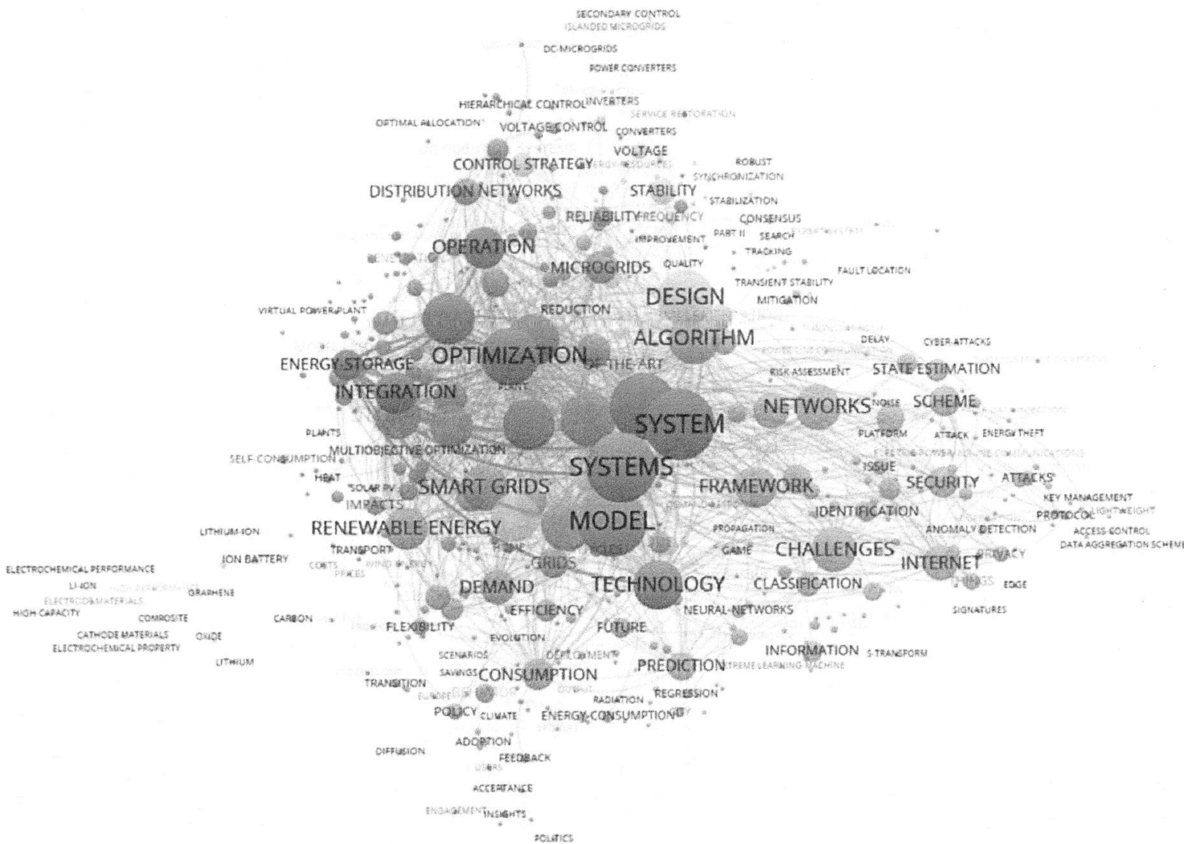

Figure 5.2 Keyword map.

reached a maturity level for the field can be read as an indication that smart grids are preparing to undergo significant advances and innovations focusing on critical issues such as power system management, integration of the latest technologies and the development of advanced communication systems. Future research indicates that optimization of smart grids will be prioritized for many purposes, including energy efficiency and sustainability, using advanced management and control systems. Data-driven decision-making will become central, with big data analytics and artificial intelligence driving insights for efficient grid operations. It can also be said that the importance of solutions that will ensure the robustness of network architectures and cybersecurity measures, network durability and protection against cyber threats will increase. The keywords with high aggregate constraints presented here form the backbone of smart grid technology research. These terms reflect the most critical and influential areas of focus in the field. "POWER SYSTEMS" and "SMART GRID TECHNOLOGY" highlight the primary emphasis on power system management and advances in smart grid technology. "COMMUNICATION TECHNOLOGY" refers to the fundamental role of communication technologies in ensuring grid operations, while "MULTI-PURPOSE OPTIMIZATION" highlights the optimization challenges in meeting multiple objectives in smart grids. While "MANAGEMENT SYSTEM" points out the vital role of management systems in the supervision of grid operations, "INFORMATION" underlines the central importance of data and information in smart grid applications. "SYSTEMS" represents the complex and interconnected nature of smart grid components. These keywords also cover various advanced technologies and techniques such as "PARTICLE Swarm OPTIMIZATION," "ARTIFICIAL INTELLIGENCE," and "MULTI AGENT SYSTEMS", highlighting the integration of cutting-edge solutions in smart grid research. "BIG DATA" refers to the use of comprehensive data analysis to optimize grid operations, while "NETWORKS" reflects the complementary role of network communications. These influential keywords collectively represent critical research areas that drive innovation and progress in the field of smart grid technology. A number of keywords in smart grid technology research appear to be important but not as central as others in the field. These terms cover various aspects of smart grid technology that are of current interest but may not dominate the research landscape. "HIGH CAPACITY" highlights the importance of managing significant amounts of data and electricity, while "LI-ION" highlights the central role of lithium-ion batteries in energy storage for smart grids. "ELECTROCHEMICAL PROPERTY" refers to research into the electrochemical properties of materials crucial to energy conversion and storage, while "CATHODE MATERIALS" indicates the continued focus on developing effective cathode materials for batteries. "LOAD REDISTRIBUTION ATTACKS" highlights the importance of addressing vulnerabilities in payload redistribution. "HIGH PERFORMANCE" reflects the continuous pursuit of high-performance components and systems in smart grids. "ELECTROCHEMICAL PERFORMANCE" showcases research into the electrochemical performance of materials for energy applications. "ELECTRODE MATERIALS" refers to the critical role of electrode materials in various electrochemical processes. "S-TRANSFORM" represents the use of this transformation for signal processing in smart grids. "ELECTRODE" serves as a key keyword reflecting the ubiquitous role of electrodes in electrochemical processes. "THINGS" refers to the far-reaching impact of the "Internet of Things" (IoT) on smart grid systems. "COMPOSITE" refers to the use of composite materials in different grid components. "PROTECTING DATA INTEGRITY" emphasizes the necessity of preserving data collection integrity in smart grid data management. "EDGE" represents the computing capabilities at the edge of the smart grid network. "OUTPUT" serves as a key term that specifies the desired results from smart grid operations. "IMPULSE NOISE" highlights the difficulty of managing impulsive noise in network communications. "OXIDE" refers to the use of oxide materials with various applications in electronics and energy storage. "GRAPHENE" reflects the increasing use of graphene in various components, particularly in energy storage and electrical conductivity. "LIGHTWEIGHT" highlights the importance of lightweight materials in the development of smart grid infrastructure. "SENSORS" refers to the expanding role of sensors in monitoring grid components and collecting data. "ELECTRICITY THEFT" refers to ongoing concerns about unauthorized electricity consumption and prevention strategies. "IRRADIANCE" represents a focus on capturing and exploiting solar radiation data for renewable energy production. "CYBERATTACKS" underscores the ongoing risk of cyberattacks on smart grid infrastructure and the need for robust security measures. "MPC" refers to the use of Model Predictive Control for smart grid management. "ENERGY THEFT" highlights concern about unregulated energy consumption and efforts to prevent theft. "ENERGY EFFICIENT" reflects the overall goal of optimizing energy efficiency across the smart grid. "ACCESS-CONTROL" emphasizes the importance of access control mechanisms in ensuring network security. Finally, "RADYO" refers to the use of radio communication technologies within smart grid systems.

These keywords represent various aspects of smart grid technology and contribute to the comprehensive development and security of future grid systems (Table 5.3). They reflect critical concepts related to smart grids. *Management* includes strategies for effectively planning, implementing, and monitoring smart grids. *Energy* refers to the production, transmission, and consumption of energy that forms the basis of smart grids. *Vehicles,* especially electric vehicles, have a significant impact on energy demand. *Data* represents the large amount of information generated and analyzed in smart grids. *Attacks* refers to security threats and potential risks faced by smart grids. *Injection* represents a strategy

Table 5.3 Social network indicators.

All degree	Betweenness centrality	High aggregate constraints	Low aggregate constraints
SYSTEMS	SYSTEMS	POWER-SYSTEMS	HIGH-CAPACITY
MODEL	MODEL	OF-THE-ART	LI-ION
MANAGEMENT	MANAGEMENT	COMMUNICATION TECHNOLOGY	ELECTROCHEMICAL PROPERTY
OPTIMIZATION	OPTIMIZATION	SMART GRID TECHNOLOGY	CATHODE MATERIALS
DESIGN	DESIGN	MULTIOBJECTIVE OPTIMIZATION	LOAD REDISTRIBUTION ATTACKS
POWER	POWER	MANAGEMENT-SYSTEM	HIGH-PERFORMANCE
GENERATION	ALGORITHM	INFORMATION	ELECTROCHEMICAL PERFORMANCE
ALGORITHM	ENERGY	SYSTEMS	ELECTRODE MATERIALS
STRATEGY	STRATEGY	PARTICLE SWARM OPTIMIZATION	S-TRANSFORM
ENERGY	GENERATION	DISTRIBUTED GENERATION	ELECTRODE
TECHNOLOGY	DEMAND RESPONSE	CLASSIFICATION	THINGS
STORAGE	PERFORMANCE	DISTRIBUTED ENERGY-RESOURCES	COMPOSITE
DEMAND RESPONSE	TECHNOLOGY	PROTECTION	PRESERVING DATA AGGREGATION
NETWORKS	NETWORKS	ARTIFICIAL-INTELLIGENCE	EDGE
IMPACT	STORAGE	IDENTIFICATION	OUTPUT

INTEGRATION	SMART GRIDS	ENERGY MANAGEMENT-SYSTEM	IMPULSIVE NOISE
FRAMEWORK	CHALLENGES	CONTROL STRATEGY	OXIDE
OPERATION	IMPACT	ELECTRICITY CONSUMPTION	GRAPHENE
CHALLENGES	RENEWABLE ENERGY	MULTIAGENT SYSTEMS	LIGHTWEIGHT
PERFORMANCE	FRAMEWORK	STATE	SENSORS
SMART GRIDS	OPERATION	INFRASTRUCTURE	ELECTRICITY THEFT
RENEWABLE ENERGY	INTEGRATION	NETWORKS	IRRADIANCE
DEMAND	INTERNET	MITIGATION	CYBERATTACKS
INTERNET	NETWORK	STABILITY	MPC
MICROGRIDS	DEMAND	BIG DATA	ENERGY THEFT
NETWORK	ENERGY MANAGEMENT	NETWORK	ENERGY-EFFICIENT
CONSUMPTION	MICROGRIDS	DISTRIBUTED GENERATION	ACCESS-CONTROL
LOAD	ELECTRIC VEHICLES	CONSUMPTION	RADIO

used to optimize and manage energy flow. *Placement* describes the strategic placement of energy resources for optimal positioning and energy efficiency. *Storage* emphasizes the use of energy storage systems and their importance to achieve energy balance. *Reconfiguration* refers to making smart grids dynamically adaptable and optimizing their performance. *System* broadly covers smart grid components. These terms represent key focal points in this field, reflecting critical issues such as management of smart grids, energy efficiency, security, data analysis, and renewable energy integration.

The contribution of different countries around the world in the technological development of smart grids is clearly visible. The h-index is an important criterion used in the research world to evaluate the scientific impact of countries and reveals their contributions in the field of smart grids technology. This metric determines a country's academic impact on smart grid technology research, based on data such as total number of citations, h-core where citations are centered, and number of articles published. In this context, the United States (US) maintains the global leadership of smart grid technology. The US h-index is 227, backed by 6848 articles containing 110 938 citations. This shows that the United States is at the forefront of smart grid research. The People's Republic of China stands out as another country that is constantly increasing its influence in the field of smart grid technology. China, with an h-index of 179, reflects its growth in this field, with 72 021 h-core citations and 9707 articles with a total of 255 350 citations. Canada is a strong contributor to smart grid technology research. Canada, with an h-index of 119, maintains its influence in this field with 1707 articles with 37 610 h-core citations. The United Kingdom (UK) is another country that has made significant contributions to the development of smart grid technology. UK, with an h-index of 117, reflects its work in this field with 2008 articles with 27 290 h-core citations. Australia continues its influence in the field of smart grid technology. Australia, with an h-index of 116, demonstrates its contributions in this field with 28 494 h-core citations and 1462 articles with a total of 59 818 citations. Other countries, such as Italy, Iran, India, and Spain, play important roles in the research field of smart grid technology and have impressive h-index values in this field. These h-index values, citations, and article numbers highlight the research contributions of these countries (Figure 5.3).

5.5 Discussion

They are effective tools for understanding and visualizing the relationships between terms in the smart grid field. The connections between these terms reflect the basic elements that make up the smart grid system. The term *control* represents regulation and direction in energy management processes, while

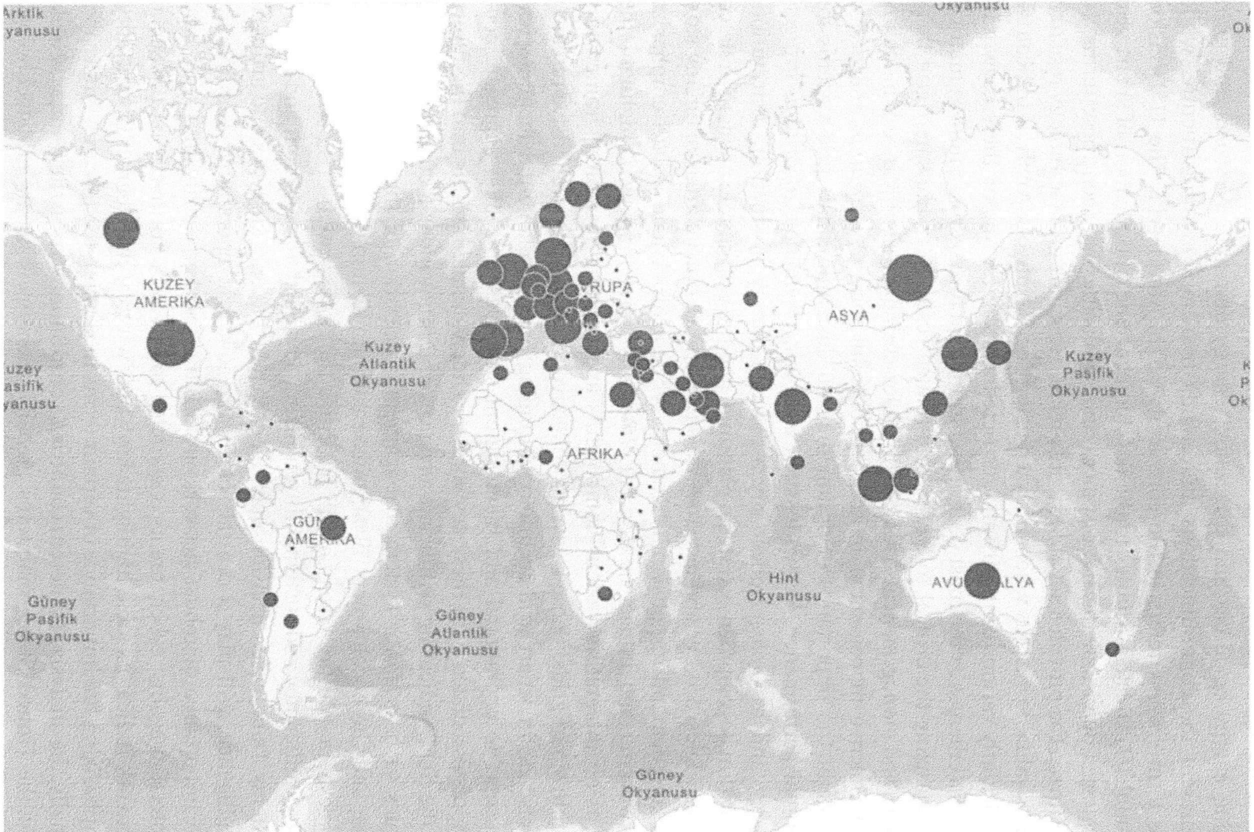

Figure 5.3 Country productivity.

response indicates the response of energy systems to changing conditions. The terms *management, optimization,* and *forecasting* are critical to effectively managing energy resources and forecasting energy demands. *Networks* and *grids* represent the communications and energy transmission infrastructure, while the term *systems* refers to the complexity of the smart grid. The IT Decision Module (ITDM) contributes to the development and optimization of smart grid strategies by visualizing the distances and similarities between these terms. In summary, the provided set of keywords related to smart grid technology research offers valuable insights when analyzed using Social Network Analysis indicators, particularly various centrality and restrictedness values. Among these keywords, *systems* emerges as the most central and influential term across multiple metrics, emphasizing the critical role of system-related research in the smart grid field. *Model* and *management* also rank high, underscoring the significance of modeling and effective management in smart grid technology. *Optimization* and *design* highlight the importance of optimizing and designing smart grid systems for optimal performance. Keywords such as *power, production, algorithm, strategy, energy,* and *technology* reflect key aspects of energy production, distribution, algorithm development, strategic planning, and the central role of energy-related technologies. *Storage* and *demand response* point to crucial areas of energy storage and demand response management, while *networks* and *impact* emphasize networking and impact assessment. *Integration* and *framework* highlight the focus on integrating various components and creating structured frameworks within smart grids. These keywords collectively depict a comprehensive research landscape in smart grid technology, ranging from fundamental technological concepts to operational and management strategies. The high aggregate constraints associated with these keywords provide valuable insights into the future of the field, indicating significant advances and innovations focusing on critical issues such as power system management, integration of the latest technologies, and the development of advanced communication systems. The analysis suggests that future research will prioritize the optimization of smart grids for energy efficiency and sustainability, utilizing advanced management and control systems. Data-driven decision-making, driven by big data analytics and artificial intelligence, will play a central role in efficient grid operations. Additionally, the importance of robust network architectures, cybersecurity measures, and protection against cyber threats is expected to increase. Furthermore, the Social Network Indicators table highlights keywords with varying degrees of centrality, emphasizing the importance of certain terms in shaping the research landscape. Finally, the analysis of different countries' contributions using the h-index reveals that the United States maintains global leadership in smart grid technology research, followed by countries like China, Canada, the United Kingdom, and Australia, among others.

References

Aksay, N.K. and Yalçın, H. (2023). Kadın Çalışmalarının Yarım Yüzyılı: Bibliyometrik Bir Analiz. *Kadin ve Demokrasi Dernegi* 9 (1): 32–65.

Borowski, P.F. (2021). Digitization, digital twins, blockchain, and industry 4.0 as elements of management process in enterprises in the energy sector. *Energies* 14 (7): 1885.

Carmichael, R., Gross, R., Hanna, R. et al. (2021). The demand response technology cluster: accelerating UK residential consumer engagement with time-of-use tariffs, electric vehicles and smart meters via digital comparison tools. *Renewable and Sustainable Energy Reviews* 139: 110701.

Daim, T.U. and Yalçın, H. (2022). *Digital Transformations: New Tools and Methods for Mining Technological Intelligence*. Edward Elgar Publishing.

Doğrul, M., Yalçın, H., and Daim, T.U. (2023). Cybersecurity technology: a landscape analysis. In: *Cybersecurity: A Technology Landscape Analysis* (ed. T.U. Daim and M. Dabic), 3–21. Springer.

Hossain, M., Madlool, N., Rahim, N. et al. (2016). Role of smart grid in renewable energy: an overview. *Renewable and Sustainable Energy Reviews* 60: 1168–1184.

Lacy, P., Cooper, T., Hayward, R., and Neuberger, L. (2010). A new era of sustainability in emerging markets? Insights from a global CEO study by the United Nations Global Compact and Accenture. *Corporate Governance* 11: 348–357.

Land, R.E. (2012). Engineering technologists are engineers. *Journal of Engineering Technology* 29 (1): 32

Mah, D.N.-y., Wu, Y.-Y., Ip, J.C.-m., and Hills, P.R. (2013). The role of the state in sustainable energy transitions: a case study of large smart grid demonstration projects in Japan. *Energy Policy* 63: 726–737.

Moura, P.S., López, G.L., Moreno, J.I., and De Almeida, A.T. (2013). The role of smart grids to foster energy efficiency. *Energy Efficiency* 6: 621–639.

Sarsenbayeva, Z., van Berkel, N., Hettiachchi, D. et al. (2023). Mapping 20 years of accessibility research in HCI: a co-word analysis. *International Journal of Human-Computer Studies* 103018.

Tseng, F.-M., Liu, J., Gil, E.I.N.P., and Lu, L.Y. (2023). Identifying and monitoring emerging blockchain technologies using patent analysis. *World Patent Information* 75: 102236.

Tu, M.-Y., Chien, T.-W., Lin, C.-Y., and Chou, W. (2023). Using coword analysis and chord diagrams to examine the effect of nutritional counseling and support (DCNS) on patients with oral and oropharyngeal cancer. *Medicine* 102 (10).

Tuballa, M.L. and Abundo, M.L. (2016). A review of the development of smart grid technologies. *Renewable and Sustainable Energy Reviews* 59: 710–725.

Wang, P., Huang, J.Y., Ding, Y. et al. (2011). Demand side load management of smart grids using intelligent trading/metering/billing system. In: *2011 IEEE Trondheim PowerTech*. IEEE.

6

Technology Intelligence: Cryptocurrencies and Emerging Technologies

Ahmet Ergurum[1], Mursel Dogrul[2], Haydar Yalçın[3], and Tugrul U. Daim[4]

[1] *University of Wisconsin-Milwaukee, Milwaukee, WI 53201, USA*
[2] *National Defense University, Istanbul, Turkey*
[3] *Department of Business Administration, Ege University, Izmir, 35800, Turkey*
[4] *Mark O. Hatfield Cybersecurity and Cyber Defense Policy Center, Department of Engineering and Technology Management, Portland State University, Portland, OR 97201, USA*

6.1 Introduction

Today, we can say that we are in an age where technology is rapidly evolving, and digital transformation affects every field. With this change, many sectors, from financial systems to supply chain management, from healthcare to the energy sector, are introduced to cryptocurrencies and emerging technologies. Cryptocurrencies, especially led by Bitcoin, have revolutionized the financial world. These digital assets, which enable decentralized, transparent, and secure value transfer, are shaking up traditional financial systems. However, cryptocurrencies are not only limited to the financial world, but they are also intertwined with the technological world. In this respect, blockchain technology offers an innovative structure that forms the basis of cryptocurrencies. This technology offers a new perspective on data storage, transfer, and security. However, blockchain stands out as the trigger of transformation not only in the financial world, but also in the supply chain, healthcare, energy, and many other sectors. Emerging technologies offer many opportunities beyond cryptocurrencies. Artificial intelligence and data analytics are important in helping us understand complex data sets and predict future trends in crypto markets. The internet of things (IoT), integrated with blockchain, offers solutions in many areas from supply chain management to product tracking. In this study, the topic areas that emerged in research in the field of crypto currencies were determined and determinations were made about the topics that emerged in cryptocurrency research. For this purpose, text mining, social network analysis, statistical power laws and topic modeling and Latent Dirichlet Allocation (LDA)

Future-Oriented Technology Assessment: A Manager's Guide with Case Applications, First Edition.
Edited by Haydar Yalçın and Tugrul U. Daim.

were used. Kleinberg's burst analysis was used to determine the emergence of concepts, their continuity and end periods. Cryptocurrencies have a wide range of impacts in terms of social, technological, economic, environmental, political, and value judgments. From a social perspective, while it may facilitate global payments by increasing financial access, it also raises risks such as security concerns and money laundering. Technologically, while blockchain technology has the potential to support secure and transparent transactions, problems such as inadequacy of technological infrastructure and security threats should also be considered. Economically, while increased participation in the financial system creates a positive impact, factors such as price fluctuations and lack of regulation can increase the risk of economic instability. Environmentally, energy conservation and the use of green technology can provide positive effects, while the energy intensity of crypto-mining operations can cause environmental impacts. In the field of politics, while the innovations it has brought to the financial system and the ease of cross-border payments have a positive impact, there is a risk of lack of regulation and encouraging illegal activities. In terms of value judgments, while financial freedom and support for decentralized systems offer a positive perspective, distrust of the traditional financial system and the risk of loss of value should not be ignored. Considering this complexity, it is important to follow the evolution and impacts of virtual currencies and take a balanced approach. In our study, the emerging topics regarding these sub-dimensions were analyzed.

Cryptocurrency, a groundbreaking form of digital currency, emerged in the early twenty-first century as a direct response to the limitations of traditional financial systems (Fauzi et al. 2020; Kadoo and Sodi 2023; Zahid 2023). The goal was to create a currency that operates independently of governmental or monetary authority control. Electronic payments could then be executed securely without revealing the identities of the transacting parties (Kayal and Rohilla 2021). This decentralized monetary system, enabling encrypted transactions (García-Corral et al. 2022), finds its roots in the cryptographic and cypherpunk movements, challenging the dominance of centralized financial institutions and redefining monetary transactions (Berg et al. 2018; Danezis and Meiklejohn 2016; Grassman et al. 2021). Satoshi Nakamoto's pivotal 2008 work, "Bitcoin: A Peer-to-Peer Electronic Cash System," marked a revolutionary moment in the history of cryptocurrency (Chohan 2022; Nakamoto 2008). The subsequent release of the open-source Bitcoin software and the successful generation of the initial Bitcoin blockchain section marked significant milestones. Initially gaining traction for its anonymous transactions and lack of governmental oversight, Bitcoin revolutionized decentralized currency, leveraging blockchain technology (Böhme et al. 2015; Campbell-Verduyn 2017; Ghosh et al. 2020; Miraz and Ali 2018; Nerurkar et al. 2021; Tredinnick 2019).

Nakamoto's pseudonymous identity and the collaborative, open-source approach underscored the fundamental principles of decentralization and anonymity at the core of the cryptocurrency domain (Jaoude and Saad 2019; Moore 2013; Taskinsoy 2019). The evolution expanded beyond Bitcoin to include a variety of alternative digital currencies, or altcoins (Nguyen et al. 2019; Siripurapu and Berman 2023; White 2014). Launched in subsequent years, these altcoins aimed to overcome limitations in Bitcoin's architecture or introduce innovative features (Kumari et al. 2023; Werbach 2018).

Post-Bitcoin, there was a surge in cryptographic projects and the development of various consensus mechanisms like proof-of-stake (PoS) and delegated proof-of-stake (DPoS) to improve scalability, energy efficiency, and transaction speed (Borge et al. 2017; Kaur et al. 2021; Yang et al. 2019). The development of cryptocurrencies is closely tied to technological breakthroughs and socioeconomic factors (Bhimani et al. 2022; Nadeem et al. 2021; Sagheer et al. 2022). Over time, the narrative expanded beyond a digital currency system, encompassing broader applications of blockchain technology in areas such as supply chain management, healthcare, voting systems, and decentralized finance (DeFi) (European Systemic Risk Board 2023; Far et al. 2023; IMF Blog 2021). The introduction of smart contracts, notably on platforms like Ethereum, extended the functionalities of blockchain beyond basic financial transactions (Javaid et al. 2022; Khan et al. 2021; Taherdoost 2023). This advancement enabled the creation of self-executing contracts with programmable conditions, leading to the emergence of decentralized applications (dApps) and the tokenization of assets (Bogner et al. 2016; Cuccuru 2017). However, the evolution of cryptocurrencies has faced significant engineering challenges and disputes. Issues include security breaches, legislative uncertainties, market volatility, scalability challenges, and environmental concerns related to energy-intensive mining methods (Gabuthy 2023; Gurgun et al. 2022; Weaver 2022; Yu et al. 2022). Regulatory responses worldwide have varied, ranging from outright bans in certain countries to cautious integration and oversight of cryptocurrencies in others (George et al. 2023). These regulatory actions have profoundly impacted the global landscape of digital currencies and their acceptance. This research provides a comprehensive thematic analysis of cryptocurrency from an engineering perspective, delving into academic literature and published research. Through rigorous academic inquiry, the article examines the complex historical context core engineering concepts, technological advancements, societal and economic impacts, regulatory challenges, and transformative outcomes resulting from scientific discoveries. It meticulously scrutinizes the engineering evolution of cryptocurrencies, highlighting key shifts in thinking and crucial periods that have influenced their growth.

6.2 Data and Method

We conducted an online search of Web of Science (WoS) indexes to access scientific documents on cryptocurrency research. The obtained document information was recorded in a relational database after going through a series of processes (such as cleaning, making it uniform) to be ready for analysis. The dataset spans from 1996 to 2024 and is sourced from 7437 different outlets, including journals and books. It comprises a total of 37 000 documents, with an impressive annual growth rate of 15.16%. The average age of the documents is 2.21 years, indicating a relatively recent and dynamic collection. On average, each document has received 14.71 citations, underscoring the impact and relevance of the content. The dataset contains a substantial number of references, totaling 647 986. In terms of document contents, there are 8332 instances of Keywords Plus (ID) and 46 088 instances of Author's Keywords (DE), providing valuable insights into the thematic elements of the documents. The dataset involves a significant number of authors, with a total of 57 864 individuals contributing to the documents. Among these, 2900 authors have single-authored documents. Collaboration is evident, as there are 3804 single-authored documents, and, on average, each document has 3.65 co-authors. Notably, international co-authorships constitute 31.5% of the collaborations, highlighting a global and collaborative nature in the production of these documents. Overall, this dataset presents a rich and dynamic collection of scholarly works, reflecting substantial growth, collaboration, and international engagement over the specified timespan.

6.2.1 Cluster Analysis

The cryptocurrency research network is segmented into 18 clusters, each offering insights into distinct aspects of the field. Cluster 0, focusing on the cryptocurrency market since 2017, explores trends and dynamics in trading. Cluster 1, which emerged in 2018, delves into the energy consumption associated with cryptocurrency mining and blockchain technologies. In Cluster 2, formed in 2017, researchers examine behavioral intentions related to cryptocurrency adoption. Federated learning takes the spotlight in Cluster 4 (2018), exploring its integration within the cryptocurrency and blockchain context. Payment channel technologies are the focus of Cluster 5 (2018). Cluster 6, dating back to 2016, explores the role of central banks in the cryptocurrency landscape. Self-sovereign identity is the theme of Cluster 7 (2017). Smart grid integration with blockchain is explored in Cluster 8 (2018), while Cluster 9 (2018) investigates the application of machine learning in the cryptocurrency domain. Cluster 10 (2017) delves into the use of blockchain in cloud storage solutions. Various consensus algorithms take center stage in Cluster 11 (2017). The application of blockchain in the real estate sector

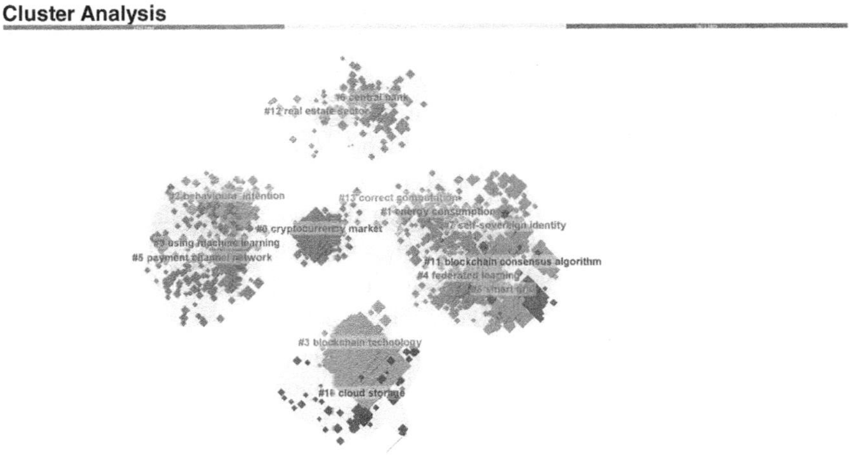

Figure 6.1 Clusters on cryptocurrency research.

is the core of Cluster 12 (2013). Clusters 13–21 focus on niche topics such as correct computation, sharing frameworks, technological convergence, blockchain consensus mechanisms, and federated blockchain-based architecture, each providing specialized insights into emerging technologies within the cryptocurrency research landscape (Figure 6.1).

6.2.2 Burst Analysis and Emerging Topics

Kleinberg's burst analysis is a powerful method used in the field of text mining and information extraction. This analysis is used to evaluate whether a particular concept or word in text data has increased significantly over a certain period. Kleinberg developed this method specifically for the purpose of identifying and analyzing the breaking points of events developing in social media, news articles or other textual data (Kleinberg 2002). This analysis is often used to understand how a particular issue or event impacts society, identify important trends, or detect emergencies. Kleinberg's burst analysis uses mathematical models and statistical analysis to identify significant increases in the frequency of a given term over time. The areas of use of this method are very wide and play an important role in fields such as social sciences, marketing, crisis management, and news media.

Kleinberg's burst analysis is considered a powerful tool for extracting meaningful information from large amounts of text data and understanding shifts in the popularity of a particular topic or word. It was used in our study to see the changes in the popularity of a topic or word related to cryptocurrencies. The burst detection analysis results from Kleinberg highlight significant trends and focal points within the cryptocurrency landscape from 1995 to 2023. Noteworthy entities and themes, along with their corresponding burst strengths, are identified.

Since 1995, digital currency has emerged as a consistent and influential topic, marked by a burst strength of 34.541. Virtual currency, closely related, exhibits sustained attention with a burst strength of 27.5044 over the same period. Money laundering (burst strength: 6.5997) and Silk Road (burst strength: 5.9431) indicate bursts of attention, likely reflecting the evolving regulatory and legal landscape. The year 2015 witnesses a surge in attention to various emerging technologies and concepts. Smart contracts dominated the scene with a burst strength of 55.3485, followed by related terms such as smart contract (47.6971) and economics (34.7528). Inefficiency (31.8104) and distributed ledger (28.3769) signify concerns and exploration in optimizing blockchain technologies. The terms cryptocurrency, distributed systems, and distributed ledgers also experienced bursts in attention. Diving deeper into specific technological components, terms like distributed ledger technology, distributed consensus, block chain, and permissioned blockchain showcase ongoing interest and research. The sharing economy, access control, and key management scheme underscore security and control aspects.

Burst analysis also unveils a focus on the intersection of blockchain with other cutting-edge technologies. IoT, digital economy, and electronic money indicate a keen exploration of the role of blockchain in shaping broader technological landscapes. Cryptographic concepts such as Byzantine fault tolerance, hash function, and directed acyclic graph underline the emphasis on secure and efficient cryptographic protocols. Payment systems, mining pool, and consensus mechanism highlight the intricate dynamics of cryptocurrency networks. Social and ethical dimensions are evident in bursts related to dark web, proof of work, and distributed ledger technologies. Business models, Hyperledger composer, and initial coin offering illustrate the evolving economic and financial paradigms within the cryptocurrency realm. This burst detection analysis provides a temporal lens, revealing the dynamic evolution of topics within the cryptocurrency domain and emphasizing the ongoing interdisciplinary exploration of technology, economics, security, and societal implications (Figure 6.2).

Prominent research focuses by year look like this:

- **2016** Bitcoin and ripples: This year, terms such as Bitcoin and its related topics, ripples, and Hurst exponent were important.
  Co-movement and granger causality: Focused on issues such as market efficiency and risk transmission.
- **2017** Long-term memory in Bitcoin: This year, the focus was on issues such as long-term addiction in relation to Bitcoin.

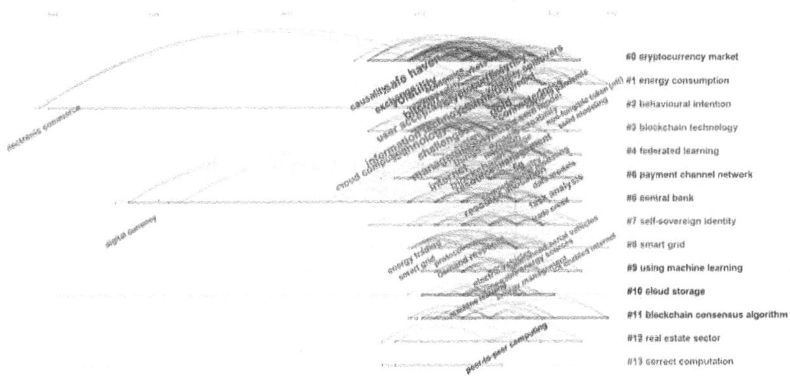

Figure 6.2 Timeline analysis cryptocurrency research.

Hurst and other measures: Focused on analyzing the performance of financial assets.

- **2018** Cryptocurrencies and risk: Along with Bitcoin, the returns and risk management of cryptocurrencies were at the forefront.
  Asymmetric volatility: Studies on asymmetric volatility in financial markets.
- **2019** Market efficiency and hypothesis: Focused on financial market efficiency and efficient market hypothesis.
  Asset pricing and efficient markets: Studies on volatility, asset pricing and efficient markets have come to the fore.
- **2020** Bitcoin and market efficiency: Focused on Bitcoin's market efficiency.
  Asset pricing and efficient markets: A year of continued interest and advanced research.
- **2021** Bitcoin and market activity: Focused on Bitcoin's efficiency and market discovery.
  Asset pricing and COVID-19: The effects of COVID-19 on asset pricing were examined.
- **2022** COVID-19 pandemic and cryptocurrencies: This year, the interaction of the COVID-19 pandemic and cryptocurrencies was significant.
  Various topics: Focused on various topics such as variance coefficient regression, hedging, crude oil market, and monetary policy.
- **2023** COVID-19 pandemic and gold: This year, the role of gold as a safe haven was examined along with the pandemic.
  Macro uncertainty and financial stress: Macro uncertainty, financial stress, and safe harbor concepts are among the prominent topics.

6.2.3 Intertopic Distance Maps and Topic Detection on Cryptocurrency Research Landscape

The intertopic distance map, which is frequently used in information extraction fields such as text mining or topic modeling, refers to the distance map between topics created using multidimensional scaling (MDS). MDS is used to visualize the relationships between items in the data set by trying to preserve the similarities or distances between the items. In the context of topic modeling, MDS is among the frequently used methods to understand the similarities or differences between documents. In this context, "intertopic distance map" can be considered as a map showing the relationships between the themes created by a topic model. In this regard, each topic is generally derived from combinations of words and concepts in the documents. Our study, MDS, aims to create a map representing the similarities or distances between these topics in cryptocurrency research documents and turn them into visuals that can help readers understand the topic similarities of the documents. Such maps are useful for analyzing large text collections, especially for understanding and visualizing thematic relationships between documents. We used an adaptation of the saliency measure for terms within the context of topics in a given corpus, as introduced by (Chuang et al. 2012).

The dynamic world of cryptocurrencies is constantly evolving with developing technologies and financial transformations. In this process of evolution, the fundamental concepts and focuses of the cryptocurrency world are expressed in a number of distinct terms. While the terms *data*, *chain,* and *management*, which stand out according to the topic detection method carried out, emphasize the data processing and management processes that form the basis of blockchain technology, the terms *technology* and *scheme* allow us to understand the technical infrastructure of the cryptocurrency world. The terms *contracts*, *networks,* and *analysis* refer to the legal frameworks, network structures, and analytical processes of cryptocurrencies, while the terms *detection, encryption*, and *security* emphasize security-oriented issues. The terms *acceptance* and *authentication*, which represent important concepts related to the adoption and proliferation of cryptocurrencies, emphasize the importance of these digital assets being accepted by society. The terms *returns, markets,* and *volatility* draw attention to the financial aspects of cryptocurrencies, referring to investment returns, market dynamics, and price fluctuations. These terms reflect the complexity and diversity of the cryptocurrency universe, offering a perspective on the future role and interactions of digital assets (Figure 6.3).

6.2.4 Topic 1: Cryptocurrency Universe and Blockchain Technology

Each map seems to provide a snapshot of the most significant terms associated with a particular topic, and the percentage of tokens gives an idea of the overall

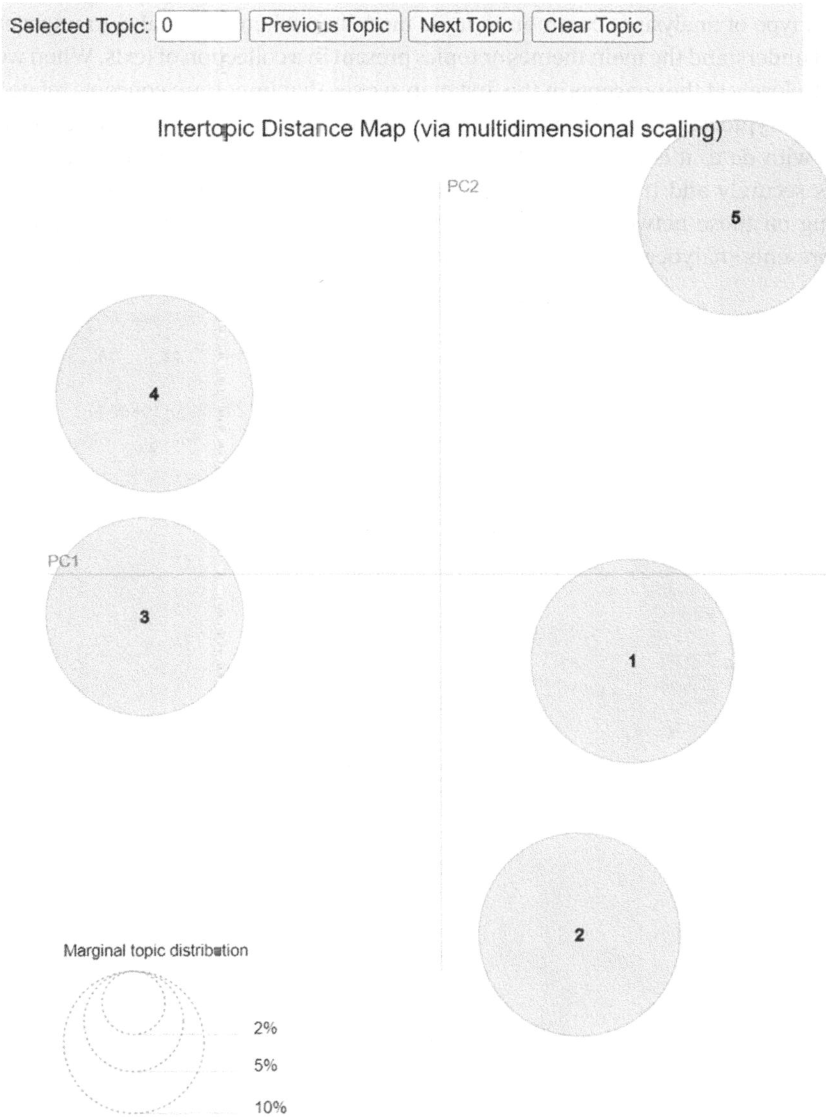

Figure 6.3 All topics.

importance or prevalence of those terms within the respective topics (Figure 6.4). This type of analysis is common in topic modeling, where the goal is to uncover and understand the main themes or topics present in a collection of texts. When we look closely at the concepts in the first map, we see that important concepts related to the cryptocurrency universe and blockchain technology come to the fore. Starting with *data*, it is emphasized that blockchain technology processes large data sets securely and transparently. Transfer transactions of cryptocurrencies operating on these networks are handled under the concept of *networks*. *Analytics* represents analytical methods used to understand the cryptocurrency markets and

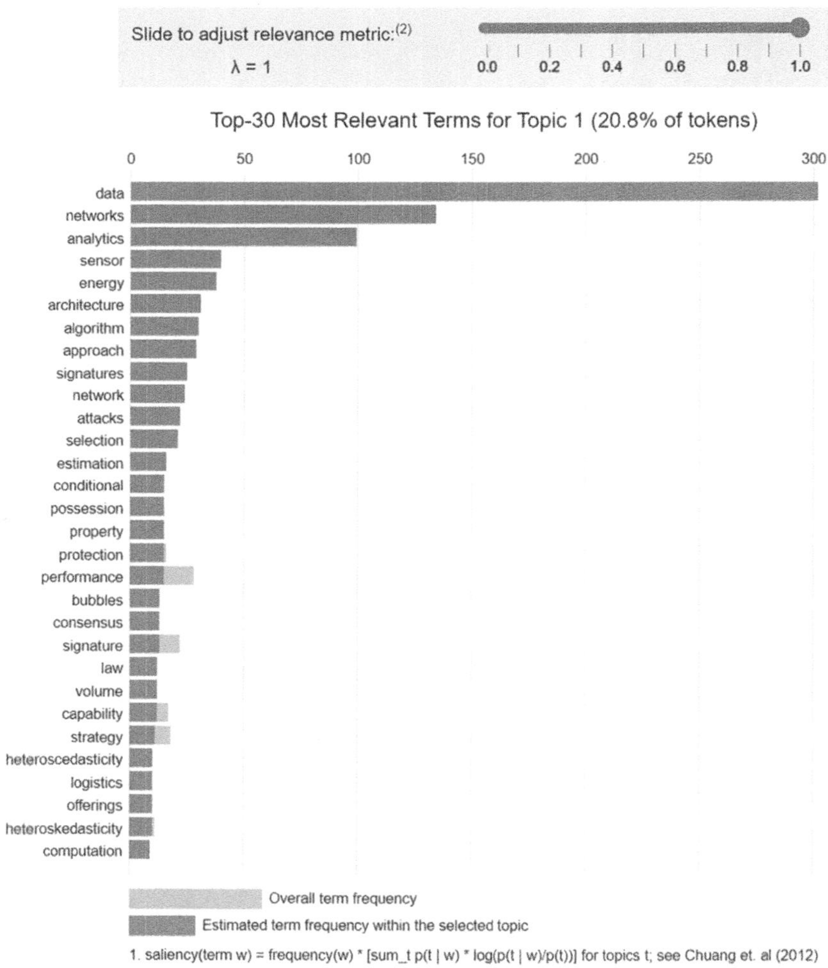

Figure 6.4 Topic 1.

make predictions about their future. The concept of *sensor* refers to the sensitivity of cryptocurrency values to market sentiments and news, while *energy* highlights the importance of energy-intensive cryptocurrency mining processes and environmental impacts. *Algorithm* refers to the algorithms used in cryptocurrency mining and price predictions, while *security* indicates that the security-focused blockchain technology prevents fraud by securely recording transactions. *Consensus* emphasizes the use of consensus protocols by cryptocurrency networks to confirm transactions, while *law* refers to an area where the legal framework is still evolving. Finally, *computation* involves complex mathematical calculations for processing and transferring cryptocurrencies. The combination of these terms is very important as it creates a framework that covers the technical, security, and economic aspects of the cryptocurrency world.

6.2.5 Topic 2: Blockchain-Based Supply Chain Management and Security Optimization

The list of the 30 most relevant terms for Topic 2, accounting for 20.5% of the tokens in the corpus, includes a thematic concentration on issues related to chain management, authentication, and optimization (Figure 6.5). The term *chain* can be thought of as indicative of participation in blockchain technology or a broader chain of processes. *Management* refers to the effective organization and oversight of processes and resources, possibly implying a strategic management perspective. The term *authentication* indicates a focus on security, particularly in terms of implementing authentication protocols. The inclusion of terms such as *scheme, protocol,* and *agreement* emphasizes a systematic and organized approach. Additionally, terms such as *traceability* and *aggregation* imply a concern for tracking the history of products or processes and collecting data in a consolidated manner. Terms such as *demand, production,* and *payment* can mean intersections with areas related to transactions, demand forecasting, and financial aspects. The inclusion of terms such as *emissions, finance,* and *integration* likely points to a broader perspective encompassing sustainability considerations, financial management, and systems integration challenges. To summarize, Topic 2 appears to revolve around a number of interconnected themes, including supply chain dynamics, security through authentication protocols, optimization of processes, and sustainability and finance considerations in the context of a systematic and well-managed framework.

6.2.6 Topic 3: Emerging Trends in Internet Science and Security

Topic 3, which includes the 30 most relevant terms representing 19.8% of the tokens in the corpus, consists of concepts centered around the dynamic

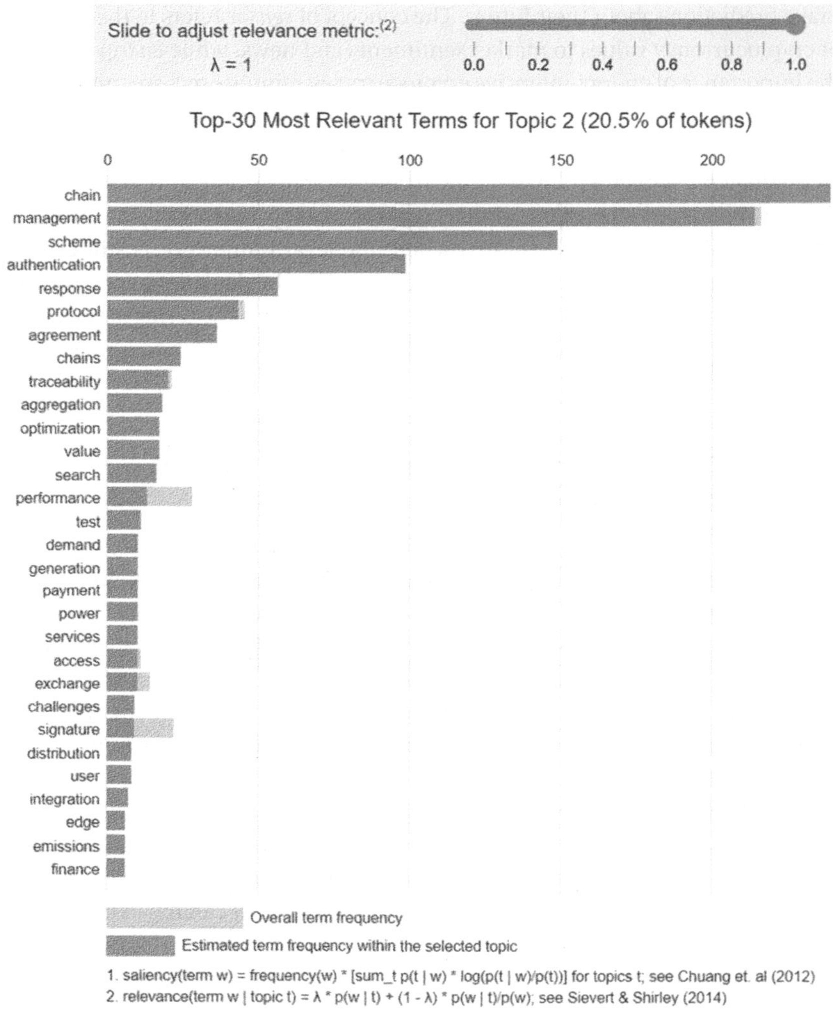

Figure 6.5 Topic 2.

intersection of the internet, technology, and their multifaceted impacts (Figure 6.6). In this context, we can define it as "Emerging Trends in Internet Science and Security." *Internet* refers to the comprehensive investigation of various issues related to the large, interconnected network. The term *adoption* suggests focusing on the widespread adoption and integration of internet-related technologies, while *model* refers to the development of conceptual frameworks

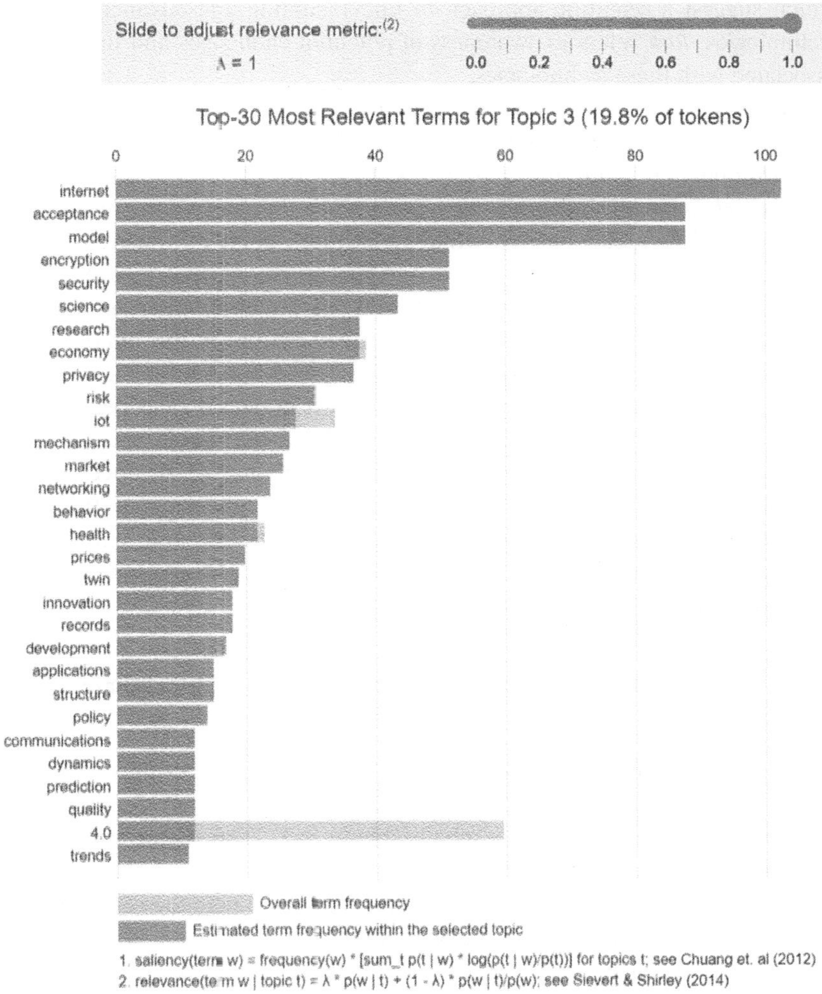

Figure 6.6 Topic 3.

and theoretical structures in the field. As terms like *encryption, security,* and *privacy* indicate, security, and privacy are top of mind. The theme underlines the necessity of protecting information and systems in the age of rapid technological advancement. *IoT* adds another layer, highlighting the interconnectivity of devices and the role of IoT in shaping the internet landscape. The thematic scope expands further with terms such as *economy*, which refers to the study of the economic impact of internet-related developments, and *science* and *research*, which suggest a scientific approach to understanding and advancing internet technologies. *Risk* refers to awareness of potential challenges and uncertainties associated with these technologies.

Various social dimensions come into focus, including *behaviour,* which may be relevant to the study of user behavior in the digital environment, and *health,* which proposes to study the impact of internet technologies on health-related aspects. *Development* and *innovation* underscore the continuous development of internet-related technologies, while *applications* point to practical uses and applications of internet science. *Trends* and *forecast* emphasize a forward-looking perspective, acknowledging the ever-changing nature of the internet space. The theme addresses specific areas such as *market* dynamics, the interconnected nature of *networking* and *policy* frameworks governing internet technologies. Terms like *logs* and *dynamics* indicate an interest in data management and the dynamic nature of internet-related trends. Overall, the theme encompasses a holistic examination of the internet, covering technological advances, societal impacts, economic considerations, and ongoing trends shaping the digital landscape. The emphasis on security, privacy, and the dynamic nature of the field reflects a comprehensive understanding of the complexities inherent in the evolving field of internet science.

6.2.7 Topic 4: Innovations in Financial Technology and Blockchain Solutions

Topic 4, which includes the 30 most relevant terms accounting for 19.7% of the tokens in the corpus, revolves around a thematic focus on the intersection of technology, finance, and blockchain (Figure 6.7). The theme can be aptly described as "Innovations in Financial Technology and Blockchain Solutions." Underlying this theme is the term *technology*, which refers to the comprehensive investigation of technological developments with a special emphasis on their impact on financial systems. *Contracts* and *returns* indicate a strong interest in contract and investment aspects, possibly in the field of smart contracts and financial returns. The inclusion of "Industry 4.0" points to the overarching concept of the fourth industrial revolution and indicates a forward-looking perspective on technological developments. *Blockchain* stands out as a pivotal term that highlights the significant focus on blockchain technology and its transformative potential within financial and technological environments. This theme dives into the intricacies of distributed ledger technologies, and *ledger* further highlights the role of ledgers in these innovations. *Adoption* and *diffusion* suggest consideration of the adoption of new technologies, possibly within financial systems, and the potential ripple effects of these adoptions. The term *control* perhaps implies the study of the mechanisms that govern technology in the context of governance

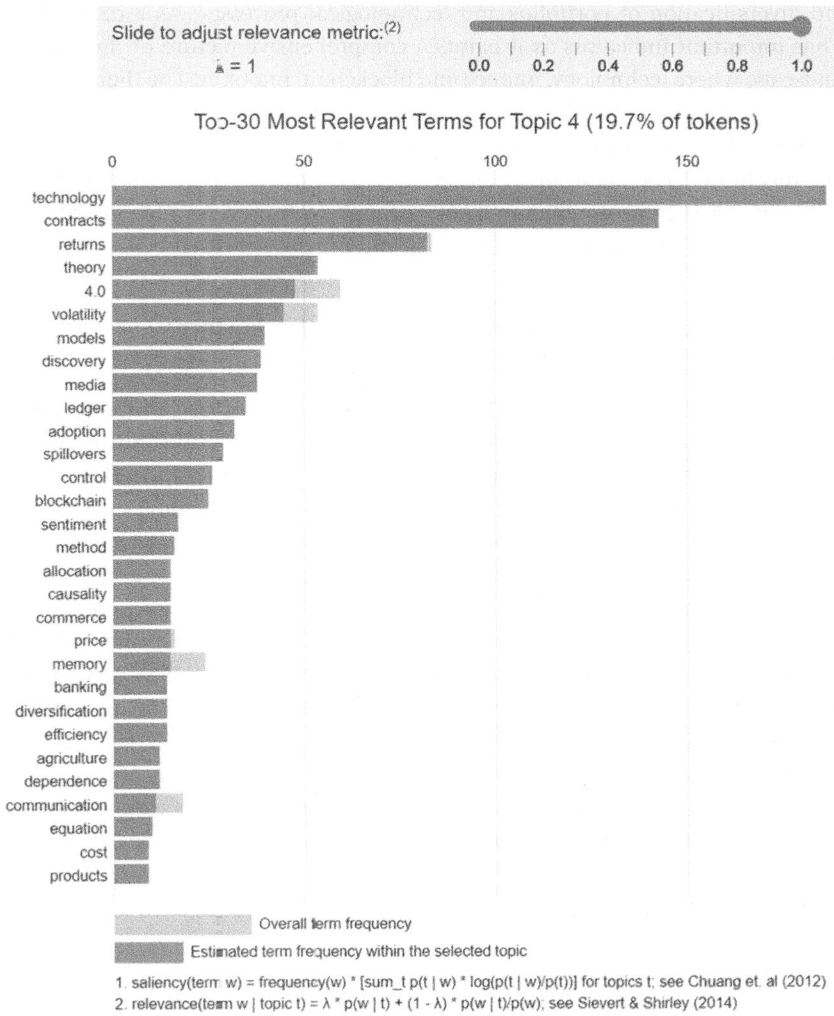

Figure 6.7 Topic 4.

frameworks. *Sentiment* potentially refers to sentiment analysis in financial markets, while *models* and *methodology* suggest a methodological approach, possibly in the context of financial or technological analysis. *Causation* refers to the study of causal relationships, especially in technological or financial fields. The theme expands its scope to various sectors including *commerce, banking, agriculture,* and *products,* indicating a wide exploration of technological solutions in different sectors. *Diversification* and *efficiency* emphasize the optimization and diversification of portfolios and technological processes. As a result, Topic 4 has important indicators as it paints a comprehensive picture of the evolving landscape where technology, finance and blockchain intersect. The theme reflects the exploration of innovative solutions, adoption models, and the transformative potential of blockchain technologies in the broader context of the fourth industrial revolution. The inclusion of terms such as *contracts, rebates,* and *blockchain* highlights the theme's depth in understanding the complexities and opportunities that arise at the intersection of technology and finance.

6.2.8 Topic 5: Advanced Systems Analysis and Governance in Current Issues

Covering the 30 most relevant terms and accounting for 19.2% of the tokens in the corpus, Topic 5 delves into a theme identified under the title "Advanced Systems Analysis and Governance in Current Issues" (Figure 6.8). Unlike the previous four topics in terms of its focus, it can be defined as a topic that involves the complex examination of various topics. *Analysis* is the basic concept. Following closely behind is *detection,* which shows a strong interest in mechanisms for identifying vulnerabilities or irregularities in systems. The term *sanctuary* implies the investigation of safe or protected environments, while the repeated mention of *systems* emphasizes a comprehensive examination of complex structures. The theme expands its reach into various areas, such as *markets,* which shows the analysis of economic systems, and *vehicles,* which proposes to focus on transportation technologies. *Contract* and *governance* denote the investigation of contractual agreements and governance structures, respectively, and underscore a focus on organizational and legal frameworks. The *framework* also emphasizes the use of structured approaches in analysis and governance. *Uncertainty* and *reality* refer to subtle considerations, possibly involving risk analysis and practical applications within complex systems.

The theme extends to urban contexts, with terms such as *city* and *road* indicating a focus on urban planning and transportation infrastructure. *Volatility, futures,* and *advantage* suggest the study of dynamic elements within systems, including economic fluctuations and strategies for gaining advantage in competitive environments. *Decomposition* refers to an analytical approach that involves breaking complex systems into understandable components. *Storage* and *identification* refer to issues related to data storage systems and identification of entities within complex frameworks. *Revocation* suggests focusing on mechanisms for withdrawing or replacing specific elements in governance structures. The theme expands to include technological aspects, with terms such as *weather* possibly referring to the study of weather systems or technologies. *Agenda* refers to a planned set of

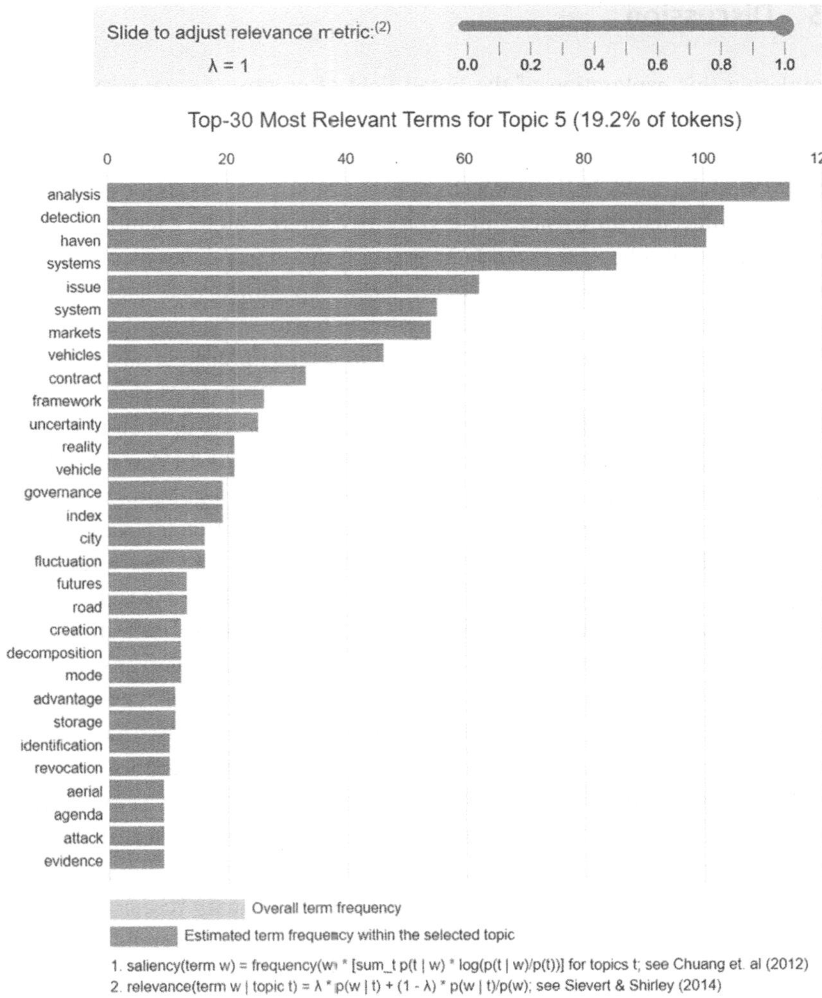

Figure 6.8 Topic 5.

activities or issues to be considered, while *attack* refers to the investigation of security issues within systems. In summary, Topic 5 provides a detailed examination of current issues, leveraging advanced systems analysis, detection mechanisms, and governance structures. The breadth of the theme covers economic, organizational, and technological systems, emphasizing a comprehensive and analytical approach to addressing the complexities inherent in contemporary challenges.

6.3 Discussion

Completing this exploration of the broad field of cryptocurrency research, our journey through 18 thematic clusters, burst analysis descriptions, temporal insights, and topic-specific insights reveals a dynamic tapestry of knowledge in this ever-evolving field. In this respect, each cluster provides a unique perspective on the multifaceted dimensions of cryptocurrency research. Kleinberg's burst analysis served as a beacon to guide us through the ups and downs of attention in the cryptocurrency world from 1995 to 2023. Exploding assets such as digital currency, virtual currency, and smart contracts have marked pivotal moments reflecting dynamic interaction, regulatory changes, technological advances, and societal changes. The temporal lens extending from 2016 to 2023 showed us the transformation of the industry from Bitcoin and fluctuations in 2016 to its deep impact in 2022. The intertopic distance maps we achieve with LDA visually bring together the thematic texture of cryptocurrency research; It highlights key aspects such as *data, chain,* and *governance,* and also goes into the intricate details of technology, legal frameworks, and financial considerations. In topic-specific analysis, we journeyed through the cryptocurrency universe, examining blockchain technology, supply chain management, internet science, financial innovations, and advanced systems analysis. Each topic represents an aspect of this ever-expanding field and provides a comprehensive guide to researchers, policymakers, and industry professionals. Ultimately, this study serves as a compass that helps us understand cryptocurrency dynamics. This method can be used not only as a valuable tool for dealing with today's complexities but also for envisioning future trajectories shaped by technological advances, economic changes, security considerations, and societal outcomes. It will undoubtedly continue to illuminate new paths and cryptocurrencies through ongoing interdisciplinary research and will make significant contributions to the evolving narrative of the currency world.

References

Berg, C., Davidson, S., and Potts, J. (2018). *Some Public Economics of Blockchain Technology*. Social Science Research Network.

Bhimani, A., Hausken, K., and Arif, S. (2022). Do national development factors affect cryptocurrency adoption? *Technological Forecasting and Social Change* 181: 121739. https://doi.org/10.1016/j.techfore.2022.121739.

Bogner, A., Chanson, M., and Meeuw, A. (2016). A decentralised sharing app running a smart contract on the Ethereum blockchain. In: *Proceedings of the 6th International Conference on the Internet of Things*.

Böhme, R., Christin, N., Edelman, B., and Moore, T. (2015). Bitcoin: economics, technology, and governance. *The Journal of Economic Perspectives* 29 (2): 213–238. JSTOR. http://www.jstor.org/stable/24292130.

Borge, M., Kokoris-Kogias, E., Jovanovic, P. et al. (2017). Proof of personhood: redemocratizing permissionless cryptocurrencies. In: *2017 IEEE European Symposium*. IEEE.

Campbell-Verduyn, M. (2017). *Bitcoin and beyond: Cryptocurrencies, Blockchains, and Global Governance*, 1e. Routledge https://doi.org/10.4324/9781315211909.

Chohan, U.W. (2022). A history of Bitcoin. SSRN scholarly paper 3047875. http://dx.doi.org/10.2139/ssrn.3047875.

Chuang, J., Manning, C.D., and Heer, J. (2012). Termite: visualization techniques for assessing textual topic models. In: *Proceedings of the International Working Conference on Advanced Visual Interfaces*. AVI.

Cuccuru, P. (2017). Beyond bitcoin: an early overview on smart contracts. *International Journal of Law and Information Technology* 25 (3): 179–195.

Danezis, G. and Meiklejohn, S. (2016). Centrally banked cryptocurrencies. arXiv. arXiv:1505.06895

European Systemic Risk Board (2023). *Crypto-Assets and Decentralised Finance: Systemic Implications and Policy Options*. European System of Financial Advisers https://data.europa.eu/doi/10.2849/131265.

Far, S.B., Rad, A.I., and Asaar, M R. (2023). Blockchain and its derived technologies shape the future generation of digital businesses: a focus on decentralized finance and the Metaverse. *Data Science and Management* 6 (3): 183–197. https://doi.org/10.1016/j.dsm.2023.06.002.

Fauzi, M.A., Paiman, N., and Othman, Z. (2020). Bitcoin and cryptocurrency: challenges, opportunities and future works. *Journal of Asian Finance, Economics and Business* 7 (8): 695–704.

Gabuthy, Y. (2023). Blockchain-based dispute resolution: insights and challenges. *Games* 14 (3): https://doi.org/10.3390/g14030034.

García-Corral, F.J., Cordero-García, J.A., de Pablo-Valenciano, J., and Uribe-Toril, J. (2022). A bibliometric review of cryptocurrencies: how have they grown? *Financial Innovation* 8 (1): 2. https://doi.org/10.1186/s40854-021-00306-5.

George, K., Boyle, M. J., and Kvilhaug, S. (2023). Cryptocurrency regulations around the world. Investopedia. https://www.investopedia.com/cryptocurrency-regulations-around-the-world-5202122 (accessed July 31, 2023).

Ghosh, A., Gupta, S., Dua, A., and Kumar, N. (2020). Security of cryptocurrencies in blockchain technology: state-of-art, challenges and future prospects. *Journal of Network and Computer Applications* 163: 102635–102635.

Grassman, R., Bracamonte, V., Davis, M., and Sato, M. (2021). Attitudes to cryptocurrencies: a comparative study between Sweden and Japan. *The Review of Socionetwork Strategies* 15 (1): 169–194. https://doi.org/10.1007/s12626-021-00069-6.

Gurgun, A.P., Genc, M.I., Koc, K., and Arditi, D. (2022). Exploring the barriers against using cryptocurrencies in managing construction supply chain processes. *Buildings* 12 (3): https://doi.org/10.3390/buildings12030357.

IMF Blog (2021). Cryptos on the rise 2022. https://www.thomsonreuters.com/en/reports/cryptos-on-the-rise-2022.html (accessed December 2022).

Jaoude, J.B. and Saad, R.G. (2019). Blockchain applications – usage in different domains. *IEEE Access* 7: 45360–45381.

Javaid, M., Haleem, A., Singh, R.P. et al. (2022). A review of blockchain technology applications for financial services. *BenchCouncil Transactions on Benchmarks, Standards and Evaluations* 2 (3): 100073. https://doi.org/10.1016/j.tbench.2022.100073.

Kadoo, S. and Sodi, K. (2023). An analysis of cryptocurrency, Bitcoin and the future. *International Journal of Advanced Research in Science, Communication and Technology* 287–292.

Kaur, S., Chaturvedi, S., Sharma, A.I., and Kar, J. (2021). A research survey on applications of consensus protocols in blockchain. *Security and Communication Networks* 2021: 1–22.

Kayal, P. and Rohilla, P. (2021). Bitcoin in the economics and finance literature: a survey. *Sn Business & Economics* 1 (7): 88. https://doi.org/10.1007/s43546-021-00090-5.

Khan, S.N., Loukil, F., Ghedira-Guegan, C. et al. (2021). Blockchain smart contracts: applications, challenges, and future trends. *Peer-to-Peer Networking and Applications* 14 (5): 2901–2925. https://doi.org/10.1007/s12083-021-01127-0.

Kleinberg, J. (2002). Bursty and hierarchical structure in streams. In: *Proceedings of the Eighth ACM SIGKDD International Conference on Knowledge Discovery and Data Mining*.

Kumari, V., Bala, P.K., and Chakraborty, S. (2023). An empirical study of user adoption of cryptocurrency using blockchain technology: analysing role of success factors like technology awareness and financial literacy. *Journal of Theoretical and Applied Electronic Commerce Research* 18 (3): 1580–1600. https://doi.org/10.3390/jtaer18030080.

Miraz, M.H. and Ali, M. (2018). Applications of blockchain technology beyond cryptocurrency. *Annals of Emerging Technologies in Computing* 2 (1): 1–6.

Moore, T.M. (2013). The promise and perils of digital currencies. *International Journal of Critical Infrastructure Protection*.

Nadeem, M.A., Liu, Z., Pitafi, A.H. et al. (2021). Investigating the adoption factors of cryptocurrencies – a case of bitcoin: empirical evidence from China. *SAGE Open* 11 (1): 2158244021998704. https://doi.org/10.1177/2158244021998704.

Nakamoto, S. (2008). *Bitcoin: A Peer-to-Peer Electronic Cash System*. https://bitcoin.org/bitcoin.pdf.

Nerurkar, P., Patel, D., Busnel, Y. et al. (2021). Dissecting bitcoin blockchain: empirical analysis of bitcoin network (2009–2020). *Journal of Network and Computer Applications* 177: 102940–102940.

Nguyen, T., Nguyen, B.P., Nguyen, T.N., and Nguyen, Q.V. (2019). Bitcoin return: impacts from the introduction of new altcoins. *Research in International Business and Finance* 48: 420–425.

Sagheer, N., Khan, K.I., Fahd, S. et al. (2022). Factors affecting adaptability of cryptocurrency: an application of technology acceptance model. *Frontiers in Psychology* 13: 903473. https://doi.org/10.3389/fpsyg.2022.903473.

Siripurapu, A. and Berman, N. (2023). *Cryptocurrencies, Digital Dollars, and the Future of Money*. Council on Foreign Relations https://www.cfr.org/backgrounder/cryptocurrencies-digital-dollars-and-future-money (accessed February 28, 2023).

Taherdoost, H. (2023). Smart contracts in blockchain technology: a critical review. *Information* 14 (2): https://doi.org/10.3390/info14020117.

Taskinsoy, J. (2019). Blockchain: a misunderstood digital revolution. Things you need to know about blockchain. *Social Science Research Network*.

Tredinnick, L. (2019). Cryptocurrencies and the blockchain. *Business Information Review* 36 (1): 39–44. https://doi.org/10.1177/0266382119836314.

Weaver, N. (2022). The death of cryptocurrency: the case for regulation. In: *Yale Law School Information Society Project, Ditigal Future Whitepaper Series*.

Werbach, K. (2018). Trust, but verify. *Berkeley Technology Law Journal* 33 (2): 487–550. JSTOR. https://www.jstor.org/stable/26533144.

White, L.J. (2014). The market for cryptocurrencies. *Social Science Research Network*.

Yang, F., Zhou, W., Wu, Q. et al. (2019). Delegated proof of stake with downgrade: a secure and efficient blockchain consensus algorithm with downgrade mechanism. *IEEE Access* 7: 118541–118555.

Yu, C., Yang, W., Xie, F., and He, J. (2022). Technology and security analysis of cryptocurrency based on blockchain. *Complexity* 2022: e5835457. https://doi.org/10.1155/2022/5835457.

Zahid, B. (2023). Cryptocurrency phenomenon: a paradigm shift in global finance – Prime Institute. https://primeinstitute.org/cryptocurrency-phenomenon-a-paradigm-shift-in-global-finance (accessed October 17, 2023).

7

Technology Intelligence: Geothermal Energy

Tugrul U. Daim[1], Mehdi Zamani[2], Ali B. Naeini[3], Fayez Alsoubaie[1], Hao Zhang[5], and Haydar Yalçın[6]

[1] *Mark O. Hatfield Cybersecurity and Cyber Defense Policy Center, Department of Engineering and Technology Management, Portland State University, Portland, OR 97201, USA*
[2] *Kiel University, Kiel, Germany*
[3] *Iran University of Science and Technology, Tehran, Iran*
[4] *Department of Engineering and Technology Management, Portland State University, Portland, OR 97201, USA*
[5] *Chengdu Medical College, Chengdu, Sichuan Province, China*
[6] *Department of Business Administration, Ege University, Izmir 35800, Turkey*

7.1 Introduction

The energy drawn from Earth's core is referred to as geothermal (Bagher et al. 2014). It entails heat from underground hot water and steam reservoirs that are often harvested to generate power or electricity. According to the Environmental and Energy Study Institute, it is one of the renewable sources of energy, which implies that it does not attract greenhouse emissions or carbon footprint (Environmental and Energy Study Institute 2006). The process entails converting heat energy for various purposes, including electricity, heating space, and cooking, in place of nonrenewable energy natural gas, nuclear energy, petroleum, coal, and other forms of fossil fuels. The continental plates undergo friction that results in heat, while in other instances, it is generated from thorium and uranium, among other elements, through a process known as radioactive decay (Riggs et al. 1991).

Earth's heat is often harvested through geothermal power plants and geothermal heat pumps, which capture steam used for electricity and heating water, among other roles in buildings. Wells are drilled in geothermal power plants to capture hot water and steam and pump it into Earth's surface. This explains the prevalence of power plants in areas deemed hot on the surface or slightly beneath,

Future-Oriented Technology Assessment: A Manager's Guide with Case Applications, First Edition.
Edited by Haydar Yalçın and Tugrul U. Daim.

such as geysers, hot springs, and areas known for volcanic activity. The process entails pumping hot water into the surface, which implies that pressure will drop, thus turning it into steam, which is also channeled toward a turbine (Stober and Bucher 2013). The turbine ensures the generator is powered to produce electricity. Earth's center is considered molten, which implies that the heat is intense with temperatures above 5000 °C, which heat the mantle.

According to Bagher et al., geothermal is drawn from *geo* and *therm,* which are Greek words meaning "earth" and "heat" from the earth, respectively (Bagher et al. 2014). The authors argue that American Paleo-Indians from North America were the first users of geothermal energy over the decamillennial. During the decamillennial, it is believed that humans used geothermal energy – in this case, water from hot springs for cleaning, bathing, or cooking (González-Acevedo and García-Zarate 2018). Bagher et al. assert that in the late eighteenth century, geothermal energy gained significant recognition and subsequent industrial use around Pisa, Italy (Bagher et al. 2014). Currently referred to as Larderello fields, steam was often harvested from natural vents and manmade holes where boric acid was extracted. Piero Ginori Conti, the Italian scientist, is credited for the first construction of a geothermal power plant that generated electricity in 1904, a concept that extended to the US in 1922 (González-Acevedo and García-Zarate 2018). Following 1973, the oil crisis that automatically meant the need to reduce overreliance on the nonrenewable source of energy, most countries shifted their focus to renewable sources of energy, including geothermal heat pumps and power plants. Regions constantly experiencing volcanoes and earthquakes often embrace geothermal power plant systems.

Just like biomass and hydropower, geothermal energy is considered one of the cleanest and sustainable sources of renewable energy (Bilgin 2018). They do not burn fossil fuels, as is the case in nonrenewable energy sources that have adverse effects on aquatic, human, and wildlife. The focus is on the fact that they have little or no contribution to climate change through the emission of greenhouse gas or carbon footprint. This, as (Bagher et al. 2014) write, implies that geothermal electricity, for example, is environmentally friendly. Apart from the transport system and land drilling process, other aspects are barely exposed to pollution, as is the case during the extraction of nonrenewable energy sources such as petroleum and coal. It is considered an affordable and reliable green energy solution since, more often, unlike other renewable sources such as solar and wind, the resource is readily available. In some instances, geothermal energy is a primary source of continuous heat and power for buildings and homes (Stober and Bucher 2013; Curtis et al. 2005). In considerable quantity and quality, it is used to produce electricity and encouraged globally because of its minimal impact on the environment. Collectively, it is a clean, natural, low-cost, and safe source of energy (Bilgin 2018).

Greenhouse applications, agriculture, and district heating are the direct uses of geothermal energy (Allahvirdizadeh 2020).

Office of Energy Efficiency & Renewable Energy is a renewable resource whose adoption should be encouraged because it is clean, homegrown, and sustainable, since it is drawn from the heat from Earth's core. The most common sources of geothermal energy include hydrothermal and steam vents, hot springs, and geysers. The process entails capturing steam and heat from the aforementioned sources and converting them through processes such as binary cycles to create electricity. According to Ehyaei et al., the nonrenewable energy resources that are dominant in the contemporary world are facing imminent depletion and discourage for their contribution to global warming, among other harmful effects (Ehyaei et al. 2020). This automatically shifts concerns toward renewable energy resources such as geothermal energy that register relatively low and manageable environmental impacts.

7.2 Impacts of Geothermal Energy

Figure 7.1 illustrate the critical global gaps and challenges affecting its development and utilization which includes economic barriers, technological limitations, issues related to energy fuels and sources, thermodynamic constraints, and regulatory frameworks. In addition, To unlock geothermal energy's full potential and ensure its successful integration into the global energy mix, these factors are explained in the next section.

7.2.1 Economic Impacts of Geothermal Energy

Installation of geothermal systems is competitive and cost-effective (Khanam and Daim 2021). Soltani et al. allude that the quality of resources and higher costs of capital is linked to the low deployment rates of geothermal energy (Soltani et al. 2021). The costs include surface and subsurface investments that entail the power plant and gathering process (Hackstein and Madlener 2021). Other costs include reservoir stimulation, drilling, and expertise needed to evaluate the wells needed and power plants' ability to generate a target capacity, operation and maintenance, and exploration and permitting. Just like other constructions, setting up and running a geothermal power plant requires skilled and semi-skilled labor, including reservoir and hydraulic engineers, geophysicists, and geologists, which implies that locals are guaranteed employment opportunities (Topić et al. 2014). Also, it attracts revenues from the electrical output (Hackstein and Madlener 2021).

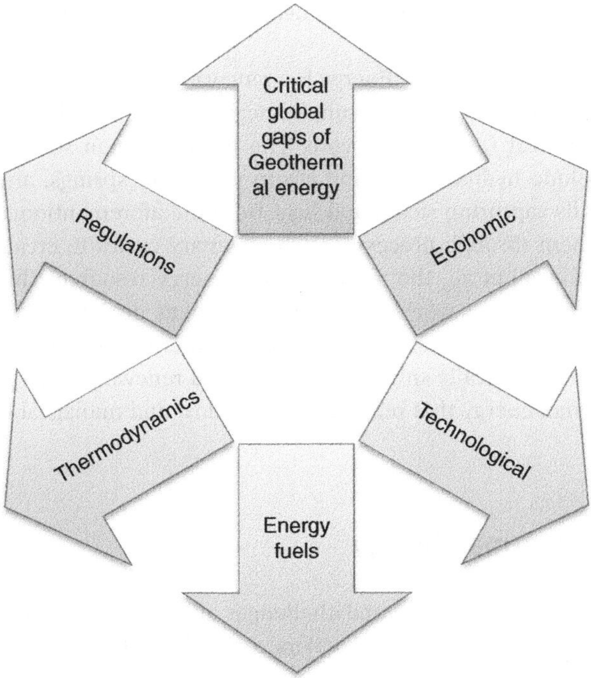

Figure 7.1 Impacts of Geothermal Energy.

7.2.2 Environmental Impacts of Geothermal Energy

Geothermal energy is a sustainable, renewable source of energy, which automatically implies that its carbon footprint is impressive. According to Dhar et al. (2020), they are a clean energy source and, as such, register relatively low water and land exploitation and emissions of hydrogen sulfide and carbon dioxide linked with global warming or climate change. On the other side, exploitation of geothermal energy attracts detrimental environmental impacts such as chemical and noise pollution, soil erosion, depletion of hot springs, and destruction of sceneries or habitat changes from land disturbance. Land subsidence is prevalent during exploitation, that also invites seismic activities. (Arnórsson 2004; Bošnjaković et al. 2019) Regardless, renewable sources of energy, in this case, geothermal, have lesser environmental impacts (Dhar et al. 2020; Bošnjaković et al. 2019; Kumar 2020).

7.2.3 Technological Impacts of Geothermal Energy

The process and technologies for converting the hydrothermal substance into electricity include binary cycle, flash, and dry steam (Aziz and Juangsa 2021;

Barbier 1997). The conversion technologies are dependent on fluid temperatures and form, which varies between steam and water, in enhanced geothermal energy.

7.2.4 Green Sustainable Science Technology

According to Pandiella et al., contemporary society increasingly targets green ecology, sustainability, and renewable energy, including geothermal energy (Pandiella-Dominique et al. 2018). The focus further entails limiting pollution from greenhouse gases drawn from fossil fuels. Reliance on binary cycle, flash, and dry steam is meant to contain overreliance on imported and fossil fuels (Aziz and Juangsa 2021).

7.2.5 Energy Fuels

Clean energy technologies, just like green sustainable science technologies, are advancing in recent times (Azad 2020). The world embraces clean energy fuels that barely attract unsustainable emission resources, environmental sustainability, and efficiency. The reliance on technologies such as binary cycles and dry steam facilitates the conversion of heat from Earth's core, which is then used as a renewable source of energy, serving as sources of electricity and in cases where production is relatively low, as sources of direct heat for buildings and homes (Michaelides 2016; Moya et al. 2018). The extracted energy, in this case, assumes either liquid or steam form.

7.2.6 Thermodynamics

Thermodynamics, in this case, overseen by geophysicists, involve the connection between temperature, heat, and energy (Heather 2020). Generating electricity or basic district heat from geothermal energy primarily involves thermodynamics, which entails the production of energy from heat or Earth's hot temperatures. This is based on the fact that Earth's surface and crust are considered extremely hot and subjected to binary cycle, dry steam, and flash steam technologies to generate energy, heat buildings, and source of electricity, among other human uses. In this regard, Kanoglu and Bolatturk allude that "the first law of thermodynamics is the application of the conservation of energy principle to heat and thermodynamic processes" (Kanoglu and Bolatturk 2016). It further entails the exergy method, which entails the second law efficiency of the thermodynamic process and energy-intensive systems (Amrollahi et al. 2011).

7.2.7 Regulation

As Chen posits, countries such as Taiwan do not have clear legislation, regulation, or legal framework that defines the adoption of geothermal energy despite its potential (Chen 2015). In cases where laws and procedures exist, concerns of complexity impede private companies' desire to invest in the generation of geothermal power. In the United States, as highlighted by Tim Spisak, State Director for New Mexico, Oklahoma, Texas and Kansas. Bureau of Land Management (Spisak 2019), the Geothermal Steam Act of 1970 is a law that recognizes geothermal energy and encourages leasing of public resources under the forest service and other related entities for geothermal power-related constructions and developments. There is a need for research and consultation to establish the different environmental legislations, including international treaties and environmental laws that might impede or enhance geothermal projects, especially for multinational entities (Arias 2016; Kaneko 2016; Sustersic et al. 2010). Consequently, they avoid legal penalties and sanctions.

7.2.8 Critical Global Gap of Geothermal Energy

Currently, governments and non-governmental organizations are working toward reducing reliance on fossil fuels, which increases climate change (Hu et al. 2021). Countries such as China that have a considerably higher carbon footprint invest in geothermal resources as a renewable energy source. However, the process, especially general geothermal power, has been affected by the low-temperature hydrothermal resources around target centers, limiting its use around district heating geothermal heat pumps. Countries such as Taiwan, despite their geothermal resources, are yet to exploit the source of energy, arguably because of the cost of setup and drilling technology, among other technical issues (Chen 2015). For this reason, there is a need to embrace economic incentives to private companies to encourage investment in geothermal power plant construction that generates considerable electricity. Also, legal barriers, including the lack of a clear procedure or regulations that cover the construction of geothermal plants in restricted areas, should be addressed collectively and locally or nationally (Chen 2015).

7.3 Methodology

The used research method and its steps are shown in Figure 7.2.

The steps of the research method in this study are as follows.

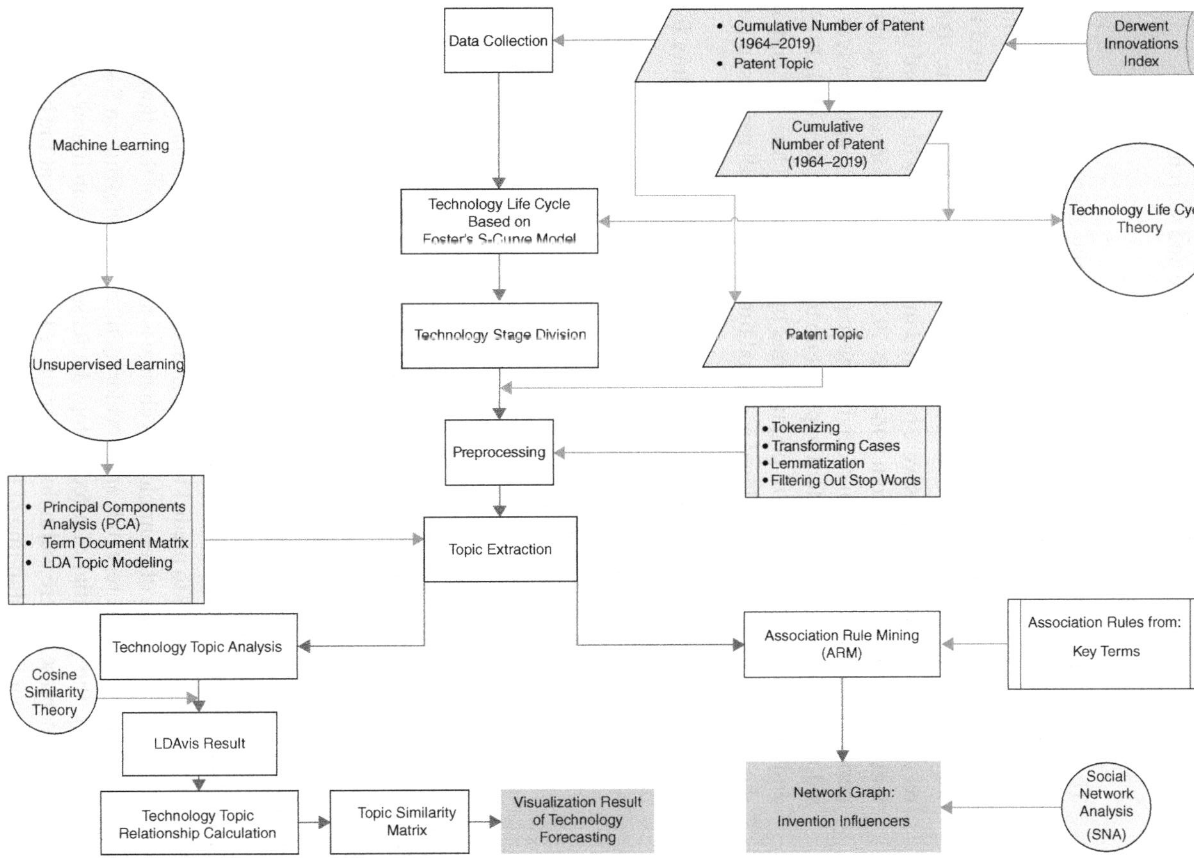

Figure 7.2 Research steps.

7.3.1 Data Collection

- ✓ Retrieving data related to geothermal energy patents in the Derwent Innovation Index database using the "topic" search route and search string:

  ```
  Search String: "Geothermal"
  ```

  Then, the patent documents in the field of geothermal energy were retrieved by referring to Derwent Innovation Index, which lists more than 82 million exclusive documents (Zhang et al. 2018). Using the "topic" search route and the built search string. Since the limit on the number of records received in the new update of the Web of Science (WOS) has increased from 500 to 1000 records, the patent documents were retrieved in 1000 pieces with record content of "full record" record from 1964 to 2019. The reason for selecting this period is that the USPTO upon receiving the request from the applicants considers the confidential status for the patent up to 18 months (Kim et al. 2019) and no patent information will be published by that time. Thus, the end of the period (i.e. 2019) is considered as a good time to review all patents in the field.
- ✓ Extracting title and abstract of patents from the retrieved documents for pre-processing
  In this section, the information retrieved from patent documents was combined and pre-processed using Bibexcel software (Persson et al. 2009), which is a screening software in the field of scientometrics and bibliometrics. A "summary" of patents was isolated and filtered after entering Excel software.

7.3.2 Extraction of Life Cycle of Geothermal Energy Technologies Based on the Number of Cumulative Patents (Foster Curve Model)

In this study, the classification of research and development activities in the field of geothermal energy was based on the technology life cycle theory, Foster S model, and the number of cumulative patents in the field.

This model is called *technological discontinuity prediction when* the obsolescence of a technology is proven by a new technology (Kasztler et al. 2012). Other authors have used similar models to predict technological discontinuity (Sahal 1985). This model states that the rate of technological change is a function of the efforts made for a technology, such as research and development costs, following an S-shaped curve. In fact, the potential of a technology increases as research and development costs increase, where technological progress begins slowly at the beginning because of a poor understanding of the principles of technology.

As technology matures, its performance increases rapidly with the accumulated costs of research and development. Finally, the potential of a technology reaches

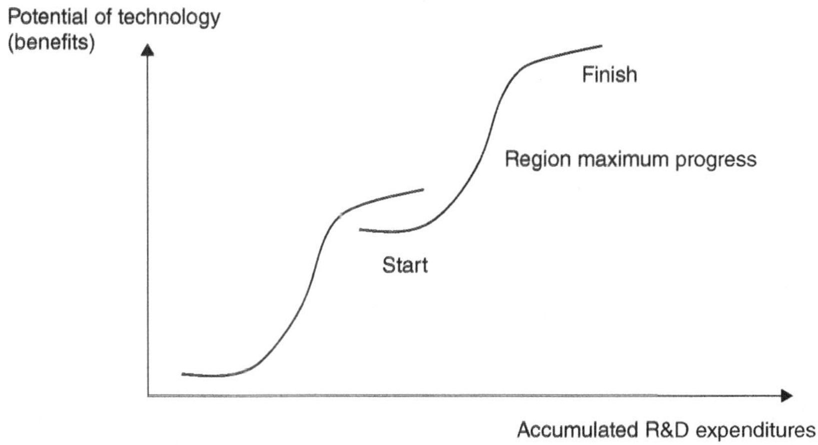

Figure 7.3 Foster S curve model.

its peak, but the company enters a phase of declining return on investment as the physical limitations of technology approach. At this step, a change in technology is necessary in order to achieve better performance (Figure 7.3) (Kasztler et al. 2012).

At this step in the described study, patents were arranged cumulatively based on the year of publication after entering the retrieved information of the first stage in Python software to identify the steps of research and development activities. Then its life cycle graph was drawn with the use of technology life cycle theory and Foster's S-curve model.

7.3.3 Mann-Kendall Test

The nonparametric test of Mann Kendall is used to analyze the data collected over time to evaluate the upward or downward trends. The more data in this test, the more likely it is that the test will find a real trend. Thus, the recommended number of measurements is at least 8–10 (Kendall 2016).

7.3.4 Sen's Slope Test

In nonparametric statistics, the Theil-Sen estimator is a method for connecting the line to the sample points on the surface (simple linear regression) by selecting the average slope of all lines through the point pair. It is also called the age slope estimator, slope selection, single median method, Kendall Robust Line-Fit Method, and Kendall–Theil Robust Line. This method is called the most popular nonparametric technique for estimating linear trends (El-Shaarawi and Piegorsch 2002; Gilbert n.d.).

At this step, Mann Kendall and Sen's Slope tests were performed to investigate the trend of the graph after drawing a life cycle graph.

7.3.5 Data Preprocessing

Text mining refers to the process of discovering knowledge from textual documents by analyzing information extracted from the text to form new hypotheses (Witten et al. 2011; IEEE 2017; Talib et al. 2016; Tan and Lee 2015; Larose and Larose 2014; Ampornphan and Tongngam 2020, 2017). Textual data involves punctuation, misspelled words, and abbreviations that should be removed or corrected before analysis (Kim et al. 2019; Singh and Kumari 2016). In addition, patents contain many terms that have such problems.

Thus, text preprocessing is considered a necessary step in patent analysis (Kim et al. 2019; Haddi et al. 2013; Zhang et al. 2008) and turns the methods such as tokenizing, transforming cases, lemmatization, filtering out stop words, and stemming more appropriate for analysis (Kim et al. 2019; Ampornphan and Tongngam 2020; Choi and Song 2018). In the "tokenizing" process, first text streams are converted into processing units known as tokens through a process.

In other words, such text streams, which are considered as character strings (e.g. sentences, phrases, and words) are converted to processing units without separators such as commas, colons, and spaces. In addition, uppercase letters are converted to lowercase letters during this process. The process of converting characters or words to lowercase or uppercase to avoid confusion between similar words is called *transforming cases*. In addition, "tokenizing" generally refers to the process of breaking text from a document into single words (tokens or terms) (Kim et al. 2019; Ampornphan and Tongngam 2020; Choi and Song 2018; Feldman and Sanger 2007; Hassler and Fliedl 2006; Stavrianou et al. 2007; Zanasi et al. 2006). The obtained words go through the process of *Lemmatization*, reducing the different types of words to "lemma" or other expressions of their origin.

By doing so, a set of different forms of a word can be analyzed as a single item in later steps (Choi and Song 2018; Hotho et al. 2005). In the next step, *stop words*, which have no meaning, are removed from the lemmatized words.

Although there is no single list of stop words, words such as articles, conjunctions, and prepositions are generally considered as stop words. In addition, common words are included in this list since they have virtually no special value in documents.

In other words, the process of removing meaningless elements such as punctuation marks, special characters, prepositions, articles, pronouns, etc. are called *filtering out stop words*. The obtained or preprocessed words are necessary for describing patent documents (Kim et al. 2019; Ampornphan and Tongngam 2020;

Choi and Song 2018; Feldman and Sanger 2007; Hotho et al. 2005). In patent analysis, the unstructured text of titles or summary of patent is pre-processed into a structured format. Then, the extracted key terms from the process for further analysis will be used in patent documents to determine the relationships between existing concepts (Ampornphan and Tongngam 2020).

In this study, the patent documents related to the time intervals of each step of the graph were separated after drawing the S-curve and a summary of the pre-processed patent documents and its keywords were extracted using the above-mentioned methods such as “tokenizing,” “transforming cases,” “lemmatization,” and “filtering out stop words.”

For this purpose, the Pandas Library (Mckinney 2011) was used to analyze the data and the Numpy Library was used to work on arrays and matrices (Harris et al. 2020). The data type in Python is shown as a string which is comprised of a set of characters. These characters can include the letters of the alphabet, numbers, and special characters.

The Python string is a set of characters surrounded by single quotes, double quotes, and triple quotes. A computer does not understand the characters since they are a combination of 0 and 1 (Sakthi Vel n.d.). The string library has been used and called in this study to identify the characters. On the other hand, the matching of the set of disciplines such as English sentences or email addresses, etc., was conducted by calling the re library (Kuching 2001) and accessing the variables by calling the sys library (SYS n.d.) in this study. In addition, the matplotlib.pyplot library (Mathplotlib n.d.) was used to draw graphs in this study. After calling the mentioned libraries, the powerful nltk library (NLTK n.d.) was used to preprocess the data, which is a leading platform including the operations of tokenizing, lemmatization, etc. In the next step, the first 100 words that exist as a pair of writing units or as three writing units in the text were extracted with a number of repetitions at least five times by calling the gensim and Phraser libraries (Gensim n.d.) and the words were stored in Corpus. Since nouns are of the most significance in the present study than adjectives, adverbs, and verbs, the names in the text were kept using the SpaCy library (SpaCy n.d.) and the words included in the other three categories were removed. Finally, the words that were placed in the noun classification were prepared as word list lemmatized for the initial modeling of latent Dirichlet allocation (LDA).

7.3.6 Latent Dirichlet Allocation (LDA) Topic Modeling

Latent semantic analysis is a thematic modeling method used for analyzing the relationships between a set of documents using statistical calculations to generate a set of concepts related to the original documents (Kim et al. 2019; Dumais 2004; Landauer and Dumais 1997).

On the other hand, probabilistic latent semantic analysis is a latent variable model that associates unobserved variables with each observation. (Kim et al. 2019; Hofmann 1999) This method has a better statistical status and presents better analysis results compared to the latent semantic analysis method (Kim et al. 2019; Hofmann 1999, 2001).

Although this method is an appropriate model for texts, it does not provide any suitable model for documents (Kim et al. 2019; Blei et al. 2003). LDA Topic Modeling is the most advanced form of latent semantic analysis method, a graphical model (Zhang et al. 2021) based on three-level Hierarchical Bayesian Model.

While the Latent Semantic Analysis (LSA), Probabilistic Latent Semantic Analysis (PLSA), and Latent Dirichlet Allocation (LDA), LDA Topic Modeling is based on the fact that documents are from a random mixture of latent topics which can be described by word distribution. In other words, the LDA uses parameters to calculate the common distribution of a topic mixture and to obtain the probability of a corpus or topic (Kim et al. 2019; Blei et al. 2003). This model has been used in different fields such as text mining and information processing. This model uses the "Word Bag Model" approach and regards each document as a possible distribution, which is comprised of a set of topics. In fact, each topic is a possible distribution of a large number of words. In other words, a "Document – Theme – Word" includes three Bayesian layers (Zhang et al. 2021; Lee et al. 2015). At this step, with 10 topics (number of latent topics, K is a free parameter in the model (Chang et al. 2009)), Chunksize = 10 000 (number of documents processed in documents at a time (Gensim n.d.)), Random State = 10 000, and Passes = 100 (number of repetitions during the processing process (Gensim n.d.)), the first LDA Topic Modeling was developed. The higher the number of documents per process and the number of iterations in the process, the higher the accuracy of the model (Fountain and Doyle 2012). Since the minimum coherence between the words was 0.4, the minimum coherence was achieved in all three steps by removing some words and re-modeling.

Finally, the final model was extracted by calling the tqdm library (da Costa-Luis n.d.), the number of 10 000 repetitions and achieving the optimal number of topics in each step. In addition, the dominant topics and weight of each in each patent text and words in each topic were identified and their word clouds (Mueller 2020) were drawn.

7.3.7 Principal Component Analysis (PCA)

This method is an unsupervised learning method in machine learning and data science. In other words, it is a mathematical technique, which converts a set of correlated variables into a set of noncorrelated variables known as principal

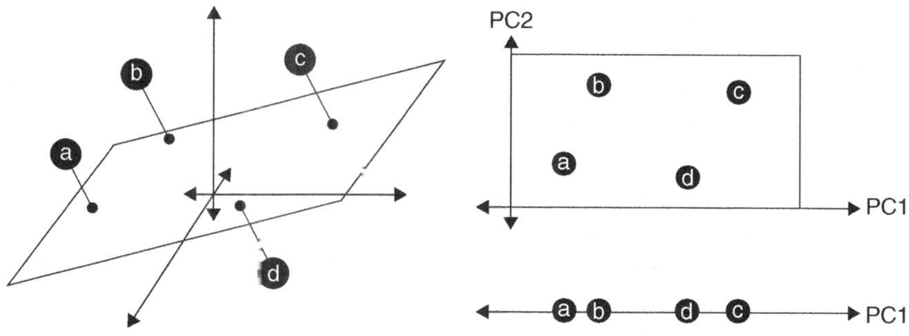

(a) An example of the original features in a 3D space.

(b) Principal components in 2D and 1D space.

Figure 7.4 An example of principal component analysis (PCA) and constructed principal component PC1 and PC2 that create different dimensional space.

components (Sarker 2021; Hotelling 1933; Pearson 1901) as shown in Figure 7.4. Figure 7.9a displays the main features in 3D space and Figure 7.9b shows the main components created by PC1 and PC2 on a two-dimensional screen and a one-dimensional line with the main PC1 component. Technically, principal component analysis (PCA) defines a complete change with the highest eigenvalues of the covariance matrix and then uses them to generate data into a new subspace of equal or less dimensions. In general, this algorithm identifies the main components and helps us to analyze a series of features which are more valuable instead of evaluating all the features. (Sarker 2021; Pedregosa et al. 2011)

In the next step, the "Term-Document" matrix was created using the words obtained from the preprocessing step and the PCA method. The "Term-Document" matrix shows the number of repetitions of words in a text (Kim et al. 2019). Since the data are textual (patents) and cannot be categorized definitively, the words and terms were clustered using the LDA Topic Modeling of the above-mentioned words and terms.

7.3.8 LDAvis Software Package

Finally, the LDAvis software package was used to visualize the results of LDA modeling. This method identifies each subject with a circle and assigns a number to each circle which is a number specific. In addition, topics are sorted according to the amount of relevant textual data. Furthermore, the distance between the circles shows the extent of the difference between the different subjects, while the size of the circles indicates the technology strength (Zhang et al. 2021; Sievert and Shirley 2014). LDAvis described each technology topic to describe the overall state of development of geothermal energy technologies. The word clouds extracted from the previous step lacks the concept of distance between words. In order to change the nature of the relationship between words, it is necessary to change the position of words from cloud to Euclidean distance. In statistics and probability theory, Jensen – Shannon Divergence measures the similarity between two probability distributions. This divergence is based on the Kullback–Leibler divergence, the second root of which is a criterion often known as the Jensen – Shannon distance. On the other hand, the pyLDAvis library reduces the properties of concepts to two properties (dimensions) using the PCA algorithm, which leads to a simpler analysis of concepts.

In this study, each term represents a document (in the LDA method, it represents a topic which consists of a set of words). In order to evaluate the development process of geothermal energy technologies, the topics in each step were described from two perspectives (using the results of reducing features to two features) "continuity" and "strength" of research and development activities. The continuity of technology refers to the state of survival of a topic. In other words, the issue of technology and its evolved topics appeared in several stages. Technology strength refers to the number of related technology research and development activities on a particular topic in the development stage, namely the number of related patents represented by the circle size in the results.

7.3.9 Text Similarity Calculation

It refers to comparing the similarity of two or more words, text, or document in such a way that the amount of similarity quantitatively (Zhang et al. 2021; Gomaa and Fahmy 2013). According to Lin, the greater the similarity and the smaller the difference, the greater the similarity, and vice versa. The maximum similarity occurs when the entities are exactly the same. The similarity theorem is as follows (Zhang et al. 2021; Lin 1998):

$$Sim(A, B) = Log\ P(common(A, B))Log\ P(description(A, B)) \tag{7.1}$$

In the above-mentioned formula, common (A, B) represents the common information A and B, description (A, B) refers to the total descriptive information A and B, and the similarity index is directly related to both. The greater the similarity of the text, the greater the correlation. However, high correlation does not necessarily mean high similarity. There are many methods to calculate text similarity. In this study, the cosine similarity method is used, which is evaluated by calculating the cosine of the fraction of two vectors. The maximum and minimum values of this index are positive one and negative one, respectively. Thus, when two vectors have a point and are in the same direction, the cosine similarity value is one. On the other hand, when the angle between two vectors is 90°, the value of cosine similarity is zero and when two vectors are in opposite directions, the value of negative cosine is one. This result is independent of the lengths of the vectors and only relates to their direction. The cosine similarity is normally used in positive distances. Therefore, the given value is between 0 and 1. In data retrieval, each term is given a different dimension and a dimension is shown by a vector whose value corresponds to the frequency of occurrence of the term in the document.

Thus, the similarity of cosines can create similarities between the two documents in terms of their topic. The cosine between two vectors can be calculated using the Euclidean point equation (Zhang et al. 2021; Lin 1998).

$$a.b = \|a\| \|b\| \cos\theta \tag{7.2}$$

Based on the two attribute vectors A and B, the similarity index is calculated as follows (Zhang et al. 2021; Lin 1998; Trappey et al. 2019).

$$Similarity = \cos\theta = \frac{A.B}{\|A\| \, \|B\|} = \frac{\sum_{i=1}^{n} A_i \times B_i}{\sqrt{\sum_{i=1}^{n} (A_i)^2} \times \sqrt{\sum_{i=1}^{n} (B_i)^2}} \tag{7.3}$$

The cosine similarity index was used to calculate the relationship between different technology topics and then identify the evolutionary relationships between them at different stages. In other words, evolutionary relationships between different topics were represented by lines in the results. In order to extract the relationship between the topics in different stages, the coordinates of the topics (x and y) obtained from the illustrated results of the previous step were entered into an Excel file.

The degree of similarity between the topics was identified by forming a matrix between the first and second period (similarity between topics in the first and second period) and a matrix between the second and third period (similarity between topics in the second and third period) and transferring it to Orange software (Orange Visual Programming 2017). It should be noted that the minimum similarity was identified to be 0.4 (Zhang et al. 2021).

7.3.10 Association Rule Mining

Association rule learning is a rule-based machine learning method to discover interesting relationships between variables in big data. This method is used for identifying strong rules discovered in a database (Platetsky-Shapiro 1991). Agrawal et al. introduced association rules to discover the rule between products in large transactions recorded by point-of-sale systems in supermarkets. Today, association rules are used in many applications, including the Internet of Things (IoT), medical diagnosis, etc. In addition, this method can be used to predict the set of items in a group and their relationship with each other. This algorithm uses support, confidence, and lift parameters to describe the rules.

Support refers to the number of repetitions of items in a data set. The "Support" rule is calculated using the following formula (Ampornphan and Tongngam 2020; Sarker 2021; Agrawal et al. 1993).

$$Support : (A \rightarrow B) = P(A \cup B) \tag{7.4}$$

The "confidence" rule indicates how the rule of numbers (an if-then term) is often true. This conditional rule is calculated as follows.

$$Confidence : (A \rightarrow B) = P(B \mid A) \tag{7.5}$$

The "lift" rule is calculated as the probability of a set of items based on the probability of unique items in the data set as follows.

$$Lift : (A \rightarrow B) = \frac{P(B \mid A)}{P(B)} \tag{7.6}$$

If the "lift" rule is higher than 1, it means that the two events are interdependent and such rules are highly useful for predicting the results in future datasets (Witten et al. 2011; Larose and Larose 2014; Ampornphan and Tongngam 2020, 2017; Agrawal and Srikant 1994). In patent analysis, the key terms extracted from each technology cluster are analyzed to determine the association rules. The association of the text derived from technical terms (technical_term #1 → technical_term #2) determines that "if technical_term #1 is extended, technical_term #2 will be also extended." Latent relationships discovered through association rules help us summarize patent documents. For instance, text association rules derived from key terms reflect patent concepts in this study (Ampornphan and Tongngam 2020).

In this study, the mlxtend.frequent_patterns and association_rules libraries (Raschka n.d.) were used to determine the relationships between concepts in each cluster.

At this step, the relationship between concepts (inventions) was identified and analyzed using the dominant topics with by forming co-occurrence matrix and association rules mining algorithm (min support = 0.01 and min confidence = 0.02) and then compared with the results of the evolution of topics.

7.3.11 Social Network Analysis

Social network analysis refers to the study of relationships and associations between actors such as individuals, groups, organizations, and processes. There are many indicators for analyzing and illustrating social relations, the main criteria of which include degree centrality (DC), betweenness centrality (BC), closeness centrality (CC), and eigenvector centrality (EC) (Ampornphan and Tongngam 2020; Farooq et al. 2018; Yang et al. 2013; Liu et al. 2019; Lee et al. 2019; Singh et al. 2020).

Degree Centrality Index: A connection center in a network (Hub) that has the most effect on the network. The node that connects many edges has the most effect on a social network. A vertex v of the graph G = (V, E) is calculated as follows.

$$C_D(v) = \deg(v) \tag{7.7}$$

Betweenness Centrality Index: The shortest link or path by which a single node bridges another node. The high value of this index shows full control or a significant role between the other two nodes in the social network. The intermediate center of vertex v of the graph $G = (V, E)$ is calculated as follows.

$$C_B = \sum_{x \neq y \varepsilon V} \sigma_{xy}(v)/\sigma_{xy} \tag{7.8}$$

σ_{xy} represents the total number of shortest paths from node x to node y, and $\sigma_{xy}(v)$ represents the number of paths passing through v.

Closeness Centrality Index: The average distance (or average of the shortest path) from each node to any other node in the network is measured using this index. The high value of this index shows the extensive communication of individuals in a social network. The closeness centrality of the vertex v of the graph $G = (V, E)$ can be calculated as follows.

$$C_C(v) = \frac{1}{\sum_j d(i,j)} \tag{7.9}$$

It should be noted that $d(i, j)$ represents the distance between vertex i and vertex j.

Eigenvector Centrality Index: This index refers to the relative scores assigned to all network nodes. The score of each node is measured by the relationships with other influential nodes. Eigenvector score means connecting a node to a large number of nodes that have high scores.

Eigenvector center is used for measuring the significance of all network nodes. In order to find the Eigenvector centrality point of the graph = (V, E) $|V|$, if $B = b\_$ (v, t) is considered an adjacency matrix, if v is connected to vertex t $b\_$ $(v, t) = 1$; otherwise $b\_$ $(v, t) = 0$.

Relative centrality, x, the score of vertex v can be calculated as follows.

$$X_v = \frac{1}{\lambda} \sum_{t \varepsilon N(v)} x_t = \frac{1}{\lambda} \sum_{t \varepsilon G} b_{v,t} x_{t2} \tag{7.10}$$

$N(v)$ refers to a set of the neighbors v and λ is a constant number.

At the last step of this study, the association of technologies with each other in each cluster and the extraction of future technologies and inventions were addressed by inserting dominant topics into Sci2 software (Xiaohua et al. 2013) and forming co-occurrence matrix and calculating social network analysis indices using Gephi software (Bastian et al. 2009).

7.4 Results of Data Analysis

7.4.1 Data Collection Initial Results

In this study, 13 965 patents were retrieved on September 30, 2021, by referring to the "Derwent Innovation Index" database using the "Topic" search route and the "Geothermal" search string. Since the patent status is considered confidential for 18 months upon, 1991 documents related to the years 2020 and 2021 were removed.

After reviewing, 148 documents that had no "abstract" were removed from the database and finally 11 826 documents were prepared for analysis.

7.4.2 Mann-Kendall Test Result

To conduct the Mann-Kendall test, the number of patent documents based on the year of publication was tested using the Python software. Figure 7.5 displays that the number of patent documents issued from 1964 to 1974 was the lowest, while this trend was relatively upward with fluctuations during 1975–2003.

Meanwhile, it took an upward trend from 2004 to 2014. After decreasing in 2015, it continued its upward trend until 2019. In addition, the results of the Mann-Kendall test indicated that the graph had an upward trend at the level of 99% confidence (Table 7.1).

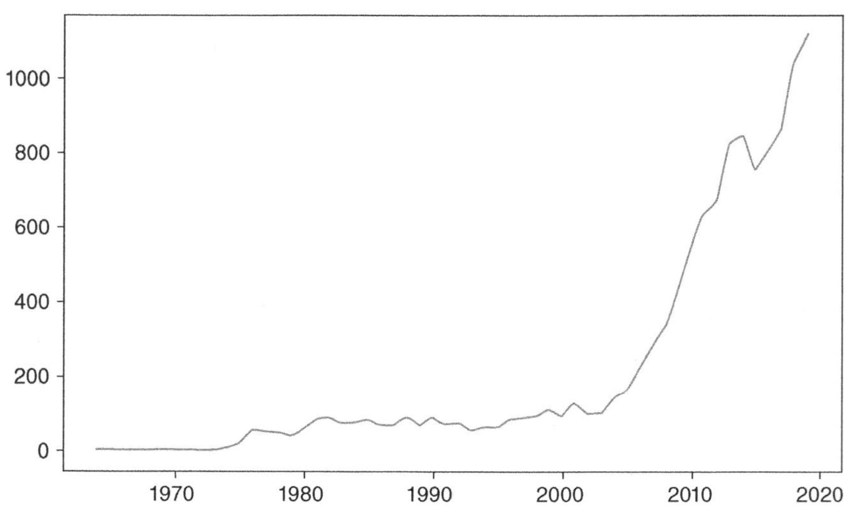

Figure 7.5 Graph of the number of patent documents issued based on the year of publication during 1964–2019.

Table 7.1 Number of patent documents issued by year of publication during 1964–2019.

Year	Number of patent
1964	1
1965	1
1966	1
1969	1
1970	4
1972	1
1973	1
1974	6
1975	19
1976	56
1977	49
1978	48
1979	38
1980	58
1981	84
1982	85
1983	74
1984	76
1985	82
1986	66
1987	69
1988	90
1989	68
1990	87
1991	69
1992	72
1993	53
1994	63
1995	63
1996	84
1997	87
1998	91

(Continued)

Table 7.1 (Continued)

Year	Number of patent
1999	111
2000	93
2001	128
2002	98
2003	101
2004	139
2005	162
2006	211
2007	275
2008	330
2009	427
2010	548
2011	635
2012	670
2013	823
2014	846
2015	752
2016	804
2017	863
2018	1044
2019	1119

In Sahal (1985):

```
import pymannkendall as mk
```

In Kendall (2016):

```
result = mk.original_test(df['NoP'])
```

In El-Shaarawi and Piegorsch (2002):

```
print (result)
Mann_Kendall_Test(trend='increasing', h=True,
p=0.0, z=8.937203548250775, Tau=0.8454281567489115,
```

```
s=1165.0, var_s=16963.0, slope=7.530555555555555,
intercept=-111.79444444444442)
```

In Gilbert (n.d.):

```
#trend : tell the trend is incrasing, decreasing,
     stationary or no Trend
#h : true if Trend is presant; Flase if the trend
     is absent
#p-value : <= 0.05 significance level
#Z-Score : Normalized test statistics --> Z = (x-mu)/
     Sigma Z<= 1   Z= 0   Z => 1
#S-core : Mann-kendal score
```

7.4.3 Sen's Slope Test Results

After reviewing the results of the Mann-Kendall test, the Sen's Slope test, which is considered the most popular nonparametric technique for estimating the linear trend was performed to evaluate the graph chart of the number of patents issued by year more accurately.

As shown in the results, the chart trend has been reported using this test to increase at the 95% confidence level.

```
from scipy.stats.mstats import theilslopes
```

In IEEE (2017):

```
theilslopes(x,y)
```

Out IEEE (2017):

```
(0.08409506398537477,
 1985.9360146252286,
 0.05347593582887700 4,
 0.15053763440860216)
```

In IEEE (2017):

```
#alpha : float Confidence degree between 0 and 1.
Default is 95% confidence. Note that alpha is
symmetric around 0.5, i.e. both 0.1 and 0.9 are
interpreted as "find the 90% confidence interval".

#medslope : float / Theil slope. (theilslope z-value :
```

```
tell the trend is incrasing, decreasing, stationary or
no Trend).

#medintercept : float / Intercept of the Theil line,
as median(y) - medslope*median(x).

#lo_slope : float / Lower bound of the confidence
interval on medslope.

#up_slope : float / Upper bound of the confidence
interval on medslope.
```

7.4.4 Results of Dividing Technology Steps

At this step, the number of patent documents was calculated cumulatively in each year (Table 7.2) and then the life cycle S-curve was drawn. (Appendix 7.B). The results of the graph indicate that the first patent for geothermal energy was filed in 1964. On the other hand, the development of technology in the field of geothermal energy was in its emerging stage until 1976 and has been on an upward trend since 1977. This development had a slight rise until 2006 and then found a sharp upward trend from 2007 to 2019.

In other words, the development of technology in the field of geothermal energy is in its early stages of growth. On the other hand, calculating the turning point of the chart (where the life cycle chart reaches from the primary growth stage to the secondary growth stage) indicating the year 2127 shows that technology development is fully in its emerging state without considering other environmental factors and only citing patent release rate. In the present study, the initial growth stage was divided into two stages of slow growth and rapid growth for more detailed analysis. Thus, the emerging stage is called the period 1964–1976, the slow growth stage is called the period 1977–2006, and the rapid growth stage is called the period 2007–2019 (Figures 7.6–7.8).

7.4.5 Results of Data Pre-processing

After removing the stop words, the words that had no semantic effect in creating topic clusters were removed. This was conducted in several steps and the coherence score of words was observed at least 0.4. First, the final word coherence score of emerging stage was 0.47, the final coherence score of slow growth stage

Table 7.2 Number of patent documents issued cumulatively by year of publication during 1964–2019.

Year	Cum
1964	1
1965	2
1966	3
1969	4
1970	8
1972	9
1973	10
1974	16
1975	35
1976	91
1977	140
1978	188
1979	226
1980	284
1981	368
1982	453
1983	527
1984	603
1985	685
1986	751
1987	820
1988	910
1989	978
1990	1 065
1991	1 134
1992	1 206
1993	1 259
1994	1 322
1995	1 385
1996	1 469

(Continued)

Table 7.2 (Continued)

Year	Cum
1997	1 556
1998	1 647
1999	1 758
2000	1 851
2001	1 979
2002	2 077
2003	2 178
2004	2 317
2005	2 479
2006	2 690
2007	2 965
2008	3 295
2009	3 722
2010	4 270
2011	4 905
2012	5 575
2013	6 398
2014	7 244
2015	7 996
2016	8 800
2017	9 663
2018	10 707
2019	11 826

was 0.4 and the coherence score of rapid growth stage was approximately 0.44 (0.4399).

The words without effect at each stage are shown separately in Tables 7.3–7.5.

7.4.6 Results of Coherence Score of Topic

Determining the number of topics K is the most important parameter in the process of analyzing the LDA Topic Modeling. In the technology evaluation process, it is better not to have more than 10 topic clusters to prevent the topics from scattering if there are too many topics. In this study, the number of topic

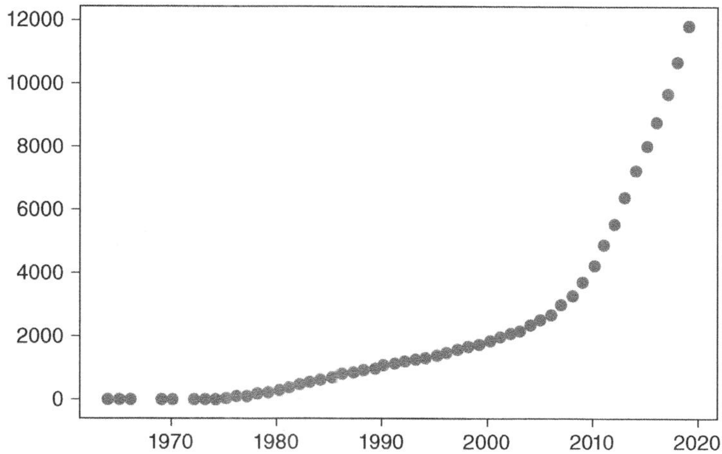

Figure 7.6 Graph of the cumulative number of patents issued based on the year of publication during 1964–2019.

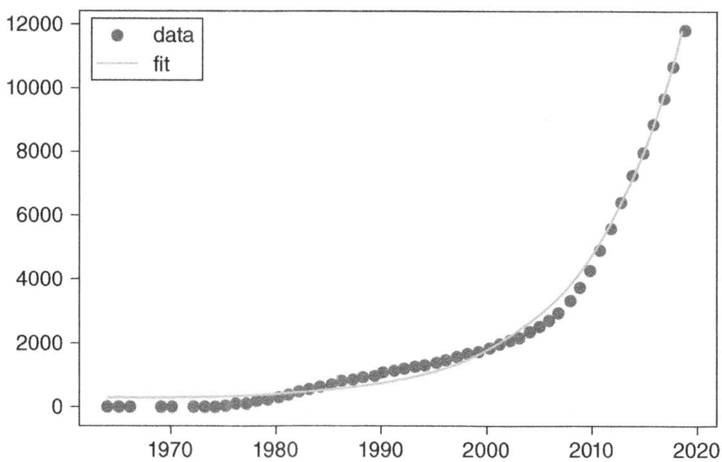

Figure 7.7 S-Curve graph of the number of patents issued based on year of publication during 1964–2019.

clusters from 2 to 10 was tested and the optimal number of topics in all three stages was obtained after the coherence test (Zhang et al. 2021).

The C_v index, based on the sliding window, is the division of a set of top words with an indirect confirmation measure using Normalized Pointwise Mutual Information (NPMI) and Cosine Similarity. In this study, C_v index was used for achieving the optimal number of topics (topic clusters) (Mifrah and Benlahmar 2020).

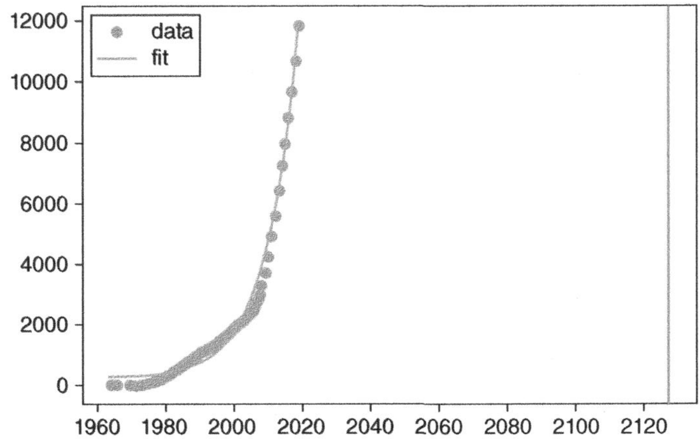

Figure 7.8 S-Curve graph of the number of patents issued based on year of publication during 1964–2019 and milestone (transition from primary to secondary growth).

Table 7.3 Words without semantic effect in creating topic clusters – emerging stage.

Stage, energy, source, fluid, magnitude, air, heating, heat, organism, depth, method, area, growth, water, temperature, temp, earth, power, surface, position, particle, insol, layer, system, flow, pressure, body, rock, waste, site, bed, degree, device, sea, column, tip, drop, return, feed, difference, micro, bearing, loop, oil, current, scale, output, sub, pref, ground, weight, gas, line, trend, plant, process, station, point, end, part, section, object, superer, level, unit, location, salt, rate, sequence, zone, side, production, sample, agent, communication, change, presence, detection, portion, response, radio, concrete, sodium, liquid, elevation, cross, example, region, desert, require, structure, exrs, network, variation, mean, value, number, proal, apparatus, medium

Table 7.4 Words without semantic effect in creating topic clusters – slow-growth stage.

Pressure, plant, fluid, advantage, stream, metal, water, pref, acid, process, method, oil, gas, material, particle, temperature, temp, layer, steel, time, depth, use, end, device, salt, acid, unit, group, scale, flow, section, surface, element, portion, body, tool, heating, heat, source, ground, line, production, earth, rock, air, side, apparatus, output, station, field, industry, performance, component, density, phase, column, length, diameter, assembly, agent, formula, treatment, stage, return, rate, zone, prodn, contg, soln, point, ground, member, content, feed, compound, waste, sample, substance, machine, product, description, figure_show, strength, property, value, equipment, area

Table 7.5 Words without semantic effect in creating topic clusters – rapid-growth stage.

Water, device, heating, heat, end, gas, system, fluid, method, plant, process, production, treatment, temperature, description, oil, tool, power, pressure, energy, layer, material, plate, surface, metal, structure, source, unit, air, portion, assembly, element, line, view, body, part, side, mechanism, ground, type, machine, output, advantage, rock, section, return, waste, liquid, operation, base, frame, wall, degree, agent, acid, day, flow, drive, apparatus, use, space, box, block, support, cover, loop, region, field, icle, earth, time, density, property, value, network, communication, component, zone, step, iron, segment, string, stream, film, panel, wood, efficiency, product, control, wellbore, show, core, bottom, substrate, head, guide, information, parameter, position, board, test, ler, mass, application, connection, construction, proof, effect, path, area, computer, measurement, cycle, station, cost, installation, model, ball, salt, compound, concentration, face, shell, load, level, point, rate, hvac, cross, equipment, comprise, scale, utility, steel, room, channel, screen, reaction, claim, column, direction, sub, sea, indepent, detection, crop, greenho, ing, build, display, data, determin, http://work.com, stance, stage, depth

The optimal number of topic clusters was five clusters at a coherence rate of 0.47 in the first stage (emerging stage), nine clusters at a coherence rate of 0.24 in the second stage (slow growth stage), and six clusters at a coherence rate of 0.28 in the third stage. The reason for the decrease in the coherence of clusters, especially in the second and third stages, is the high number of iterations ($N = 10\,000$). The higher the number of iterations, the higher the accuracy of the model and the lower the coherence of clusters.

The graph of the optimal number of topic clusters and the results of coherence test in all three stages of emerging, slow growth, and rapid growth are shown in Figures 7.9–7.11, as well as Tables 7.6–7.8, respectively.

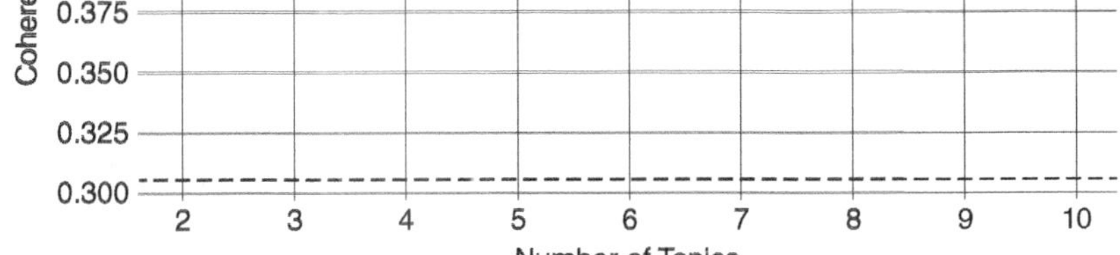

Figure 7.9 The optimal number of topic clusters in the emerging stage based on the coherence index.

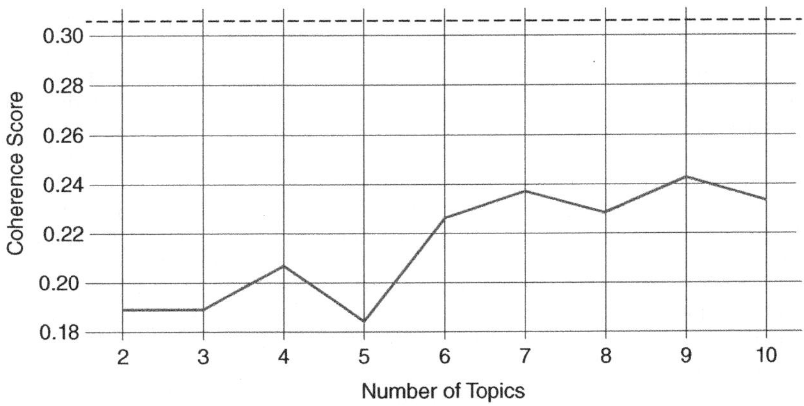

Figure 7.10 The optimal number of topic clusters in the slow-growth stage based on the coherence index.

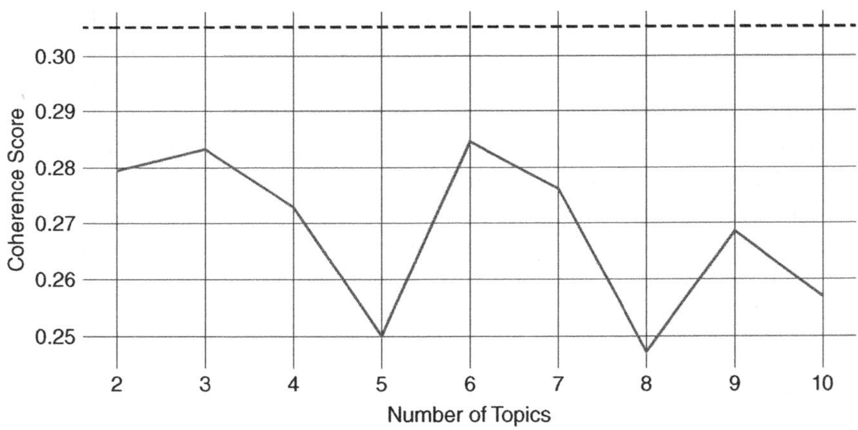

Figure 7.11 The optimal number of topic clusters in the rapid-growth stage based on the coherence index.

7.4.7 Latent Dirichlet Allocation (LDA) Topic Modeling with the Number of Optimal Topic Clusters

At this step, the LDA Topic Modeling was constructed using the optimal number of thematic clusters. The topic clusters of each stage along with the probability of the presence of each word in each cluster can be observed in Tables 7.9–7.11.

Table 7.6 Results of coherence score of topic clusters in emerging stage.

Number of topics	Coherence score
5	0.4779
4	0.4767
3	0.4765
7	0.4757
8	0.4666
10	0.4656
9	0.4584
2	0.4518
6	0.4461

Table 7.7 Results of coherence score of topic clusters in slow-growth stage.

Number of topics	Coherence score
9	0.2424
7	0.2368
10	0.2327
8	0.2286
6	0.2261
4	0.2064
3	0.1894
2	0.1881
5	0.1839

In order to better analyze and name each cluster, the display limit for the number of words in each topic increased to 15 words (Tables 7.12–7.14).

Then, word clouds were drawn at each stage for the better display of topic words and clusters as well as dominant topics (Figures 7.12–7.14).

Table 7.8 Results of coherence score of topic clusters in rapid-growth stage.

Number of topics	Coherence score
6	0.2844
3	0.2831
2	0.2793
7	0.276
4	0.2728
9	0.2685
10	0.2567
5	0.2501
8	0.247

Table 7.9 Topic clusters – emerging stage.

```
[(0,
  '0.061*"pump" + 0.048*"steam" + 0.046*"turbine" +
   0.034*"pipe" + '
  '0.025*"generator" + 0.025*"solution" +
   0.023*"transfer" + '
  '0.023*"evaporator" + 0.015*"circuit" + 0.013*"outlet"'),
 (1,
  '0.059*"bore" + 0.035*"steam" +
   0.025*"generator" + 0.022*"well" + '
  '0.018*"cavity" + 0.015*"voltage" + 0.015*"junction" +
   0.013*"gradient" + '
  '0.013*"wave" + 0.013*"carrier"'),
 (2,
  '0.024*"form" + 0.021*"separator" + 0.014*"vessel" +
   0.011*"mineral" + '
  '0.011*"deposit" + 0.011*"force" +
   0.011*"expansion" + 0.011*"hydrocarbon" + '
  '0.011*"conductivity" + 0.011*"signal"'),
 (3,
  '0.063*"brine" + 0.035*"vapor" + 0.026*"well" +
   0.026*"chamber" + '
  '0.022*"form" + 0.018*"flash" + 0.017*"contact" +
   0.015*"steam" + '
  '0.014*"expansion" + 0.012*"generator"'),
```

Table 7.9 (Continued)

```
(4,
'0.045*"shaft" + 0.021*"form" + 0.019*"steam" +
0.018*"turbine" + '
'0.015*"chamber" + 0.015*"drill" + 0.015*"well" +
0.013*"pipe" + '
'0.013*"deposit" + 0.013*"rubble"')]
```

Table 7.10 Topic clusters – slow-growth stage.

```
[(0,
'0.136*"drill" + 0.018*'sensor" + 0.017*"bit" +
0.015*"well" + '
'0.014*"measurement" + 0.012*"signal" + 0.012*"datum" +
0.010*"formation" + '
'0.008*"probe" + 0.006*'downdrill"'),
(1,
'0.071*"brine" + 0.027*'steam" + 0.022*"liquid" +
0.017*"form" + '
'0.016*"vessel" + 0.014*"formation" +
0.012*"membrane" + 0.011*"resin" + '
'0.011*"reaction" + 0.011*"recovery"'),
(2,
'0.038*"cement" + 0.025*"well" + 0.020*"composition" +
0.017*"polymer" + '
'0.015*"slurry" + 0.013*"formation" + 0.010*"ion" +
0.009*"ratio" + '
'0.008*"cpd" + 0.008*"corrosion"'),
(3,
'0.037*"seal" + 0.023*"tube" + 0.022*"string" +
0.020*"ring" + 0.018*"drill" '
'+ 0.016*"housing" + 0.015*"casing" + 0.015*"sleeve" +
0.014*"pipe" + '
'0.011*"force"'),
(4,
'0.030*"hydrogen" + 0.028*"solution" + 0.023*"filter" +
0.011*"boron" + '
'0.011*"mineral" + 0.010*"iron" + 0.009*"drawing" +
0.008*"chemical" + '
'0.008*"panel" + 0.008*"barrier"'),
```

(Continued)

Table 7.10 (Continued)

```
(5,
 '0.137*"pipe" + 0.043*"exchanger" + 0.030*"tank" +
  0.026*"pump" + '
 '0.020*"building" + 0.020*"storage" + 0.018*"supply" +
  0.016*"conditioning" '
 '+ 0.015*"drawing" + 0.014*"circulation"'),
(6,
 '0.083*"steam" + 0.066*"tube" + 0.056*"turbine" +
  0.025*"chamber" + '
 '0.025*"vapor" + 0.020*"generator" + 0.015*"liquid" +
  0.013*"outlet" + '
 '0.013*"rotor" + 0.012*"separator"'),
(7,
 '0.037*"generator" + 0.030*"valve" + 0.024*"pump" +
  0.020*"control" + '
 '0.020*"circuit" + 0.020*"supply" + 0.019*"exchanger" +
  0.018*"drawing" + '
 '0.014*"cycle" + 0.012*"efficiency"'),
(8,
 '0.018*"alloy" + 0.013*"resistance" + 0.012*"chemical" +
  0.012*"block" + '
 '0.009*"impurity" + 0.008*"matrix" + 0.008*"cell" +
  0.007*"deg" + '
 '0.007*"electrode" + 0.007*"ion"')]
```

Table 7.11 Topic clusters – rapid-growth stage.

```
[(0,
  '0.136*"drill" + 0.018*"sensor" + 0.017*"bit" +
   0.015*"well" + '
  '0.014*"measurement" + 0.012*"signal" + 0.012*"datum" +
   0.010*"formation" + '
  '0.008*"probe" + 0.006*"downdrill"'),
 (1,
  '0.071*"brine" + 0.027*"steam" + 0.022*"liquid" +
   0.017*"form" + '
  '0.016*"vessel" + 0.014*"formation" +
   0.012*"membrane" + 0.011*"resin" + '
  '0.011*"reaction" + 0.011*"recovery"'),
```

Table 7.11 (Continued)

```
(2,
 '0.038*"cement" + 0.025*"well" + 0.020*"composition" +
  0.017*"polymer" + '
 '0.015*"slurry" + 0.013*"formation" + 0.010*"ion" +
  0.009*"ratio" + '
 '0.008*"cpd" + 0.008*"corrosion"'),
(3,
 '0.037*"seal" + 0.023*"tube" + 0.022*"string" + 0.020*"ring" +
  0.018*"drill" '
 '+ 0.016*"housing" + 0.015*"casing" + 0.015*"sleeve" +
  0.014*"pipe" + '
 '0.011*"force"'),
(4,
 '0.030*"hydrogen" + 0.028*"solution" + 0.023*"filter" +
  0.011*"boron" + '
 '0.011*"mineral" + 0.010*"iron" + 0.009*"drawing" +
  0.008*"chemical" + '
 '0.008*"panel" + 0.008*"barrier"'),
(5,
 '0.137*"pipe" + 0.043*"exchanger" + 0.030*"tank" +
  0.026*"pump" + '
 '0.020*"building" + 0.020*"storage" + 0.018*"supply" +
  0.016*"conditioning" '
 '+ 0.015*"drawing" + 0.014*"circulation"'),
(6,
 '0.083*"steam" + 0.066*"tube" + 0.056*"turbine" +
  0.025*"chamber" + '
 '0.025*"vapor" + 0.020*"generator" + 0.015*"liquid" +
  0.013*"outlet" + '
 '0.013*"rotor" + 0.012*"separator"'),
(7,
 '0.037*"generator" + 0.030*"valve" + 0.024*"pump" +
  0.020*"control" + '
 '0.020*"circuit" + 0.020*"supply" + 0.019*"exchanger" +
  0.018*"drawing" + '
 '0.014*"cycle" + 0.012*"efficiency"'),
(8,
 '0.018*"alloy" + 0.013*"resistance" + 0.012*"chemical" +
  0.012*"block" + '
 '0.009*"impurity" + 0.008*"matrix" + 0.008*"cell" +
  0.007*"deg" + '
 '0.007*"electrode" + 0.007*"ion"')]
```

Table 7.12 Top clusters of emerging stage.

Topic	Terms per topic
Topic 1	Pump, steam, turbine, pipe, generator, solution, transfer, evaporator, circuit, outlet, tube, supply, combination, utilization, gene
Topic 2	Bore, steam, generator, well, cavity, voltage, junction, gradient, wave, carrier, tube, supply, signal, resin, explosion
Topic 3	Form, separator, vessel, mineral, deposit, force, expansion, hydrocarbon, conductivity, signal, pump, spring, produce, use, apparatus
Topic 4	Brine, vapor, well, chamber, form, flash, contact, steam, expansion, generator, gravel, casing, basin, bore, generating
Topic 5	Shaft, form, steam, turbine, chamber, drill, well, pipe, deposit, rubble, conduit, material, passage, velocity, duct

Table 7.13 Top clusters of the slow growth stage.

Topic	Terms per topic
Topic 1	Drill, sensor, bit, well, measurement, signal, datum, formation, probe, downdrill, condition, string, deposit, contact, interval
Topic 2	Brine, steam, liquid, form, vessel, formation, membrane, resin, reaction, recovery, vapor, exchanger, state, deposition, flash
Topic 3	Cement, well, composition, polymer, slurry, formation, ion, ratio, cpd, corrosion, copolymer, inhibitor, loss, oxide, property
Topic 4	Seal, tube, string, ring, drill, housing, casing, sleeve, pipe, force, expansion, rod, thread, valve, position
Topic 5	Hydrogen, solution, filter, boron, mineral, iron, drawing, chemical, panel, barrier, formation, type, adsorption, sulfur, hydrogen_sulfide
Topic 6	Pipe, exchanger, tank, pump, building, storage, supply, conditioning, drawing, circulation, carrier, installation, transfer, pile, cost
Topic 7	Steam, tube, turbine, chamber, vapor, generator, liquid, outlet, rotor, separator, conduit, inlet, exhaust, nozzle, shaft
Topic 8	Generator, valve, pump, control, circuit, supply, exchanger, drawing, cycle, efficiency, operation, lithium, electricity, wind, language
Topic 9	Alloy, resistance, chemical, block, impurity, matrix, cell, deg, electrode, ion, film, volume, sheet, plate, corrosion

Table 7.14 Top clusters of the rapid-growth stage.

Topic	Terms per topic
Topic 1	Pipe, tube, drill, sleeve, probe, rod, groove, shaft, cavity, fix, pile, connect, cylinder, rotation, drawing
Topic 2	Pipe, pump, exchanger, tank, circulation, supply, age, valve, outlet, inlet, drawing, tube, utilization, conditioner, conditionerer
Topic 3	Filter, injection, formation, solution, brine, well, tank, sand, pris, spray, mixture, extraction, carbon_dioxide, reservoir, hydrocarbon
Topic 4	Generator, steam, turbine, wind, drawing, age, supply, engine, electricity, evaporator, transfer, carbon_dioxide, hydrogen, fuel, circuit
Topic 5	Floor, module, conductor, cable, insulation, wire, battery, carbon, cell, pris, cement, oxide, posite, thermoelectric, aluminum
Topic 6	Drill, sensor, bit, formation, drawing, monitor, signal, well, downdrill, blade, conditioner, hydrocarbon, enable, performation, follow

Figure 7.12 Word clouds of top clusters at the emerging stage.

7.4.8 Analysis of Research and Development Status of Geothermal Energy Technologies

In the process of displaying topic clusters as word clouds, the distance between words is not clear. Since the Euclidean distance between words in each cluster shows the similarity of concepts to each other in the present study, the word clouds were transferred to Euclidean distance at this stage. (Figures 7.15–7.17)

At this step, the relationship between topic clusters in three periods was measured in pairs (first period with the second period; second period with the third period) using the cosine similarity index (Tables 7.15 and 7.16).

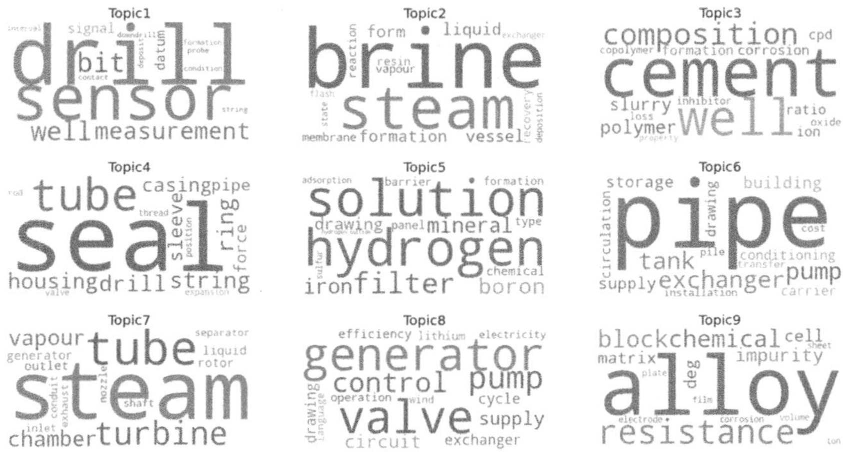

Figure 7.13 Word clouds of top clusters at the slow-growth stage.

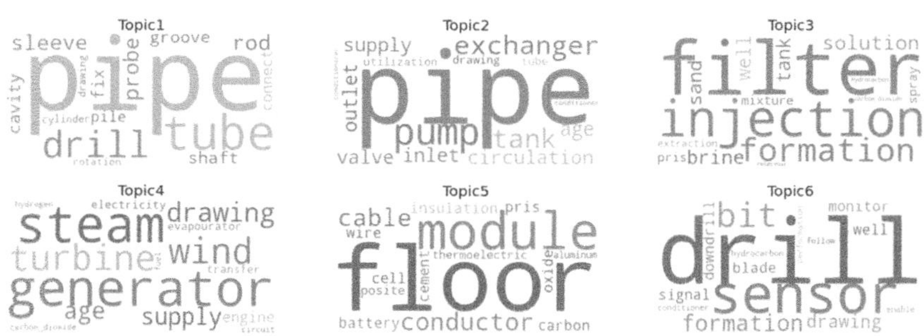

Figure 7.14 Word clouds of top clusters at the rapid-growth stage.

Figure 7.18 indicates the evolutionary process of topic clusters during the three periods of emerging, slow growth, and rapid growth. The lines between the topics in the graph indicate the evolution of the second topic from the first topic, and the relationship between the different colors shows the relationship between the different topics. In this figure, the size of each circle shows the technological strength of each topic and is directly related to the number of words in the topic cluster. On the other hand, technology continuity refers to the survival of a topic, i.e. technology has evolved and appeared in several stages. In addition, the evolutionary relationships between different topics are shown with lines of different colors. In other words, the number of related technology research and development activities in a particular topic indicates the technology strength, which is equivalent to the number of related patents in the form of circle size in this study.

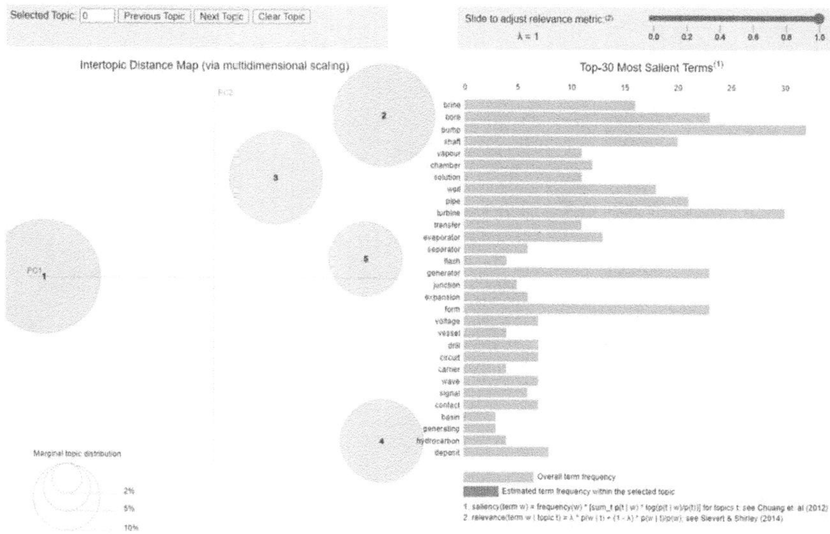

Figure 7.15 Topic clusters based on principal component analysis (PCA) in Euclidean distance- emerging stage.

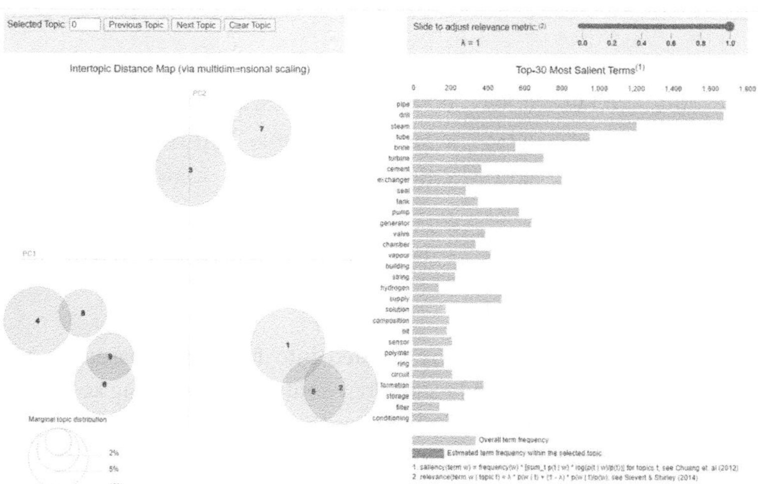

Figure 7.16 Topic clusters based on principal component analysis (PCA) in Euclidean distance – slow-growth stage.

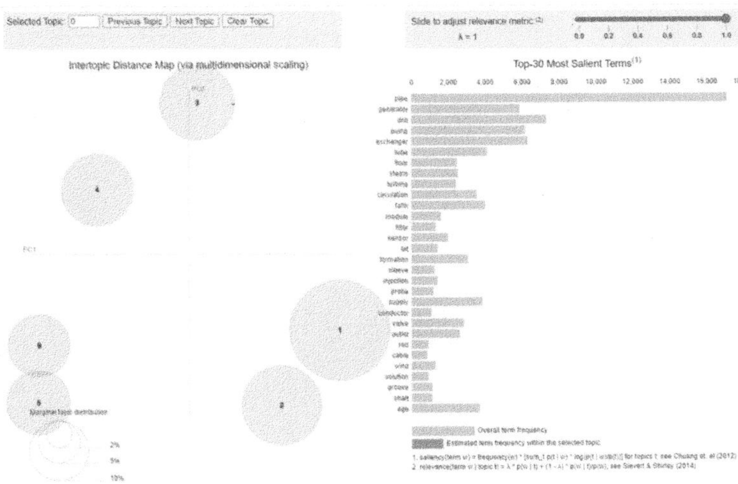

Figure 7.17 Topic clusters based on principal component analysis (PCA) in Euclidean distance – rapid-growth stage.

Table 7.15 Similarities between topic clusters of emerging and slow-growth stages.

		Topic number – period 2								
		1	2	3	4	5	6	7	8	9
Topic number – period 1	1	101	0.139	0.206	0.293	0.427	0.522	0.618	0.935	1
	2	0.073	0.106	0.166	0.246	0.37	0.46	0.551	0.857	0.92
	3	0.044	0.07	0.12	0.189	0.301	0.383	0.468	0.758	0.819
	4	0.014	0.03	0.066	0.12	0.212	0.284	0.359	0.622	0.679
	5	0	0.002	0.017	0.049	0.114	0.169	0.228	0.453	0.503

In this study, technology topics were divided into four types based on two indicators of technology strength and technology continuity, including key topics, hot topics, breakthrough topics, and blank topics (Zhang et al. 2021).

Key issues refer to issues that have a high level of both technology strength and technology continuity and have a significant effect on technology development. Hot topics are high-tech topics but have no continuity. In fact, such topics include the technologies that have recently been considered, and their technical capabilities are impressive. It should be noted that hot topics are likely to become

Table 7.16 Similarities between topic clusters of slow-growth stage and rapid-growth stage.

Topic number – period 2	Topic number – period 3					
	1	2	3	4	5	6
1	0.094	0.301	0.421	0.549	0.811	0.936
2	0.054	0.226	0.333	0.448	0.690	0.808
3	0.020	0.148	0.236	0.336	0.553	0.661
4	0.002	0.078	0.146	0.226	0.411	0.506
5	0.012	0.019	0.058	0.113	0.254	0.331
6	0.051	0.001	0.016	0.049	0.152	0.214
7	0.115	0.010	0.000	0.012	0.079	0.125
8	0.405	0.157	0.086	0.039	0.001	0.002
9	0.512	0.228	0.141	0.079	0.012	0.001

the most effective technology topics in the future. Breakthrough topics have low technology strength but high continuity. There are two possible scenarios for such topics. The first is that they play no significant role in the future of the field due to low strength. On the other hand, based on the second scenario, such topics may become hot due to a change in a specific factor and lead to significant progress in that area. Finally, blank topics are at a low level in terms of technology strength and continuity. In other words, they do not create much value in terms of research and development in the field.

The conducted classification can be used as an appropriate reference to advance the goals and appropriate investment for research and development in different branches. In this study, these two indicators (technology strength and technology continuity) and finally the four above-mentioned categories were used for research and development-oriented policy. Long-term investment is recommended for key topics. In case of hot topics, it is suggested to increase investment in the early stages since research and development on these topics will lead to a window of opportunity in the future. Breakthrough issues should be regarded carefully with a more accurate assessment, and investment should be conducted by considering several factors and more appropriate indicators. Eventually, over-investment in blank topics should be avoided as they will create almost no added value to advance technological goals.

Based on Figure 7.18, the only key topic in Topic 1 geothermal energy is in the third stage. This topic is shown in dashed lines because it is first considered as the

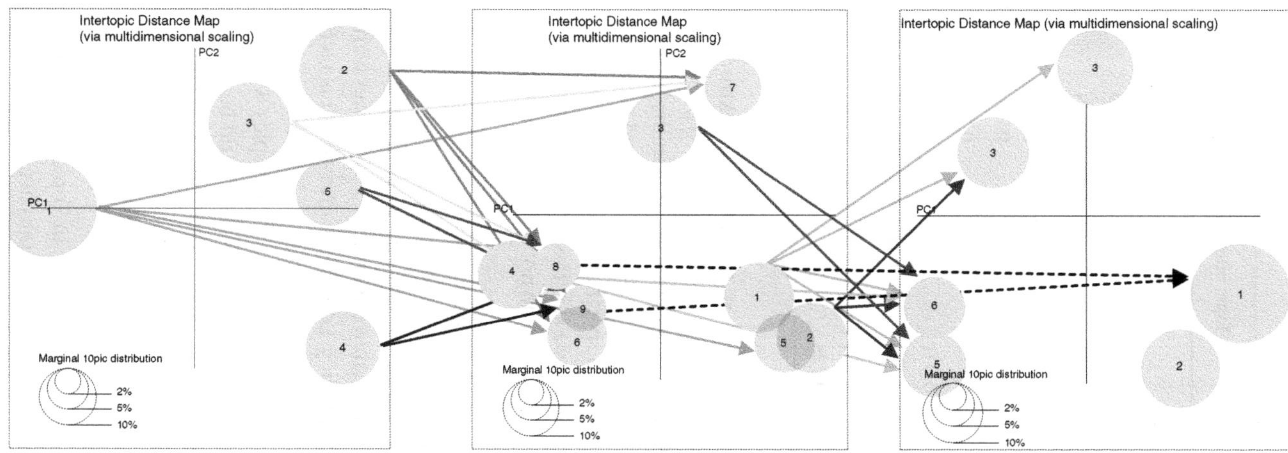

Figure 7.18 Evolution of topic clusters during emerging, slow-growth, and rapid-growth stages.

only key topic and second evolved from several topics in the second and first stages. This topic has evolved jointly from Topic 8 and Topic 9 in the second stage and from all five topics in the first stage. Topic 1 has gone through all three stages with the evolution of different topics and has a high continuity. Despite the fact that Topic 8 and Topic 9 have low technology strength in stage 2, they are considered as the main link of the topics during the three periods. Based on social networks, it is considered as a bridge. On the other hand, high-tech topics have evolved from the first stage and have high technological strength. This topic and related topics are the core of the field of geothermal energy and research and development on technologies related to its topics should be among the investment priorities. Topic 3 and Topic 4 both have high technology strength and evolved from Topic 1 in the second stage and Topic 1 and Topic 2 in the second stage, respectively. However, they have a weak continuity and are not related to the topics of the first period. Thus, these two topics are considered hot topics.

These two topics should be viewed hopefully and invested as an opportunity window during the early stages to implement appropriate research and development policies. On the other hand, Topic 2 has high technology strength but has no continuity. This topic can be a hot topic, but the risk of investment on it is higher since there are two approaches to it. Firstly, this topic has no continuity and there may not be sufficient tacit knowledge about it. Thus, investing in it can have a high risk. Second, it may become the most effective technology in the future due to its newness and high power of technology. Thus, careful analysis of investment in technologies should be conducted in this field from the perspective of research and development.

Topic 5 and Topic 6 evolved jointly from Topic 1, Topic 2, Topic 3, and Topic 4 in Stage 2. Both topics have low strength but had no continuity in all periods. Thus, they are regarded as blank topics. It should be noted that investment policies should be cautious (but with a lower priority than hot topics) due to the evolution of these two topics since the evolution of these two topics originated from several topics in Stage 2 and in policy evaluation. Such a topic should also be regarded in research and development. Furthermore, there is no breakthrough topic (low technology strength and high continuity) in the third period. Based on the analysis of the above-mentioned technologies in Topic 1, it should be considered by experts and specialists in research and development in the field of geothermal energy more than other topics. As mentioned, the technologies of this Topic are considered as the core of this field. Unlike the technologies in Topic 1, the technologies in Topic 5 and Topic 6 should not be among the priorities of research and development experts. In addition, research and development policies on Topic 2, Topic 3, and Topic 4 technologies should be conducted carefully by considering multiple indicators and more accurate evaluations than other technologies categorized in this study.

7.4.9 Breakthrough Inventions and Technologies of Geothermal Energy

7.4.9.1 Results of Association Rules Mining

In this study, the relationship between inventions and technologies, as well as its interaction was measured using the dominant topics extracted from the LDA Topic Modeling in each patent document during all three stages (Tables 7.17–7.19) and defining the minimum Support and Confidence in the association rule search algorithm at 0.01 and 0.02, respectively (Tables 7.20–7.22).

Table 7.17 Dominant topics extracted from the Latent Dirichlet Allocation (LDA) topic modeling in each document along with their weight – emerging stage.

Document_ No	Dominant_ Topic	Topic_ Perc_ Contrib	Keywords	Text
1	3	0.96895	brine, vapor, well, chamber, form, flash, con...	[brine, brine, concentration, basin, brine, ba...
2	3	0.95518	brine, vapor, well, chamber, form, flash, con...	[deposit, caso, borebore, bore, liquor, contac...
3	1	0.9522	bore, steam, generator, well, cavity, voltage, ...	[order, steam, generator, bore, stratum, bore, ...
4	2	0.9267	form, separator, vessel, mineral, deposit, for...	[purpose, prevent, equipment, corrosion, cost, ...
5	1	0.96893	bore, steam, generator, well, cavity, voltage, ...	[recovery, distribution, rubble, cone, cavity, ...
...	...	...	...	...
87	4	0.95515	shaft, form, steam, turbine, chamber, drill, w...	[chamber, shaft, height, mountain, turbine, sh...
88	4	0.96343	shaft, form, steam, turbine, chamber, drill, w...	[exploration, deposit, mineral, existence, ele...
89	4	0.95763	shaft, form, steam, turbine, chamber, drill, w...	[duct, meter, rise, duct, passage, vane, react...
90	1	0.96614	bore, steam, generator, well, cavity, voltage, ...	[generator, comprise, rejection, generator, di...
91	0	0.93768	pump, steam, turbine, pipe, generator, solutio...	[valley, rejection, mountain, top, supplying, ...

Table 7.18 Dominant topics extracted from the Latent Dirichlet Allocation (LDA) topic modeling in each document along with their weight – slow growth stage.

Document_ No	Dominant_ Topic	Topic_ Perc_ Contrib	Keywords	Text
1	8	0.822928	alloy, resistance, chemical, block, impurity, …	[charge, detonation, cartridge, cartridge, rea…
2	6	0.375183	steam, tube, turbine, chamber, vapor, generat…	[comprise, steam, collection, vessel, bottom, …
3	1	0.749602	brine, steam, liquid, form, vessel, formation,…	[access, brine, brine, appt, pressurize, liqui…
4	1	0.940729	brine, steam, liquid, form, vessel, formation,…	[state, liquid, formation, degree, injection, …
5	0	0.822151	drill, sensor, bit, well, measurement, signal,…	[consist, well, bottom, well]
…	…	…	…	…
2595	7	0.452302	generator, valve, pump, control, circuit, supp…	[generator, configuration, array, generator, w…
2596	0	0.975964	drill, sensor, bit, well, measurement, signal,…	[drill, bit, protrusion, fallback, insert, ins…
2597	3	0.668971	seal, tube, string, ring, drill, housing, casi…	[welding, position, plate, butt, face, plate, …
2598	7	0.642286	generator, valve, pump, control, circuit, supp…	[vessel, form, liquid, bing, liquid, ambient, …
2599	2	0.904603	cement, well, composition, polymer, slurry, fo…	[mechanization, drill, tillage, drill, mechani…

Table 7.19 Dominant topics extracted from the Latent Dirichlet Allocation (LDA) topic modeling in each document along with their weight – rapid-growth stage.

Document_ No	Dominant_ Topic	Topic_ Perc_ Contrib	Keywords	Text
1	1	0.967709	pipe, pump, exchanger, tank, circulation, supp…	[feed, pipe, pump, reduction, tube, circulatio…
2	1	0.841408	pipe, pump, exchanger, tank, circulation, supp…	[exchanger, increase, conductorion, convection…

Table 7.19 (Continued)

Document_ No	Dominant_ Topic	Topic_ Perc_ Contrib	Keywords	Text
3	0	0.73909	pipe, tube, drill, sleeve, probe, rod, groove,…	[groove, pipe, cement, cement, shape, cement, …
4	1	0.964928	pipe, pump, exchanger, tank, circulation, supp…	[pump, pris, inter, integd, duct, geothermal, …
5	1	0.751214	pipe, pump, exchanger, tank, circulation, supp…	[utilization, conditioner, performation, condi…
…	…	…	…	…
9132	4	0.910352	floor, module, conductor, cable, insulation, w…	[analyz, module, audition, module, olfactory, …
9133	5	0.358632	drill, sensor, bit, formation, drawing, monito…	[well, site, perforation, battery, driv, shaft…
9134	1	0.843686	pipe, pump, exchanger, tank, circulation, supp…	[vinyl, circulation, age, tank, circulation, m…
9135	2	0.711156	filter, injection, formation, solution, brine,…	[hav, index, ratio, young, modulus, loss, volu…
9136	3	0.901454	generator, steam, turbine, wind, drawing, age,…	[teacher, concept, conversion, electricity, ce…

Table 7.20 Examples of association rules mining between key terms, support, confidence, and lifting them – emerging stage.

Row	Antecedents	Consequents	Support	Confidence	Lift
1	({' steam'})	({' generator'})	0.63	0.72	1.15
2	({' generator'})	({' steam'})	0.63	1.00	1.15
3	({' well'})	({' steam'})	0.57	1.00	1.15
4	({' steam'})	({' well'})	0.57	0.66	1.15
5	({' pipe'})	({' steam'})	0.54	1.00	1.15
6	({' steam'})	({' pipe'})	0.54	0.62	1.15
7	({' pipe'})	({' turbine'})	0.54	1.00	1.86
8	({' turbine'})	({' pipe'})	0.54	1.00	1.86
9	({' steam'})	({' turbine'})	0.54	0.62	1.15
10	({' turbine'})	({' steam'})	0.54	1.00	1.15

Table 7.20 (Continued)

Row	Antecedents	Consequents	Support	Confidence	Lift
11	({' pipe', ' steam})	({' turbine'})	0.54	1.00	1.86
12	({' pipe', ' turbine'})	({' steam'})	0.54	1.00	1.15
13	({' steam', ' turbine'})	({' pipe'})	0.54	1.00	1.86
14	({' pipe'})	({' steam', ' turbine'})	0.54	1.00	1.86
15	({' steam'})	({' pipe', ' turbine'})	0.54	0.62	1.15
16	({' turbine'})	({' pipe', ' steam'})	0.54	1.00	1.86
17	({' form'})	({' chamber'})	0.41	1.00	2.46
18	({' chamber'})	({' form'})	0.41	1.00	2.46
19	({' steam'})	({' chamber'})	0.41	0.47	1.15
20	({' chamber'})	({' steam'})	0.41	1.00	1.15
...	...	...	...	...	...

Tables 7.20–7.22 indicate the examples of association between terms (inventions related to terms) in all three stages based on the maximum amount of support. There are three common criteria for describing a relationship as discussed below:

1) Rules with high support value show the popularity of technologies and inventions. For example, the technologies related to inventions (generator, steam/pipe, turbine/form, and chamber) are among the most popular technologies in the field of geothermal energy in the first period. In other words, generator-related technologies and inventions were widely developed from steam-related technologies in the early period. In addition, there is a similar connection between pipe and turbine, as well as chamber and form. In the second period, the technologies and inventions related to the exchanger and pump can be mentioned among the high-value communications. In other words, pump-related technologies have been widely developed from exchanger-related technologies. In the third period, circulation-related technologies used the development of exchanger-related technologies.
2) High-reliability rules implied the probability of technology and invention. For instance, if technologies related to pipe and steam are developed in the first period, it is likely that technologies related to turbine will be developed. Furthermore, there is this probability for steam and chamber related technologies. In the second period, if formation-related technologies are developed, there is a probability of developing well-related technologies accordingly. In addition, the probability of developing pump-related technologies in case of developing exchanger and drawing technologies is at a high level. On

Table 7.21 Examples of association rules mining between key terms, support, confidence, and lifting them – slow-growth stage.

Row	Antecedents	Consequents	Support	Confidence	Lift
1	({' well'})	({' formation'})	0.28	1.00	2.56
2	({' formation'})	({' well'})	0.28	0.71	2.56
3	({' exchanger'})	({' drawing'})	0.26	1.00	3.22
4	({' drawing'})	({' exchanger'})	0.26	0.84	3.22
5	({' drawing'})	({' pump'})	0.26	0.84	3.22
6	({' pump'})	({' drawing'})	0.26	1.00	3.22
7	({' drawing'})	({' supply'})	0.26	0.84	3.22
8	({' supply'})	({' drawing'})	0.26	1.00	3.22
9	({' exchanger'})	({' pump'})	0.26	1.00	3.84
10	({' pump'})	({' exchanger'})	0.26	1.00	3.84
11	({' exchanger'})	({' supply'})	0.26	1.00	3.84
12	({' supply'})	({' exchanger'})	0.26	1.00	3.84
13	({' pump'})	({' supply'})	0.26	1.00	3.84
14	({' supply'})	({' pump'})	0.26	1.00	3.84
15	({' exchanger', ' drawing'})	({' pump'})	0.26	1.00	3.84
16	({' exchanger', ' pump'})	({' drawing'})	0.26	1.00	3.22
17	({' drawing', ' pump'})	({' exchanger'})	0.26	1.00	3.84
18	({' exchanger'})	({' drawing', ' pump'})	0.26	1.00	3.84
19	({' drawing'})	({' exchanger', ' pump'})	0.26	0.84	3.22
20	({' pump'})	({' exchanger', ' drawing'})	0.26	1.00	3.84
…	…	…	…	…	…

the other hand, the development of exchanger-related technologies is likely to depend on the development of circulation-related technologies.

3) The rules of "lift" further imply a strong relationship between technology and the development of inventions. The values of this index in the samples given in Tables 7.20–7.22 are more than 1, which means that technologies and inventions are highly related to each other. In other words, the potentially greater relationship between technologies and related inventions based on this index makes it easier to project the extent of technology development.

7.4.9.2 Results of Analysis for Indicators Related to Social Networks

Association rules were analyzed in the previous section using indicators related to social networks. At this step, the dominant topics extracted from the LDA

Table 7.22 Examples of association rules mining between key terms, support, confidence, and lifting them – rapid-growth stage.

Row	Antecedents	Consequents	Support	Confidence	Lift
1	({' age'})	({' supply'})	0.53	1.00	1.88
2	({' supply'})	({' age'})	0.53	1.00	1.88
3	({' age'})	({' circulation'})	0.35	0.67	1.88
4	({' circulation'})	({' age'})	0.35	1.00	1.88
5	({' age'})	({' exchanger'})	0.35	0.67	1.88
6	({' exchanger'})	({' age'})	0.35	1.00	1.88
7	({' age'})	({' inlet'})	0.35	0.67	1.88
8	({' inlet'})	({' age'})	0.35	1.00	1.88
9	({' outlet'})	({' age'})	0.35	1.00	1.88
10	({' age'})	({' outlet'})	0.35	0.67	1.88
11	({' age'})	({' pump'})	0.35	0.67	1.88
12	({' pump'})	({' age'})	0.35	1.00	1.88
13	({' age'})	({' tank'})	0.35	0.67	1.54
14	({' tank'})	({' age'})	0.35	0.82	1.54
15	({' valve'})	({' age'})	0.35	1.00	1.88
16	({' age'})	({' valve'})	0.35	0.67	1.88
17	({' age'})	({'pipe'})	0.35	0.67	1.29
18	({'pipe'})	({' age'})	0.35	0.68	1.29
19	({' exchanger'})	({' circulation'})	0.35	1.00	2.82
20	({' circulation'})	({' exchanger'})	0.35	1.00	2.82
…	…	…	…	…	…

Topic Modeling were transferred to Sci2 software to form co-occurrence matrix and then to Gephi software for illustration. First, the Force Atlas 2 algorithm – which removes the nodes with the most connections to each other but aligns the nodes connected to the poles in the surrounding clusters (Altuntas and Gok 2020) – was applied on co-occurring networks of emerging, rapid-growth, and slow-growth stages. By performing statistical calculations and calculating the degree centrality, betweenness centrality, closeness centrality and Eigenvector centrality, co-occurrence networks between words in all periods were drawn (Figures 7.19–7.22). In addition, Tables 7.23–7.26 show the highest number of indices for words in all three periods.

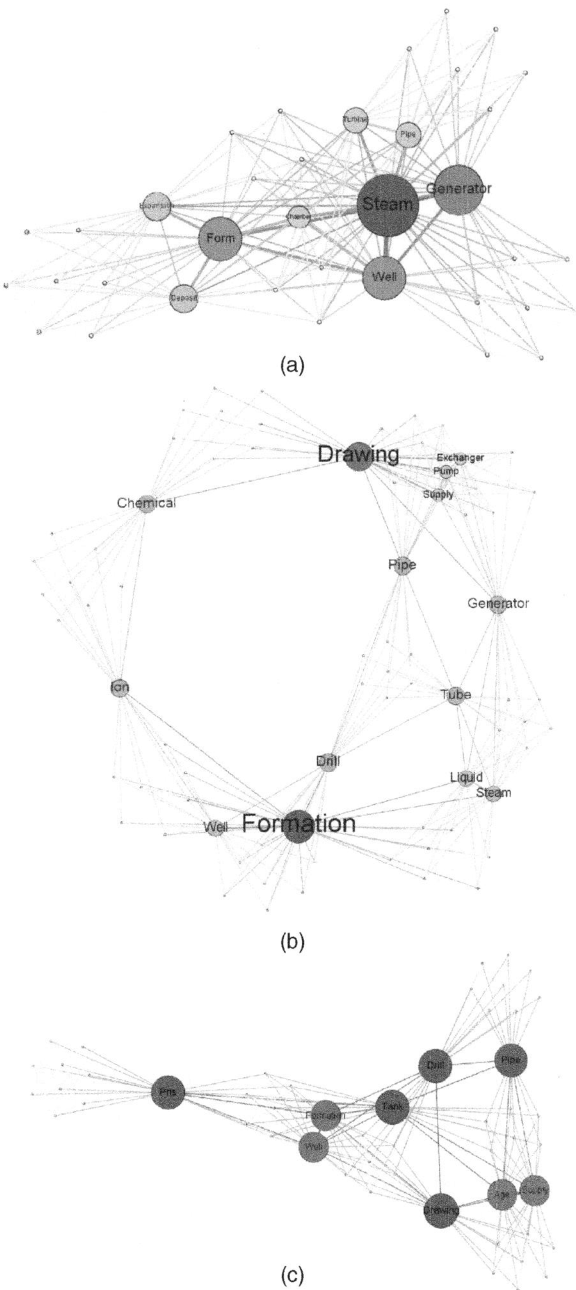

Figure 7.19 Co-occurrence network derived from association rules based on the degree centrality index – emerging stage (a), slow growth (b), and rapid growth (c).

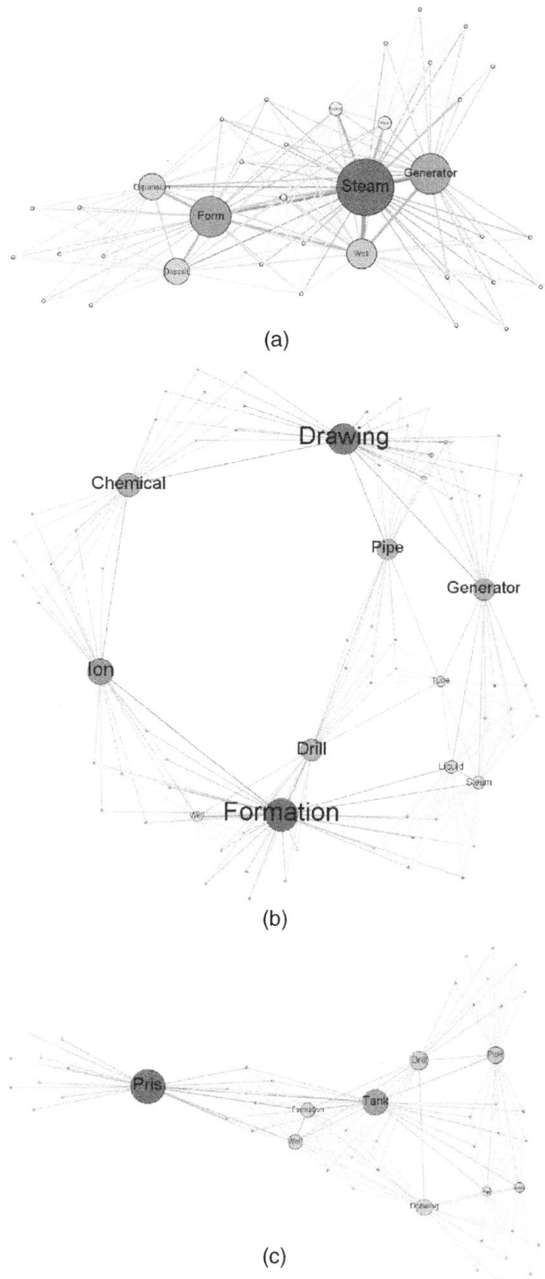

Figure 7.20 Co-occurrence network derived from association rules based on the betweenness centrality index – emerging stage (a), slow growth (b), and rapid growth (c).

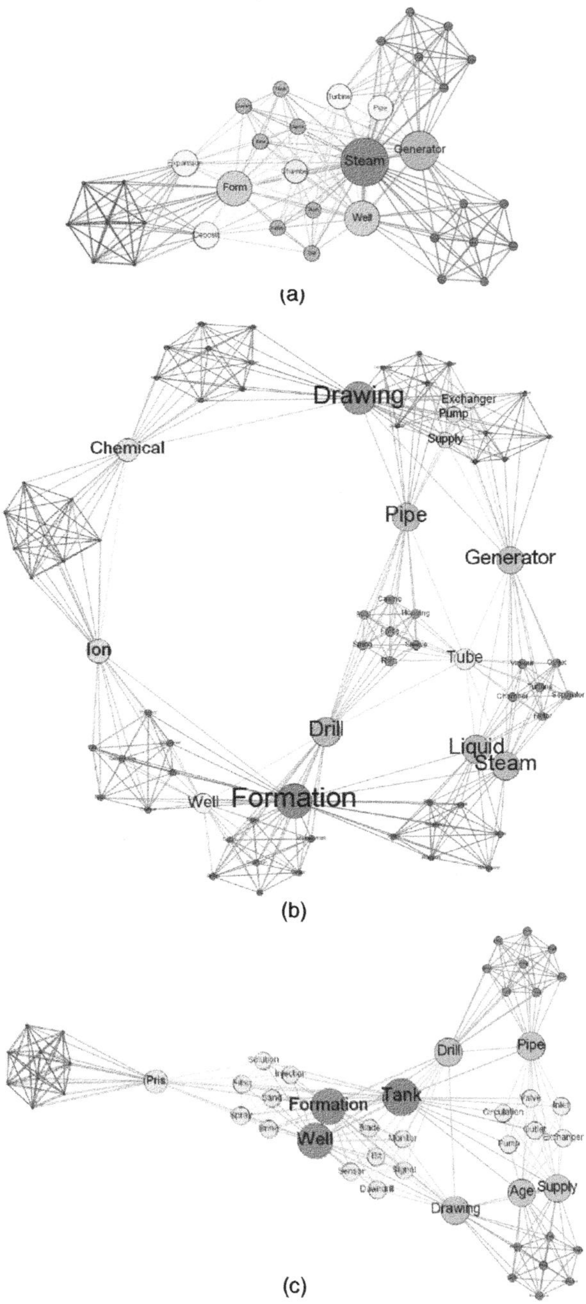

Figure 7.21 Co-occurrence network derived from association rules based on the closeness centrality index – emerging stage (a), slow growth (b) and rapid growth (c).

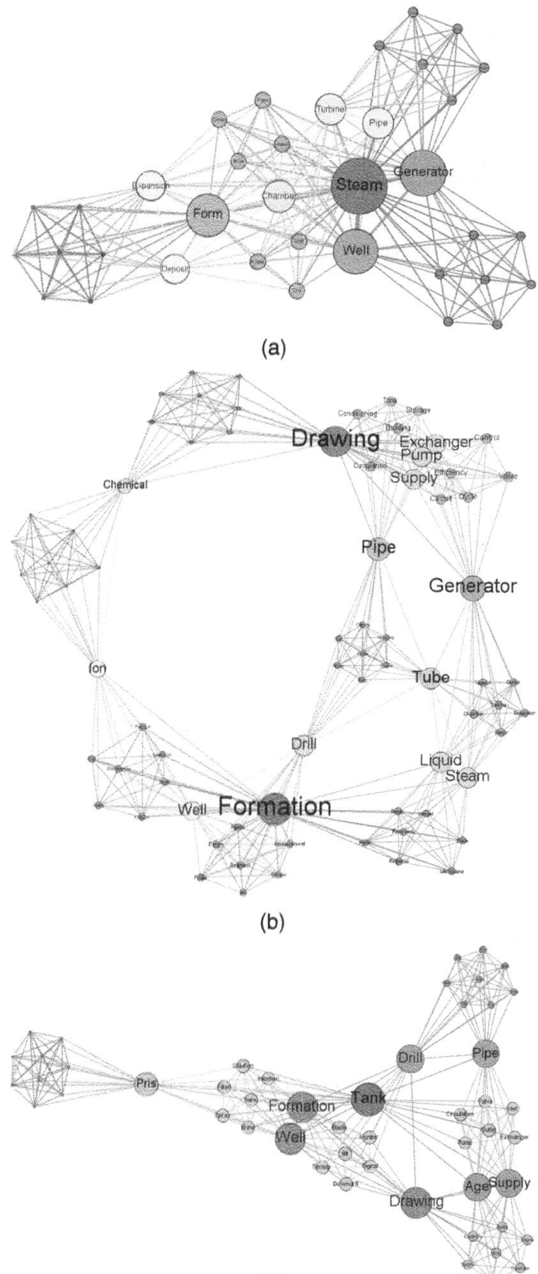

Figure 7.22 Co-occurrence network derived from association rules based on the Eigenvector centrality index – emerging stage (a), slow growth (b), and rapid growth (c).

Table 7.23 Words extracted from association rules based on the highest degree centrality index – emerging stage (a), slow growth (b), and rapid growth (c).

	Emerging stage		Slow-growth stage		Rapid-growth stage	
Row	**Label**	**Degree**	**Label**	**Degree**	**Label**	**Degree**
1	Steam	28	Formation	26	Drawing	18
2	Generator	24	Drawing	24	Pipe	18
3	Form	22	Ion	18	Tank	18
4	Well	22	Generator	18	Drill	18
5	Expansion	17	Chemical	18	Pris	18
6	Deposit	17	Tube	18	Supply	17
7	Turbine	16	Drill	18	Age	17
8	Pipe	16	Pipe	18	Well	17
9	Chamber	15	Liquid	17	Formation	17
10	Brine	9	Steam	17	Valve	9
11	Vapor	9	Well	17	Inlet	9
12	Flash	9	Exchanger	15	Exchanger	9
13	Contact	9	Supply	15	Outlet	9
14	Wave	9	Pump	15	Circulation	9
15	Gradient	9	Electrode	9	Sand	9
16	Cavity	9	Cell	9	Brine	9
17	Voltage	9	Resistance	9	Pump	9
18	Junction	9	Matrix	9	Tube	9
19	Carrier	9	Recovery	9	Cavity	9
20	Bore	9	Signal	9	Fix	9

In Figures 7.19–7.22, nodes represent the words and edges represent their co-occurrence in the topic clusters of the LDA Topic Modeling. In Figure 7.19, the size of each node depends on the degree centrality. The higher the degree of centrality, the larger the node. The values of the "lift" index of association rules represent the edges which connect the nodes in the network. The size of each edge depends on the value of this index. The higher the above-mentioned index, the larger the edge size. In the above figures, the most popular and influential technologies and inventions are directly related to the size of nodes. Nodes play a significant role in promoting new technologies and inventions in their

Table 7.24 Words extracted from association rules based on the highest betweenness centrality index – emerging stage (a), slow growth (b), and rapid growth (c).

	Emerging stage		Slow-growth stage		Rapid-growth stage	
Row	Label	Betweenness centrality	Label	Betweenness centrality	Label	Betweenness centrality
1	Steam	122.85	Formation	603.37	Pris	369.00
2	Form	86.02	Drawing	578.43	Tank	272.07
3	Generator	85.02	Ion	471.24	Drill	190.53
4	Well	60.35	Chemical	415.91	Drawing	167.07
5	Expansion	53.15	Generator	393.73	Pipe	166.53
6	Deposit	51.68	Drill	384.79	Well	145.45
7	Turbine	22.48	Pipe	375.03	Formation	145.45
8	Pipe	22.48	Liquid	207.65	Supply	85.45
9	Chamber	6.98	Steam	207.65	Age	85.45
10	Brine	0.00	Tube	163.19	Exchanger	0.00
11	Vapor	0.00	Well	156.20	Outlet	0.00
12	Flash	0.00	Exchanger	36.94	Valve	0.00
13	Contact	0.00	Supply	36.94	Pump	0.00
14	Wave	0.00	Pump	36.94	Circulation	0.00
15	Gradient	0.00	Resistance	0.00	Inlet	0.00
16	Cavity	0.00	Matrix	0.00	Tube	0.00
17	Voltage	0.00	Electrode	0.00	Cavity	0.00
18	Junction	0.00	Impurity	0.00	Fix	0.00
19	Carrier	0.00	Deg	0.00	Rod	0.00
20	Bore	0.00	Block	0.00	Groove	0.00

respective fields. Based on Figure 7.19 and Table 7.23, the technologies and inventions related to steam, generator, form, well, expansion, deposit, turbine, pipe, chamber, brine in terms of degree centrality were considered as the most effective technologies and inventions in the emerging stage. However, influential technologies and inventions were related to formation, drawing, ion, generator, chemical, tube, drill, pipe, liquid, and steam in the slow-growth stage. Effective technologies and inventions in the rapid growth stage included drawing, pipe, tank, drill, pris, supply, age, well, formation, valve. The betweenness centrality index shows the relationship between inventions and technologies in the

Table 7.25 Words extracted from association rules based on the highest closeness centrality index – emerging stage (a), slow growth (b), and rapid growth (c).

	Emerging stage		Slow-growth stage		Rapid-growth stage	
Row	Label	Closeness centrality	Label	Closeness centrality	Label	Closeness centrality
1	Steam	0.83	Formation	0.51	Tank	0.61
2	Generator	0.76	Drawing	0.50	Well	0.60
3	Form	0.73	Drill	0.49	Formation	0.60
4	Well	0.73	Liquid	0.48	Pipe	0.55
5	Expansion	0.66	Steam	0.48	Drill	0.55
6	Deposit	0.66	Pipe	0.48	Drawing	0.55
7	Turbine	0.65	Generator	0.48	Supply	0.54
8	Pipe	0.65	Ion	0.46	Age	0.54
9	Chamber	0.64	Chemical	0.46	Pris	0.52
10	Brine	0.57	Tube	0.45	Exchanger	0.47
11	Vapor	0.57	Well	0.45	Outlet	0.47
12	Flash	0.57	Exchanger	0.43	Valve	0.47
13	Contact	0.57	Supply	0.43	Pump	0.47
14	Rubble	0.57	Pump	0.43	Circulation	0.47
15	Shaft	0.57	Force	0.40	Inlet	0.47
16	Drill	0.57	Ring	0.40	Sand	0.47
17	Wave	0.51	Seal	0.40	Filter	0.47
18	Gradient	0.51	String	0.40	Brine	0.47
19	Cavity	0.51	Sleeve	0.40	Injection	0.47
20	Voltage	0.51	Housing	0.40	Spray	0.47

network, leading to the transfer and sharing of knowledge. Based on Figure 7.20 and Table 7.24, the nodes of steam, form, generator, well, expansion, deposit, turbine, pipe, chamber in the first stage, formation, drawing, ion, chemical, generator, drill, pipe, liquid, steam, tube, well, exchanger, supply, pump in the second period and pris, tank, drill, drawing, pipe, well, formation, supply, age in the third period had the highest betweenness centrality index.

Figure 7.21 and Table 7.25 indicate the words with the highest degree of closeness. The technologies and inventions related to terms with the highest

Table 7.26 Words extracted from association rules based on the highest Eigenvector centrality index – emerging stage (a), slow growth (b), and rapid growth (c).

	Emerging stage		Slow-growth stage		Rapid-growth stage	
Row	**Label**	**Eigen centrality**	**Label**	**Eigen centrality**	**Label**	**Eigen centrality**
1	Steam	1.00	Formation	1.00	Tank	1.00
2	Generator	0.86	Drawing	0.98	Drawing	0.95
3	Well	0.85	Generator	0.84	Well	0.94
4	Form	0.82	Pipe	0.79	Formation	0.94
5	Chamber	0.68	Tube	0.75	Drill	0.88
6	Turbine	0.66	Drill	0.73	Supply	0.87
7	Pipe	0.66	Liquid	0.72	Age	0.87
8	Deposit	0.64	Steam	0.72	Pipe	0.87
9	Expansion	0.64	Exchanger	0.71	Pris	0.78
10	Rubble	0.45	Supply	0.71	Signal	0.54
11	Shaft	0.45	Pump	0.71	Monitor	0.54
12	Drill	0.45	Well	0.67	Downdrill	0.54
13	Brine	0.45	Ion	0.63	Bit	0.54
14	Vapor	0.45	Chemical	0.58	Blade	0.54
15	Flash	0.45	Control	0.48	Sensor	0.54
16	Contact	0.45	Efficiency	0.48	Sand	0.54
17	Transfer	0.37	Valve	0.48	Filter	0.54
18	Outlet	0.37	Circuit	0.48	Brine	0.54
19	Evaporator	0.37	Cycle	0.48	Injection	0.54
20	Circuit	0.37	Building	0.47	Spray	0.54

closeness degree are potentially used for developing new products and services. The words with the highest degree of closeness in the first period were steam, generator, form, well, expansion, deposit, turbine, pipe, chamber, brine, in the second period were the words formation, drawing, drill, liquid, steam, pipe, generator, ion, chemical, tube, and in the third period were tank, well, formation, pipe, drill, drawing, supply, age, pris, and exchanger. Finally, the words with the highest degree of eigenvector determine the most significant inventions and technologies for other nodes of each technical part.

The technologies and inventions related to steam, generator, well, form, chamber, turbine, pipe, deposit, expansion, rubble in the emerging stage, the technologies and inventions related to formation, drawing, generator, pipe, tube, drill, liquid, steam, exchanger, supply in the slow growth stage, and the technologies and inventions related to tank, drawing, well, formation, drill, supply, age, pipe, pris, signal in the rapid growth stage had the highest value of eigenvector centrality (Figure 7.22 and Table 7.26).

7.5 Discussion

Enhancing geothermal energy will require government and private investments, including introducing economic incentives to limit the economic barriers, including high exploration, setup, operation, permit, and other costs (Hackstein and Madlener 2021; Chen 2015). The focus is on increasing its presence as an alternative source of energy for direct use, including heating buildings and a source of electricity with minimal carbon footprint or greenhouse gas emissions (Allahvirdizadeh 2020; Hu et al. 2021; Varvitsioti and Tsifoutidis 2021).

Hu et al. (2021) add that the focus should be on increasing "significant investment in research and development of enhanced geothermal systems (EGS) technologies for exploiting hot dry rock HDR resources and establishment of favorable policy instruments are recommended as key technological and policy measures" to enhance the adoption and success of geothermal energy. Chen et al. (Chen 2015) add that countries such as Taiwan with considerable geothermal energy capabilities should ensure the legal issues and regulations impeding successful adoption of the aforementioned are contained. Since most of the environmental effects are subject to mitigation, there is a need to prioritize the technological impacts of geothermal energy, including green-sustainable science technology. In this regard, as Gonzalez-Acevedo and Garcia-Zarate assert (González-Acevedo and García-Zarate 2018), they should be considered the green alternative and championed in homes to limit overreliance on nonrenewable energy sources that have detrimental effects on humans, water, wildlife, and the environment in general. The goal is to contain atmospheric emissions using geothermal energy.

The international bodies, including the United Nations, should intervene and encourage or facilitate the adoption of geothermal energy, especially in developing countries with low electrical power installed (Barbier 2002; Obonyo 2021). As of 2002, countries such as Kenya and El Salvador have about 8% and 17%, respectively, on geothermal electricity. In other words, Ellmies et al. (2004) write geothermal energy is pivotal in sustainable development (Ellmies et al. 2004).

Importantly, geothermal electricity generation faces challenges, including the inadequate exploitable sites, which implies that more resources and technology should be invested to identify, explore, and construct power plants that go beyond usual direct heat use (Kabeyi and Jeremiah 2019). Kabeyi and Jeremiah (2019) also assert that "research and development into new, efficient, and cost-effective technologies will enhance safety and environmental integrity with respect to geothermal energy and electricity development." Regardless, the process is costly and requires a collective effort from private and public entities, from exploration, feasibility, drilling, development, reinjection wells, steam gathering, commissioning, construction, and production. Stages such as drilling and the high costs require investments in geophysical technology that can also minimize risks.

7.6 Conclusions

The clusters that emerged in our investigations and the adjacent and divergent positions of these clusters were determined. In this respect, it is possible to say that our study has reached important conclusions regarding geothermal energy technologies. It is stated that it has a unique aspect in terms of combining social network analysis with LDA. On the other hand, it is seen that we have important outputs in terms of revealing the evolutionary development of the related technologies in emerging, slow growth and rapid growth periods. In the examinations made, it was determined that the key topics were high-tech topics, but it was also determined that these topics did not show continuity. We think that identifying key issues, which can be defined as topics that have a high level of technology power and technology continuity and that have a significant impact on technology development, will contribute positively to policy studies on engineering and technology management.

7.7 Limitations and Future Research

When we evaluate in terms of the limitations of the study, it should be noted that the privacy and trade secret restrictions in the period until the publication of the patent data limit the timeliness of the patent data accessed on the relevant subject. On the other hand, the fact that the schemes used in patent classification studies include a wider classification effort creates a limitation on the scope since it increases the number of patent documents that can be entered into the data set used. It is thought that this limitation can be avoided thanks to the workflow proposed in the study and the combined analysis methods.

7.A Python Code of Mann-Kendall Test

```
import numpy as np
import matplotlib.pyplot as plt
import pandas as pd
```

In Kim et al. (2019):

```
df = pd.read_excel(r"D:\Documents\CV\Ph.D\Ph.D
     Thesis\Ph.D Thesis\Raw Data-30-9-2021\Book1.xlsx")
df.head()
```

Out Kim et al. (2019):

	Year	NoP
0	1964	1
1	1965	1
2	1966	1
3	1969	1
4	1970	4

In Persson et al. (2009):

```
x = list(df.Year)
y = list(df.NoP)
```

In Kasztler et al. (2012):

```
import matplotlib.pyplot as plt
plt.rc('font', size=12)
fig, ax = plt.subplots(figsize=(10, 6))
ax.plot(x,y, color='tab:orange')
```

Out Kasztler et al. (2012):

```
[<matplotlib.lines.Line2D at 0x21007ac8730>]
```

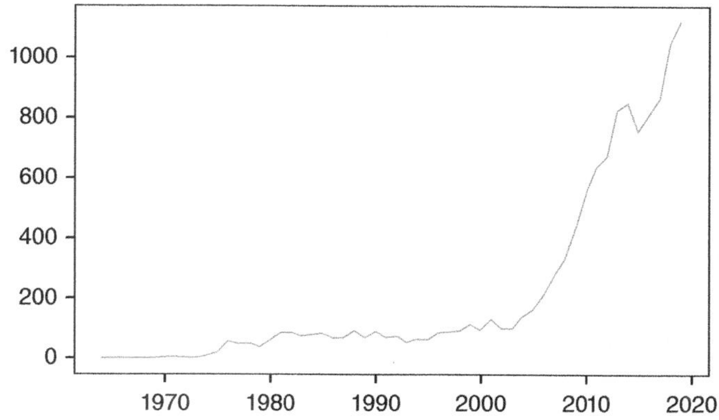

In Sahal (1985):

```
import pymannkendall as mk
```

In Kendall (2016):

```
result = mk.original_test(df['NoP'])
```

In El-Shaarawi and Piegorsch (2002):

```
print (result)
Mann_Kendall_Test(trend='increasing', h=True, p=0.0,
z=8.937203548250775, Tau=0.8454281567489115, s=1165.0,
var_s=16963.0, slope=7.530555555555555,
intercept=-111.79444444444442)
```

In Gilbert (n.d.):

```
#trend : tell the trend is incrasing, decreasing,
 stationary or no Trend
#h : true if Trend is presant; Flase if the trend
 is absent
#p-value : <= 0.05 significance level
#Z-Score : Normalized test statistics --> Z = (x-mu)/
 Sigma       Z<= 1    Z= 0    Z => 1
#S-core : Mann-kendal score
```

In Witten et al. (2011):

```
from scipy.stats.mstats import theilslopes
```

In IEEE (2017):

```
theilslopes(x,y)
```

Out IEEE (2017):

```
(0.08409506398537477,
 1985.9360146252286,
 0.053475935828877004,
 0.15053763440860216)
```

In Talib et al. (2016):

```
#alpha : float Confidence degree between 0 and 1.
Default is 95% confidence. Note that alpha is
symmetric around 0.5, i.e. both 0.1 and 0.9 are
interpreted as "find the 90% confidence interval".

#medslope : float / Theil slope. (theilslope z-value :
tell the trend is incrasing, decreasing, stationary
or no Trend).

#medintercept : float / Intercept of the Theil line,
as median(y) - medslope*median(x).

#lo_slope : float / Lower bound of the confidence
interval on medslope.

#up_slope : float / Upper bound of the confidence
interval on medslope.
```

7.B Python Code of Life Cycle S Curve

```
import numpy as np
import matplotlib.pyplot as plt
import pandas as pd
```

In Kim et al. (2019):

```
data = pd.read_excel("D:\Documents\CV\Ph.D\Ph.D
       Thesis\Ph.D Thesis\Raw Data-30-9-2021\Book1.xlsx")
```

In Persson et al. (2009):

```
data.head()
```

Out Persson et al. (2009):

	Year	Cum
0	1964	1
1	1965	2
2	1966	3
3	1969	4
4	1970	8

In Kasztler et al. (2012):

```
xdata=list(data.Year)
```

In Sahal (1985):

```
ydata=list(data.Cum)
```

In Kendall (2016):

```
plt.plot(xdata, ydata, 'o')
```

Out Kendall (2016)

```
[<matplotlib.lines.Line2D at 0x263ea83d2e0>]
```

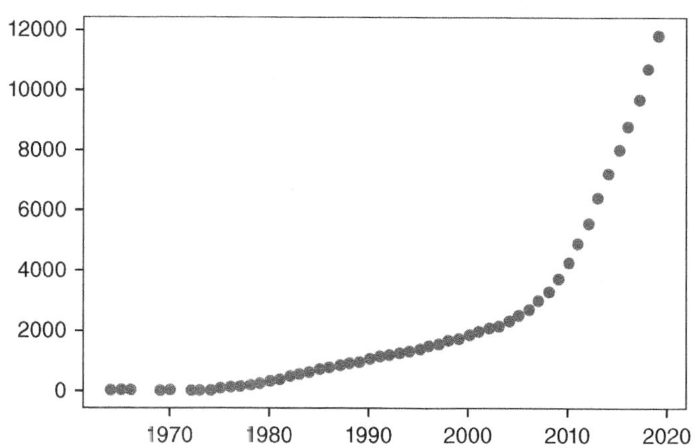

In El-Shaarawi and Piegorsch (2002):

```
from scipy.optimize import curve_fit

def sigmoid(x, L ,x0, k, b):
    y = L / (1 + np.exp(-k*(x-x0)))+b
    return (y)

p0 = [max(ydata), np.median(xdata),1,min(ydata)]

popt, pcov = curve_fit(sigmoid, xdata, ydata, p0,
maxfev=6000)

x = np.linspace(1964, 2019,53)
y = sigmoid(x, *popt)

plt.plot(xdata, ydata, 'o', label='data')
plt.plot(x,y, label='fit')
plt.legend(loc='best')
plt.axvline(x=2023)
```

Out El-Shaarawi and Piegorsch (2002):

```
<matplotlib.lines.Line2D at 0x263eab3c400>
```

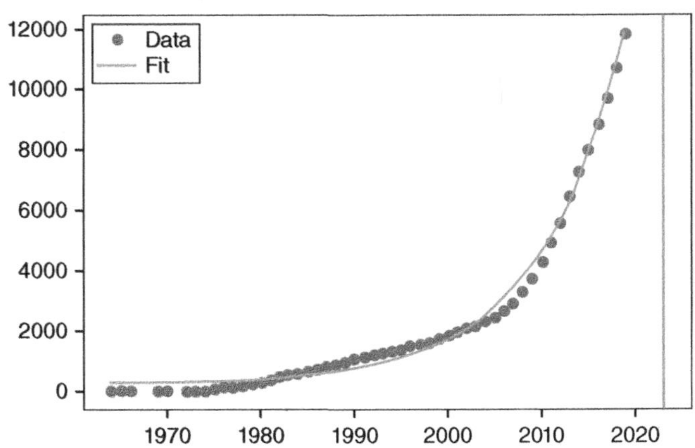

In Gilbert (n.d.):

```
popt
# L, X0, K, b
```

Out Gilbert (n.d.):

```
array([1.47263445e+09, 2.12705093e+03, 1.08707395e-01,
2.07709278e+02])
```

In Witten et al. (2011):

```
sigmoid(2023,1.47263445e+09, 2.12705093e+03,
1.08707395e-01, 2.07709278e+02)
```

Out Witten et al. (2011):

```
18227.050868621537
```

References

Agrawal, R. and Srikant R. (1994). Fast algorithms for mining association rules in large databases. In: *Proceedings of the 20th International Conference on Very Large Data Bases*, 487–499. San Francisco, CA: Morgan Kaufmann Publishers Inc. https://www.bibsonomy.org/bibtex/960c924ccbe1ff429a30f7433ec53122.

Agrawal, R., Imielinski, T., and Swami, A.N. (1993). Mining association rules between sets of items in large databases. SIGMOD Conference. *Proceedings of the 1993 ACM SIGMOD International Conference on Management of Data* 22 (2): 207–216. https://doi.org/10.1145/170036.170072.

Allahvirdizadeh, P. (2020). A review on geothermal wells: well integrity issues. *Journal of Cleaner Production* 275 (1): 124009. https://doi.org/10.1016/j.jclepro.2020.124009.

Altuntas, F. and Gok, M.S. (2020). Technological evolution of wind energy with social network analysis. *Kybernetes* doi: 10.1108/K-11-2019-0761.

Ampornphan, P. and Tongngam, S. (2017). Patent knowledge discovery using data analytics. In: *Proceedings of the 2017 International Conference on Information Technology*, 42–46. Singapore: ICIT.

Ampornphan, P. and Tongngam, S. (2020). Exploring technology influencers from patent data using association rule mining and social network analysis. *Information* 11 (6): 333. https://doi.org/10.3390/info11060333.

Amrollahi, Z., Ertesvåg, I.S., and Bolland, O. (2011). Thermodynamic analysis on post combustion CO2 capture of natural gas fired power plant. *International Journal of Greenhouse Gas Control* 5 (3): 422–426. https://doi.org/10.1016/S1876-0147(10)02034-3.

Arias, E.R. (2016). Geothermal energy in the framework of international environmental law. In: *Geothermal Power Generation*, 763–786. Woodhead Publishing https://doi.org/10.1016/B978-0-08-100337-4.00026-7.

Arnórsson, S. (2004). Environmental impact of geothermal energy utilization. *Geological Society, London, Special Publications* 236 (1): 297–336. https://doi.org/10.1144/GSL.SP.2004.236.01.18.

Azad, A.K. (2020). *Advances in Clean Energy Technologies*. Academic Press https://doi.org/10.1016/C2019-0-01802-X.

Aziz, M. and Juangsa, F.B. (2021). Dry steam power plant: thermodynamic analysis and system improvement. In: *Thermodynamic Analysis and Optimization of Geothermal Power Plants*, 97–111. Elsevier https://doi.org/10.1016/B978-0-12-821037-6.00015-9.

Bagher, A.M., Vahid, M., and Mohsen, M. (2014). Geothermal energy. *Journal of Engineering and Technology Research* 6 (8): 146–150.

Barbier, E. (1997). Nature and technology of geothermal energy: a review. *Renewable and Sustainable Energy Reviews* 1 (1–2): 1–69. https://doi.org/10.1016/S1364-0321(97)00001-4.

Barbier, E. (2002). Geothermal energy technology and current status: an overview. *Renewable and Sustainable Energy Reviews* 6 (1–2): 3–65. https://doi.org/10.1016/S1364-0321(02)00002-3.

Bastian, M., Heymann, S., and Jacomy, M. (2009). Gephi: an open source software for exploring and manipulating networks. https://www.researchgate.net/publication/221297890_Gephi_An_Open_Source_Software_for_Exploring_and_Manipulating_Networks (accessed 24 May 2024).

Bilgin, O. (2018). The importance of geothermal energy resources in Turkey. *Open Access Library Journal* 05 (02): 1–8. https://doi.org/10.4236/oalib.1104317.

Blei, D.M., Ng, A.Y., and Jordan, M.I. (2003). Latent Dirichlet allocation. *Journal of Machine Learning Research* 3: 993–1022.

Bošnjaković, M., Stojkov, M., and Jurjević, M. (2019). Environmental impact of geothermal power plants. *Tehnicki Vjesnik – Technical Gazette* 26 (5): 1515–1522. https://doi.org/10.17559/TV-20180829122640.

Chang, J., Gerrish, S., Wang, C. et al. (2009). Reading tea leaves: how humans interpret topic models. *Advances in Neural Information Processing Systems* 22.

Chen, W.-M. (2015). An analysis of current status and legal issues of geothermal power plant construction in Taiwan (published in Chinese). *Science and Technology Law Review* 27: 56–72.

Choi, D. and Song, B. (2018). Exploring technological trends in logistics: topic modeling-based patent analysis. *Sustainability* 10 (8): 2810. https://doi.org/10.3390/su10082810.

da Costa-Luis, C. (n.d.). tqdm documentation. http://tqdm.github.io/ (accessed Nov. 13, 2021).

Curtis, R., John, L., Sanner, B., and Rybach, L. (2005). Ground source heat pumps–geothermal energy for anyone, anywhere: current worldwide activity. *Proceedings World Geothermal Conger's* 24: 175.

Dhar, A., Naeth, M.A., Jennings, P.D., and Gamal El-Din, M. (2020). Geothermal energy resources: potential environmental impact and land reclamation. *Environmental Reviews* 28 (4): 415–427. https://doi.org/10.1139/er-2019-0069.

Dumais, S.T. (2004). Latent semantic analysis. *Annual Review of Information Science and Technology* 38 (1): 188–230. https://doi.org/10.1002/aris.1440380105.

Ehyaei, M.A., Ahmadi, A., Rosen, M.A., and Davarpanah, A. (2020). Thermodynamic optimization of a geothermal power plant with a genetic algorithm in two stages. *Processes* 8 (10): 1277. https://doi.org/10.3390/pr8101277.

Ellmies, R., Franken, G., Kessels, K., and Kraml, M. (2004). Geothermal energy for sustainable development. *Bulletin Für Angewandte Geologie* 2: 4–11.

El-Shaarawi, A.H. and Piegorsch, W.W. (2002). *Encyclopedia of Environmetrics*. Chichester: Wiley.

Environmental and Energy Study Institute (2006). Fact sheet: Geothermal energy.

Farooq, A., Joyia, G.J., Uzair, M., and Akram, U. (2018). Detection of influential nodes using social networks analysis based on network metrics. https://www.semanticscholar.org/paper/Detection-of-influential-nodes-using-social-based-Farooq-Joyia/ddbfc3aad890dcbebe1435fcd3e8f9d4cb641d24 (accessed 24 May 2024).

Feldman, R. and Sanger, J. (2007). *The Text Mining Handbook: Advanced Approaches in Analyzing Unstructured Data*. Cambridge: Cambridge University Press http://www.loc.gov/catdir/enhancements/fy0633/2005029330-d.html.

Fountain, S.B. and Doyle, K.E. (2012). Learning by chunking. In: *Encyclopedia of the Sciences of Learning* (ed. N.M. Seel), 1814–1817. Boston, MA: Springer https://link.springer.com/referenceworkentry/10.1007%2F978-1-4419-1428-6_1042.

Gensim (n.d.). Gensim: topic modelling for humans. https://radimrehurek.com/gensim/ (accessed: Nov. 13, 2021).

Gilbert, R.O. (n.d.). *Statistical Methods for Environmental Pollution Monitoring*. New York: Wiley.

Gomaa, W.H. and Fahmy, A.A. (2013). A survey of text similarity approaches. *International Journal of Computer Applications* 68 (13): https://doi.org/10.5120/11638-7118.

González-Acevedo, Z. and García-Zarate, M. (2018). Geothermal energy as an alternative to reduce atmospheric emissions and provide green energy. In: *Green Technologies to Improve the Environment on Earth*. IntechOpen https://doi.org/10.5772/intechopen.80759.

Hackstein, F.V. and Madlener, R. (2021). Sustainable operation of geothermal power plants: why economics matters. *Geothermal Energy* 9 (1): 10. https://doi.org/10.1186/s40517-021-00183-2.

Haddi, E., Liu, X., and Shi, Y. (2013). The role of text pre-processing in sentiment analysis. *Procedia Computer Science* 17: 26–32. https://doi.org/10.1016/j.procs.2013.05.005.

Harris, C.R., Millman, K., van der Walt, S. et al. (2020). Array programming with NumPy," (in English). *Nature* 585 (7825): 357–362. https://doi.org/10.1038/s41586-020-2649-2.

Hassler, M. and Fliedl, G. (2006). Text preparation through extended tokenization. *WIT Transactions on Information and Communication Technologies* 37: 13–21. http://www.witpress.com/elibrary/wit-transactions-on-information-and-communication-technologies/37/16699.

Heather, K. (2020). *Thermodynamics of Geothermal Energy*.

Hofmann, T. (1999). Probabilistic latent semantic analysis. *Proceedings of the Fifteenth Conference on Uncertainty in Artificial Intelligence* 289–296.

Hofmann, T. (2001). Unsupervised learning by probabilistic latent semantic analysis," (in English). *Machine Learning* 42 (1): 177–196. https://doi.org/10.1023/A:1007617005950.

Hotelling, H. (1933). Analysis of a complex of statistical variables into principal components. *Journal of Educational Psychology* 24 (7): 498–520. https://doi.org/10.1037/h0070888.

Hotho, A., Nürnberger, A., and Paass, G. (2005). A brief survey of text mining. https://www.semanticscholar.org/paper/A-Brief-Survey-of-Text-Mining-Hotho-N%C3%BCrnberger/8f74f5623c4e5c5931641a264cfd7c02097e1e22

Hu, Y., Cheng, H., and Tao, S. (2021). Opportunity and challenges in large-scale geothermal energy exploitation in China. *Critical Reviews in Environmental Science and Technology* 1–22. https://doi.org/10.1080/10643389.2021.1971004.

IEEE (2017). *Proceedings of the 2017 International Conference on Information Technology – ICIT 2017*. New York: ACM Press.

Kabeyi, B. and Jeremiah, M. (2019). Geothermal electricity generation, challenges, opportunities and recommendations. *International Journal of Advances in Scientific Research and Engineering* 5 (8): 53–95. https://doi.org/10.31695/IJASRE.2019.33408.

Kaneko, M. (2016). Geothermal energy laws in the world (2). *Journal of the Geothermal Research Society of Japan* 34 (3): 85–100.

Kanoglu, M. and Bolatturk, A. (2016). Thermodynamic analysis of geothermal power plants. In: *Alternative Energy and Shale Gas Encyclopedia*, 290–300. Wiley https://doi.org/10.1002/9781119066354.ch28.

Kasztler, A., Apilo, T., Marcin, B. et al. (2012). Practices, future requirements and building blocks of a new innovation model. https://www.semanticscholar.org/paper/Practices%2C-future-requirements-and-building-blocks-Kasztler-Tiina/3f4f0d899a6e657289637096fe22bca59bc0ece6

Kendall, M.G. (2016). Kendall analysis for the Fort Ord Site. https://www.statisticshowto.com/wp-content/uploads/2016/08/Mann-Kendall-Analysis-1.pdf.

Khanam, M. and Daim, T. (2021). A market diffusion potential (MDP) assessment model for residential energy efficient (EE) technologies in the U.S. *Renewable and Sustainable Energy Reviews* 144: https://doi.org/10.1016/j.rser.2021.110968.

Kim, K.H., Han, Y.J., Lee, S. et al. (2019). Text mining for patent analysis to forecast emerging technologies in wireless power transfer. *Sustainability* 11 (22): 6240. https://doi.org/10.3390/su11225240.

Kuchling, A.M. (2001). Regular expression HOWTO – Python 3.12.3. documentation. http://docs.python.org/3/howto/regex.html (accessed: Nov. 13, 2021).

Kumar, M. (2020). Social, economic, and environmental impacts of renewable energy resources. In: *Wind Solar Hybrid Renewable Energy System* (ed. K. Okedu, A. Tahour, and A.G. Aissaou), 1–9. IntechOpen https://doi.org/10.5772/intechopen.89494.

Landauer, T. and Dumais, S. (1997). A solution to plato's problem: the latent semantic analysis theory of acquisition, induction, and representation of knowledge. https://www.semanticscholar.org/paper/A-Solution-to-Plato%27s-Problem%3A-The-Latent-Semantic-Landauer-Dumais/68dd4b89ce1407372a29d05ca9e4e1a2e0513617

Larose, D.T. and Larose, C.D. (2014). *Discovering Knowledge in Data: An Introduction to Data Mining /*. Hoboken, NJ: Wiley.

Lee, Y.-S., Lo, R., Chen, C.-Y. et al. (2015). News topics categorization using latent Dirichlet allocation and sparse representation classifier. https://www.semanticscholar.org/paper/News-topics-categorization-using-latent-Dirichlet-Lee-Lo/1dc2276d92a7312c9ffd4b4dd790c83d4ba7d800

Lee, S., Cha, Y., Han, S., and Hyun, C. (2019). Application of association rule mining and social network analysis for understanding causality of construction defects. *Sustainability* 11 (3): 618. https://doi.org/10.3390/su11030618.

Lin, D. (1998). An information-theoretic definition of similarity. http://citeseerx.ist.psu.edu/viewdoc/summary? https://doi.org/10.1.1.55.1832

Liu, W., Tao, Y., Yang, Z., and Bi, K. (2019). Exploring and visualizing the patent collaboration network: a case study of smart grid field in China. *Sustainability* 11 (2): 465. https://doi.org/10.3390/su11020465.

Mathplotlib (n.d.). Source code for Matplotlib.pyplot (version 3.4.3). https://matplotlib.org/3.4.3/_modules/matplotlib/pyplot.html (accessed Nov. 13, 2021).

Mckinney, W. (2011). pandas: a foundational Python library for data analysis and statistics. https://www.researchgate.net/publication/265194455_pandas_a_Foundational_Python_Library_for_Data_Analysis_and_Statistics

Michaelides, E.E. (2016). Future directions and cycles for electricity production from geothermal resources. *Energy Conversion and Management* 107: 3–9. https://doi.org/10.1016/j.enconman.2015.07.057.

Mifrah, S. and Benlahmar, E.H. (2020). Topic modeling coherence: a comparative study between LDA and NMF models using covid'19 corpus. *International Journal of Advanced Trends in Computer Science and Engineering* 5756–5761. http://pesquisa.bvsalud.org/global-literature-on-novel-coronavirus-2019-ncov/resource/pt/covidwho-828717.

Moya, D., Aldás, C., and Kaparaju, P. (2018). Geothermal energy: power plant technology and direct heat applications. *Renewable and Sustainable Energy Reviews* 94: 889–901. https://doi.org/10.1016/j.rser.2018.06.047.

Mueller, A. (2020). WordCloud for Python Documentation — Wordcloud 1.8.1. *Documentation*. http://amueller.github.io/word_cloud (accessed: Nov. 13, 2021).

NLTK (n.d.). Natural language toolkit. https://www.nltk.org/ (accessed Nov. 13, 2021).

Obonyo, R. (2021, January 6). Push for renewables: How Africa is building a different energy pathway. *Africa Renewal* https://www.un.org/africarenewal/magazine/january-2021/push-renewables-how-africa-building-different-energy-pathway.

Orange Visual Programming (2017) Documentation Release 3. https://http://orange3.readthedocs.io/projects/orange-visual-programming/en/latest/widgets/unsupervised/distancematrix.html (accessed: Nov. 13 2021.033Z).

Pandiella-Dominique, A., Bautista-Puig, N., and de Filippo, D. (2018). Mapping growth and trends in the category "Green and Sustainable Science and Technology". *STI 2018 Conference Proceedings* 1038–1042.

Pearson, K. (1901). LIII. On lines and planes of closest fit to systems of points in space. *The London, Edinburgh, and Dublin Philosophical Magazine and Journal of Science* 2 (11): 559–572. https://doi.org/10.1080/14786440109462720.

Pedregosa, F., Varoquaux, G., Gramfort, A. et al. (2011). Scikit-learn: machine learning in Python. *Journal of Machine Learning Research* 12 (85): 2825–2830. https://www.jmlr.org/papers/v12/pedregosa11a.html.

Persson, O., Danell, R., and Schneider, J.W. (2009). How to use Bibexcel for various types of bibliometric analysis. In: *Celebrating Scholarly Communication Studies*. International Society for Scientometrics and Informetrics.

Platetsky-Shapiro, G. (1991). *Discover, Analysis, and Presentation of Strong Rules*. AAAI/MIT Press https://www.bibsonomy.org/bibtex/85578d6243c6d1620855dc2ef07bd7a7.

Raschka, S. (n.d.). Association rules – mlxtend. http://rasbt.github.io/mlxtend/user_guide/frequent_patterns/association_rules (accessed Nov. 13, 2021).

Riggs, S.R., Ellison, S.P., Fisher, W.L. et al. (1991). Mineral resources and geopressured-geothermal energy. In: *The Gulf of Mexico Basin*, 495–528. Geological Society of America https://doi.org/10.1130/DNAG-GNA-J.495.

Sahal, D. (1985). Technological guideposts and innovation avenues. *Research Policy* 14 (2): 61–82. https://doi.org/10.1016/0048-7333(85)90015-0.

Sakthi Vel, S. (n.d.). Strings in Python. INTRODUCTION Built-in String Methods PYTHON*Language Technology- UOK*. https://www.academia.edu/44926625/Strings_in_Python_INTRODUCTION_Built_in_String_Methods_PYTHON

Sarker, I.H. (2021). Machine learning: algorithms, real-world applications and research directions. *SN Computer Science* 2 (3): 1–21. https://doi.org/10.1007/s42979-021-00592-x.

Sievert, C. and Shirley, K.E. (2014). LDAvis: A method for visualizing and interpreting topics. https://www.researchgate.net/publication/265784473_LDAvis_A_method_for_visualizing_and_interpreting_topics

Singh, T. and Kumari, M. (2016). Role of text pre-processing in twitter sentiment analysis. *Procedia Computer Science* 89: 549–554. https://doi.org/10.1016/j.procs.2016.06.095.

Singh, S., Thapar, V., and Bagga, S. (2020). Exploring the hidden patterns of cyberbullying on social media. *Procedia Computer Science* 167: 1636–1647. https://doi.org/10.1016/j.procs.2020.03.374.

Soltani, M., Moradi Kashkooli, F., Souri, M. et al. (2021). Environmental, economic, and social impacts of geothermal energy systems. *Renewable and Sustainable Energy Reviews* 140: 110750.

SpaCy. (n.d.). SpaCy 101: Everything you need to know. https://spacy.io/usage/spacy-101 (accessed: Nov. 13 2021.996Z).

Spisak, T. (2019). Geothermal energy development. Press release, State Director for New Mexico, Oklahoma, Texas and Kansas. Bureau of Land Management. June 20, 2019. https://www.doi.gov/ocl/geothermal-energy-development.

Stavrianou, A., Andritsos, P., and Nicoloyannis, N. (2007). Overview and semantic issues of text mining. *ACM SIGMOD Record* 36 (3): 23–34. https://doi.org/10.1145/1324185.1324190.

Stober, I. and Bucher, K. (2013). *Geothermal Energy*, vol. 10. Berlin, Heidelberg: Springer-Verlag.

Sustersic, V., Babic, M., Gordic, D. et al. (2010). An overview of the regulatory framework for the geothermal energy in Europe and Serbia. *Thermal Science* 14 (suppl): 115–123. https://doi.org/10.2298/TSCI100616068S.

SYS (n.d.). System-specific parameters and functions – Python 3.10.0 documentation. http://docs.python.org/3/library/sys.html (accessed: Nov. 13 2021.468Z).

Talib, R., Kashif, M., Ayesha, S., and Fatima, F. (2016). Text mining: techniques, applications and issues. *International Journal of Advanced Computer Science and Applications* 7 (11): https://doi.org/10.14569/IJACSA.2016.071153.

Tan, M. and Lee, W.-L. (2015). Evaluation and improvement of procurement process with data analytics. *International Journal of Advanced Computer Science and Applications* 6 (8): https://doi.org/10.14569/IJACSA.2015.060809.

Topić, D., Šljivac, D., Vezmar, S. et al. (2014). Positive and negative impacts of renewable energy sources. *International Journal of Electrical and Computer Engineering Systems* 5 (2): 47–55.

Trappey, A.J.C., Chen, P.P.J., Trappey, C.V., and Ma, L. (2019). A machine learning approach for solar power technology review and patent evolution analysis. *Applied Sciences* 9 (7): 1478. https://doi.org/10.3390/app9071478.

Varvitsioti, E. and Tsifoutidis, G. (2021). A regulatory roadmap to the past, present and future of geothermal energy in Greece. *Materials Proceedings* 5 (1): 10.

Witten, I.H., Frank, E., and Hall, M.A. (2011). *Data Mining: Practical Machine Learning Tools and Techniques*, 3e. Burlington, MA: Morgan Kaufmann.

Xiaohua, Q., Guojun, L., and Ming, X. (2013). Sci2 – new mapping knowledge analysis software. *Libraly Journal* 32 (9): 79. [Online]. Available: http://www.libraryjournal.com.cn/EN/Y2013/V32/I9/79.

Yang, D.H., Kang, J.H., Park, Y.B. et al. (2013). Association rule mining and network analysis in oriental medicine. *PLoS One* 8 (3): e59241. https://doi.org/10.1371/journal.pone.0059241.

Zanasi, A., Brebbia, C.A., and Ebecken, N.F.F. (ed.) (2006). *Data Mining VII: Data, Text, and Web Mining and their Business Applications*. Southampton: WIT.

Zhang, W., Yoshida, T., and Tang, X. (2008). Text classification based on multi-word with support vector machine. *Knowledge-Based Systems* 21 (8): 879–886. https://doi.org/10.1016/j.knosys.2008.03.044.

Zhang, Z., Chen, X., and Huang, J. (2018). Research on innovation posture of automated driving technology based on patentometrics. *Technology and Investment* 09 (03): 137–153. https://doi.org/10.4236/ti.2018.93010.

Zhang, H., Daim, T., and Zhang, Y. (2021). Integrating patent analysis into technology roadmapping: a latent dirichlet allocation based technology assessment and roadmapping in the field of Blockchain. *Technological Forecasting and Social Change* 167: 120729. https://doi.org/10.1016/j.techfore.2021.120729.

8

Technology Intelligence: Heat Pump Water Heaters

Fayez Alsoubaie, Aynur Kirbac, and Tugrul U. Daim

Mark O. Hatfield Cybersecurity and Cyber Defense Policy Center, Department of Engineering and Technology Management, Portland State University, Portland, OR 97201, USA

8.1 Introduction

Today, end-use energy efficiency is critical for energy and climate change research and policy. Since the 1970s, energy efficiency has continued to attract governments' attention to mitigate climate change by reducing CO_2 emissions due to the combustion of fossil fuel and reducing industry operation cost (Trianni et al. 2018). According to Roslizar et al. (2014), one way of achieving this goal is to improve residential and commercial building energy efficiency. The building sector accounts for 27.8% of the total energy consumption, and 65% is attributed to heating, ventilation, and air condition (HVAC) systems. Additionally, even the most efficient fossil fuel heating technology accounts for significant carbon emissions. Buildings present a significant target for behavioral change, design practice, and the application of technological options (Cabeza et al. 2020). In particular, given the degree to which conventional energy sources are used in the building sector, it is a priority for policymakers to consider energy reduction approaches for the different building systems. Heat water technologies are natural resource options available for replacing current energy systems and reducing carbon emissions associated with them. The heat pump has the potential to meet the low-temperature heat required by many industrial processes and save households and businesses 33% of their cost a year (Willem et al. 2017). With the heat and cooling systems, water heat technologies provide an opportunity for businesses to save cost.

Heat pump technologies utilize thermal or electric energy to extract heat from external renewable sources (Willem et al. 2017). Heat pump water heater (HPWH) research is one of the different heat source systems that heat natural water from the sea or ground. Few studies have attempted to clarify the nature of HPWH systems or identify trends that explicate their development (Amirirad et al. 2018;

Future-Oriented Technology Assessment. A Manager's Guide with Case Applications, First Edition.
Edited by Haydar Yalçın and Tugrul U. Daim.

Shen et al. 2021). The findings of the previous research are based on subjective classification. Therefore, they fail to offer an objective basis for policy roadmap establishment. In particular, the previous investigation has classified HPWHs in terms of type. Researchers have approached heat water technologies differently.

This research seeks to fill this knowledge gap by conducting a bibliometric analysis to show the most relevant thematic areas covered in the publications in the last 10 years to show how data in a peer-reviewed journal is connected. The aim is to offer the most-effective heat pump water technologies. The study also shows the practical application of HPWH publications and identifies the most productive authors. Further, it demonstrates the classification of HPWH based on the degree (density) and the relevance degree (centrality) of most collaborating countries in developing HPWH studies and most used heat pump systems. Finally, it aims to deduce the global network of co-authors conducting patents on heat-pump water heaters.

8.2 Literature Review

8.2.1 Heat Pump Water Heater Technology

HPWHs technologies are more costly than conventional water heaters because HPWHs work by taking heat from the surrounding/ambient air and then bringing it to the water in the tank in order to heat low-pressure liquid refrigerant in the heat pump's evaporation and vaporize the liquid (Ashdown 2004; Zhang et al. 2007). Air-to-water heat pumps provide more than 50% energy savings (Ashdown 2004) and has two to three times more energy efficiency than conventional water heaters or electric resistance systems because this technology uses electricity to transfer heat from one area to another rather than generating heat directly[1] (Sachs et al. 2011). Another name for the HPWHs is hybrid water heaters since to use hot water during the peak hours this technology allows to maximize energy savings and efficiency switching control panels to high demand or hybrid to obtain sufficient hot water. Thus, householders can take a shower in different bathrooms and use washing machine and dishwasher at the same time (Trout 2021).

This technology works the reverse of the refrigerator. While the HPWH takes heat from surrounding air and transfers it to the water in the tank to heat water, the refrigerator takes heat from inside a machine and transfers it into the surrounding room. Thus, HPWHs can be used for space conditioning systems such as heating and cooling home and water heating.[2]

1 https://www.energy.gov/energysaver/heat-pump-water-heaters
2 https://www.bpa.gov/EE/Technology/EE-emerging-technologies/Projects-Reports-Archives/Pages/Integrated-Heat-Pump-Water-Heater.aspx

While the gas water heater and electric water heater are conventional water heater types that many residents 52% and 42% use these technologies in their homes, respectively, advanced technologies of water heater are ENERGY STAR Gas, Gas Thankless, Condensing Storage, Solar Thermal and HPWH (Hoeschele et al. 2012; Zhang et al. 2007).

There are some significant criteria and features of HPWH technology such as energy efficiency, energy saving, installed costs, saved energy costs, climate requirements, and lifespan.[3] With the emergence of new water heater technologies such as HPWH, there are differences in both performance and hot water distribution characteristics compared to gas and electric storage water heaters. For example, in terms of the HPWH, energy efficiency of HPWH is higher than 2.0 while the energy efficiency of gas water heater and electric water heater are maximum 0.62 and 0.95, respectively. In addition, based on the climate loads and installation details, energy savings can be more than 40% (Hoeschele et al. 2012). While the HPWH technologies provide more energy savings and lower operating costs, they have much more initial and installed cost than conventional ones. Similarly, HPWH uses less than half electricity unlike resistance water heaters (Sachs et al. 2011). In addition to the energy saving benefits of HPWH. Emphasize other advantages of HPWH such as safety in use and low running fare. While the new construction cost of HPWH can be $1000, the retrofit cost can go up $1500 (Hoeschele et al. 2012). For HPWH, the only maintenance expense is the air filter. While the lifespan of conventional electric water heaters is 8–12 years, the lifespan of HPWH is around 13–15 years, which is longer than conventional ones (Trout 2021).

On the other hand, HPWH needs to be installed in an area, which stays between 4.4 and 32.2° C and (40 and 90°F) range yearly and allow at least 28.3 m^3 (1000 ft^3) of air space near the water heater. These technologies should be installed in a hot room because they tend to cool the room air, where they are located. Thus, it can be said that HPWH can be operating efficiently in the warm climate instead of cold climate because they absorb heat from the air.[4]

Therefore, selecting the proper climate influences the performance of HPWH in terms of the hot water load. Based on gallons per day there are four different categories of hot water loads such as low (20 gpd), moderate (45 gdp), above average (65 gpd) and high (110 gpd) (Hoeschele et al. 2012). To maximize the performance of HPWH and minimize the HPWH's cost, appropriate climate preferences are significantly important because operating cost can be different based on climate and hot water load categories. For instance, while the nominal annual HPWH operating costs of low, moderate, above-average, and high categories of hot water loads

---

3 https://www.energystar.gov/products/what_makes_product_energy_star
4 https://www.energy.gov/energysaver/heat-pump-water-heaters

are $102, $194, $290, and $574, respectively, in the cold, very cold and subarctic climate, these cost are $62, $114, $163, and $295 in hot, dry, humid, and mixed climate.[5] According to Sachs et al. (2011), HPWH provides a dehumidified atmosphere in the humid climate areas for that reason the cost of saved energy of this technology is $0.03/kWh.

Energy Star[6] is one of the programs that help to protect the environment and save electricity bills and money, supported by the US Department of Energy and US Environmental Protection Agency that promote energy efficiency. Energy Star certified HPWHs in order to improve energy efficiency. According to Energy Star[7] households who use HPWH less than 55 gal their energy cost saving can grow up approximately $12 billion each year. Significantly, 140 billion pounds of annual greenhouse gas emission can be hindered. This amount is equal to 13 million vehicles' emissions. While a household (four people) can save $350 per year, they can save up to $3750 over the 13 years that is the HPWHs lifespan. A four-person family can pay for its additional cost in 2.3 years and can save 2690 annual kWh.[8]

Beyond the economic and technical advantages of HPWH technologies, in the next part, researchers can find out the HPWT's current and future technology trends, top authors, and top institutions in the arena based on bibliometrics, social network analysis (SNA), and interviews.

8.2.2 Bibliometric Analysis

For efficiency, companies need to explore hidden patterns analyzing historical patent and scientific paper data (Jeong et al. 2019) to make their technology-related decisions more accurately (Daim and Gerdsri 2009). Both patent and scientific paper data that represent technology and innovation-based priorities can be the main decision-making tool for practitioners. Thus, firms can prioritize their R&D strategies depending on the most significant and appropriate ones (Wang et al. 2020).

Technological progress can be tracked by forecasting technological opportunities. For this reason, bibliometrics analysis is an important tool that enables companies to reshape their research portfolio and define novel technology developments to increase productivity and profitability (Norling et al. 2000). It helps to forecast emerging technologies and provide early warning signals for uncertainty situations. Hence, it can be said that this analysis supports

5 https://www.energy.gov/energysaver/heat-pump-water-heaters
6 https://www.energystar.gov/products/what_makes_product_energy_star
7 https://www.energystar.gov/products/water_heaters/high_efficiency_electric_storage_water_heaters/benefits_savings
8 https://www.energystar.gov/products/what_makes_product_energy_star

the decision-making process, providing quality information doing technology scanning and monitoring (Daim and Gerdsri 2009).

Bibliometric analysis that is a quantitative analysis (Liu et al. 2020) can be conducted based on a patent database and a scientific paper database. Most of the patents are produced by companies, and scientific papers are published by academicians. There is a strong relationship between patents and scientific research. While patents identify technological opportunities (Park and Yoon 2018), scientific papers represent scientific knowledge (Meyer 2002). Bibliometric analysis provides not only the quantity of the publications and patents, but also the quality of the company's technological and scientific knowledge background (McMillan and Hamilton Iii 2000).

Bibliometrics analysis uncover and categorize the associations between published documents' variables such as authors, institutions, key words, topics, and titles (Jeong et al. 2019; Xu et al. 2021). It leads researchers to provide deep insights into knowledge-based fields identifying major journals, articles, institutions, countries, authors, keywords, and predicting past, current, and future trends (Liu et al. 2020).

In this study, some bibliometrics metrics are used – such as co-authorship analysis, co-citation analysis, and co-occurrence analysis – to identify cooperation networks and main keywords:

- *Co-citation analysis.* Co-citation analysis contributes to identify emerging and trend topics from many publications based on their similarities. "If two articles were cited simultaneously by n paper ($n = 1, 2, \ldots$), then the two articles are defined to have a relationship of co-citation in the strength of n" (Liu et al. 2021). In the co-citation network, while the articles are represented by nodes, the link between two nodes is co-cited by other papers (Liu et al. 2021).
- *Co-authorship analysis.* Co-authorship analysis illustrates the framework of the research network (Xu et al. 2021).
- *Co-occurrence analysis.* Co-occurrence analysis helps to identify trends of the specific domain counting the frequency of the words that appeared together in the papers (Setyowati and Heriyanto 2019).
- *Co-word analysis.* Co-word analysis is necessary to identify the trajectory of the research and trend topics.

Although bibliometric is applied by many researchers, it needs to be supported by other analyses like SNA. A combination of these analyses might overcome the uncertainty on the rapidly developing technology (Yoon and Lee 2012). Therefore, using bibliometrics with SNA helps to explore in detail interrelationship of authors, institutions, countries, and keywords.

8.3 Social Network Analysis

The aim of the SNA is to reveal the best experts or top institutions in the specific area. SNA figures out how the network is connected, measuring and visualizing that base on some metrics connections and interactions between actors, groups, organizations, and other connected entities (Dey and Roy 2016). SNA results can help decision makers contact domain experts to identify emerging technology trajectory or forecast future technologies (Daim et al. 2013).

In this chapter we aim to classify the keywords on the highest-ranked degree of betweenness and the degree of centrality. The main metrics of SNA are:

- *Degree*. Number of direct connections in a network for an actor (node).
- *Betweenness*. Number of shortest paths between two actors that actor X resides on.
- *Closeness*. Distance of actor X to all other players in a network.
- *Eigenvector*. A high eigenvector score is assigned to an actor that is connected to other high-scoring actors.

SNA has been used in different aspects that could change through the time (Shuying et al. 2019). Especially in complex systems, SNA is one of the best ways to contribute to identifying interactions among actors in a network.

8.4 Methodology

8.4.1 Data Collection

All data were retrieved using specific keywords and phrases. The process involved developing a list of keywords and verifying them. The next phase entailed adjusting based on the literature review to get a better result. We also applied an expert who made a further adjustment to develop a final keyword list that was highly reflective of this objective.

The VosViewer software was used to verify the keywords based on occurrence, and the total link strength. A total of 2620 keywords were obtained and the total strength of the co-occurrence links with the other keywords calculated. The keywords with the highest total link strength were selected, and the unidentified keywords were unselected. Examples of keywords include "heat," "compressor," and "heat pump water heater," are the most vital keywords connected or used mainly in patents and scientific papers. In the data collection step, Table 8.1 that shows a lexical search query to retrieve data from Web of Science (WoS), Scopus, Sumobrain, and Derwent Innovation Index were used to gather academic publications and patents on heat pump technologies from 2010 to 2020.

Table 8.1 Heat pump water heater outputs.

Installed costs	**$1000 (new) – $1500 (retrofit)**
Energy efficiency	>2.0 EF
Energy saving	≥40%
Maintenance	Air filter
Climate	Hot – warm – dry – humid – mixed
Cost of saved energy	~$0.03/kWh
Lifespan	13–15 years

Source: Strategy Guideline: Proper Water Heater Selection. Adapted from DOE; Hoeschele et al., 2015; Sachs et al., 2011.

The research filtered only articles published in English until the end of 2020. The filtering resulted in 108 articles and those cited more than 15 times were considered highly cited (Barbosa and Schneck 2015). We followed the structured approach of keywords-to-keywords verification to keyword adjustment, we applied the approach to find the top experts and the technologies (Table 8.2; Figure 8.1).

Table 8.2 Search queries.

Scopus database	**Keywords**
Results 475 *Timespan: 2010–2021* *Scholarly data*	*(TITLE-ABS-KEY ["Heat pump"] AND TITLE-ABS-KEY [gas OR sorption OR thermal AND compression OR vapor OR thermostat])*
Web of science core collection	***Keywords***
Results 1210 *Timespan: 2010–2021* *Scholarly data*	*TITLE: ("Heat pump") AND TOPIC: (gas OR sorption OR thermal AND compression OR vapor OR thermostat)*
SumoBrain database	***Keywords***
Results 1001 *Timespan: 2010–2021* *Patent data*	*TTL/"Heat pump" and (gas heat pump or sorption or thermal or compression or vapor or thermostat)*
Derwent innovation index	***Keywords***
Results 521 *Timespan: 2011–2021* *Patent data*	*TITLE: ("Heat pump") AND TOPIC: (gas OR sorption OR thermal AND compression OR vapor OR thermostat)*

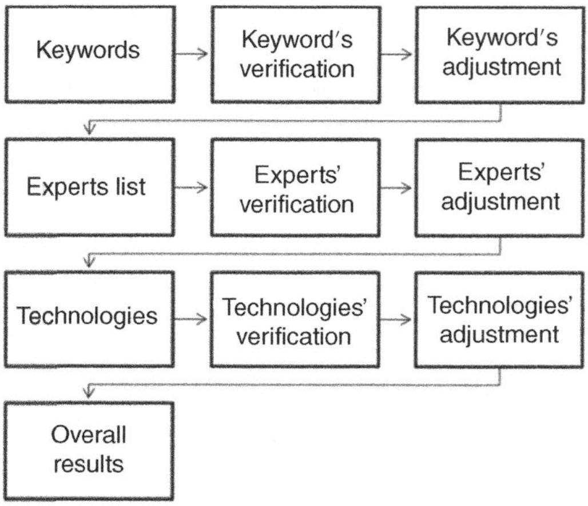

Figure 8.1 HPWH application theory practice data resources.

Bibliometric analysis was performed using quantitative, and knowledge mapping approaches. A quantitative method was applied to find answers related to keywords that have appeared in HPWH patents and what disciplines have collaborated in conducting HPWH research. Keywords are indicated from the Darwent innovation index with 521 patents. Searching peer-reviewed patents between 2011 and 2021 that included keywords "heat," "pump," "water heater," "gas," "absorption," and "thermostat" in their titles and abstracts. A total of 2620 co-occurrence of a keyword met the threshold. WoS provided a quantitative analysis technique while knowledge mapping was accomplished through RStudio, Gephi, SCI2, Citespace, VOSviever, and Pajek (Bornmann et al. 2018). With keyword network analysis, the software was used to link strength between keywords, countries, institutions, authors, and technologies. Keyword analysis provided a basis for explaining the content and revealing the association between topics. Thus, it was possible to investigate each domain of the research objective (Cho et al. 2020). Additionally, visual representation of data was obtained using RStudio, Gephi, SCI2, Citespace, and Pajek. The process produced the following findings – co-citation network of heat pump systems and application, most productive authors and countries. Only studies that indicated the study area were included for analysis.

8.4.2 Findings

The findings part is divided into three sections that include co-authorship analysis, co-word analysis, and interview results.

8.4.3 Co-authorship Network Analysis: WoS, Scopus, Sumobrain

Figure 8.2 shows a global network of co-authors' researches about HPWHs. The distance between two nodes describes the closeness of the author's scientific connections. The graph shows the analysis of the top five authors who co-authored the publication around the world.

Figure 8.3 illustrates a global network of co-author researchers of heat-pump water heaters from the WoS. The graph shows the top five authors who co-authored the publication around the world. The size of the nodes represent the citation counts. The authors (Yao, Y; Jiang, Y; Zang, H; Wang, RZ; Ogura, H) in this figure are the top specialists in the domain. Jiang, Y ranked first with nine publications and the highest-ranking betweenness of centrality 0.041559.

Figure 8.4 is a global network of co-authors conducting patents on heat-pump water heaters from Sumobrain patent data collection. The graph shows the top five inventors who co-authored the publication around the world. Park, S ranked first with 14 publications and the highest-ranking betweenness of centrality 0.004145. One of his patents is the heat pump, including at least two refrigerant injection flow paths into a scroll compressor.

8.4.4 Top Institutions, Countries, Authors, Papers, and Patent

In Tables 8.3 and 8.4, all experts, their institutions, and countries are identified based on the findings of the co-authorship network analysis and the SNA from WoS, Scopus, and Smobrain databases. Based on the identified top experts, the

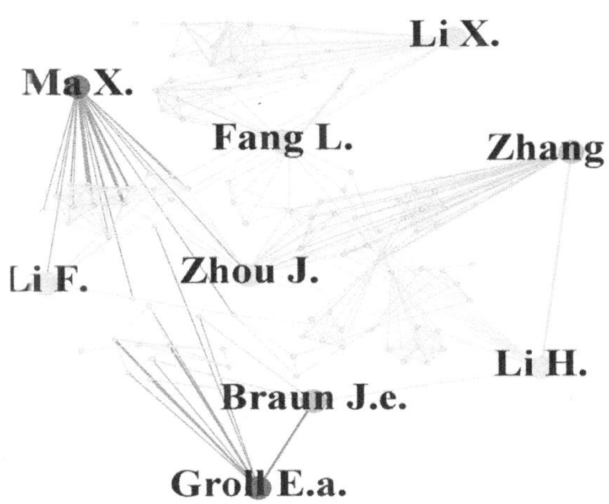

Figure 8.2 A global co-authoring network for research on heat pump water heaters (Scopus 2021).

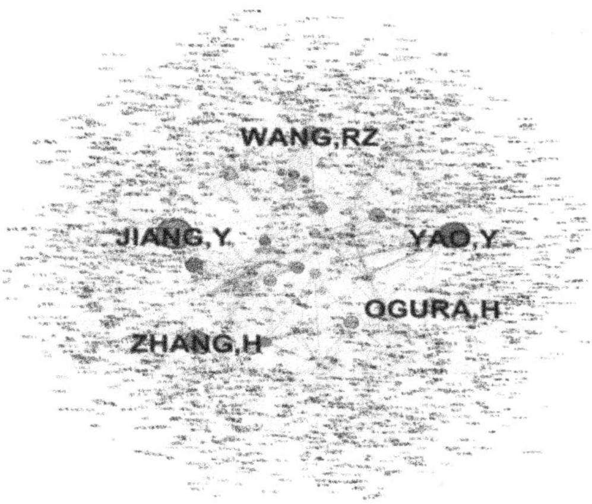

Figure 8.3 A global network of co-authors that have researched HPWH from WoS.

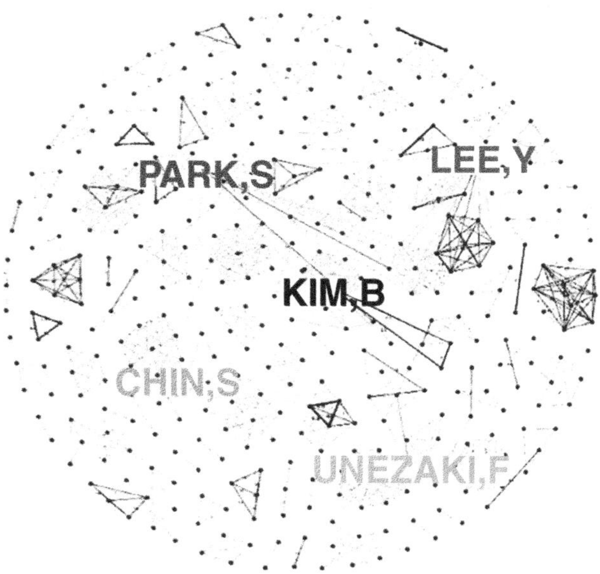

Figure 8.4 A global network analysis of co-author conducting patents on heat-pump water heaters from Sumobrain patent data (http://Sumobrain.com 2021).

Tables 8.3 and 8.4 also summarize top experts' best papers and patents, respectively, about HPWH technologies.

8.4.4.1 Scholarly Data

The highlighted articles are top papers that discuss the different HPWH technologies with special attention to cost reduction and energy efficiency. The authors of

Table 8.3 SNA top scholarly data analysis summary.

Scientific papers institutions	Countries	Authors	Main keywords
		Ma, X. Solar ejector cooling technologies	Ejector; phase change material; refrigerant; solar collector; solar ejector cooling system
University of Hull **Ranking:** Affiliation: 501	UK	*Zhou, J.* Experimental investigation of a solar driven direct-expansion heat pump system employing the novel PV/micro-channels-evaporator modules.	Coefficient of performance; direct expansion; direct-expansion heat pumps; experimental investigations
Expertise area: Biological and marine science, creative industries, and earth science		*LI, J.* Experimental research on a solar air-source heat pump system with phase change energy storage	
Purdue University Ranking: 53	USA	*Braun, J.E.* Virtual refrigerant pressure sensors for use in monitoring and fault diagnosis of vapor-compression equipment	Air conditioning; compressors; condensers (liquefiers); equipment; pressure measurement; pressure sensors; reciprocating compressors; refrigerants
Expertise area: engineering, nursing, and pharmacy		*Groll, E.A.* Past, present, and future of refrigeration: pathways to the next generation heating and cooling technologies	Cooling; heat pump; heating; history; refrigeration; working fluids
		Yazawa, K. Optimum design for thermoelectric in a sub-cooled transcritical CO_2 heat pump for data center cooling	sub-cooler; thermoelectric

(*Continued*)

Table 8.3 (Continued)

Scientific papers institutions	Countries	Authors	Main keywords
Federal Institute of Technology in Lausanne	**Switzerland**	*Favrat, D.* Multi-criteria optimization of a district cogeneration plant integrating a solid oxide fuel cell-gas turbine combined cycle, heat pumps and chillers.	Absorption; cogeneration plants; computer simulation; gas turbines; heat pump systems
Ranking: 89			
Expertise area: technology, natural sciences, and engineering		*Maréchal, F.* Performance and profitability perspectives of a CO_2 based district energy network in Geneva's City Centre	CO_2; district energy; heat pump; refrigerant; thermoeconomics; urban energy systems
		Schiffman, K. Modeling and experimental investigation of an oil-free micro compressor-turbine unit for an organic rankine cycle driven heat pump	
Holy Names University	**China**	*Li, H.* Exergetic and exergoeconomic assessment of a novel CHP system integrating biomass partial gasification with ground source heat pump	Biomass partial gasification; combined heating and power system.
Ranking:3610			
Expertise area: liberal arts and professional education		*Wang, S.* Further analysis of the influence of interstage configurations on two-stage vapor compression heat pump systems	Interstage subcooling/desuperheating; optimization; refrigerant; two-stage vapor compression
		Wang, Y. Explore energy saving operation strategy: indoor VOCs removal performance of silica gel rotor in clean-air heat pump system at low regeneration air temperature.	Air cleaning; energy saving; regeneration air temperature; ventilation; VOCs
		Li, F. Energy and exergy analyses of a solar-driven ejector-cascade heat pump cycle	Air-source heat pump; ejector; energy; exergy; solar.

Tianjin University Ranking: 281 Expertise area: computer science, applied and pure science, and engineering	**China**	*Jiang, Y.* A high-temperature hybrid absorption-compression heat pump for waste heat recovery *Wu, Z. X.* *Zhang, Y.* Experimental research of a new steam heat pump system for recovering industrial waste heat	Heat pump; low-grade waste heat; mixture working fluid
Shanghai Jiao Tong University Ranking:47 Expertise area: engineering, economics, and science	**China**	*Wang, R. Z.* Air-source heat pump for distributed steam generation: a new and sustainable solution to replace coal-fired boilers in China *Zhao, Y.* Modeling and dynamic control simulation of unitary gas engine heat pump *Ge, T.S.* Performance simulation of a joint solid desiccant heat pump and variable refrigerant flow air conditioning system in EnergyPlus	
Tsinghua University Ranking: 28 Expertise area: science and literature	**China**	*Fu, L.* Application of absorption heat pump and direct-contact total heat exchanger to advanced-recovery flue-gas waste heat for gas boiler. *Zhu, C.Y.* Vertical U-pipe flow characteristics in absorption heat pump: experimental study under vacuum conditions	

Table 8.4 SNA top analysis summary.

Patents institutions	Countries	Authors
Korea Institute of Energy Research Rank: 531 Expertise area: Renewable energy Solar energy Climate	South Korea	*Park, S.* Dehumidifier for a compressor in compression-absorption heat pump system *Lee, Y* Thermoelectric heat pump apparatus
LG Electronics Inc. Industry world rank: 4 Expertise area: Renewable energy, Solar panels	**South Korea**	*Chin, S.* Heat pump-type hot water feeding apparatus *Ha, S.* Heat pump-type hot water feeding apparatus *Ko, Y.* Heat pump including at least two refrigerant injection flow paths into a scroll compressor *Jang, Y.* Heat pump and method of controlling the same *Park, H.* Heat pump type speed heating apparatus *Choi, H.* Heat pump type hot water supply apparatus

Mitsubishi Electric Corporation Industry world rank: 2 Expertise area: Thermal power plants, gas turbine, air quality control system	**Japan**	*Unezaki, F.* Air conditioner heat pump with injection circuit and automatic control thereof *Shinomoto, Y.* AC-DC converter, method of controlling the same, motor driver, compressor driver, air-conditioner, and heat pump type water heater
Hitachi Ltd. Industry rank: 68 Expertise area: Power system control, Energy management, Renewable energy and Thermal	**Japan**	*Shibata, T.* Compressor and heat pump system *Araki, H.* Cooling apparatus, gas turbine system using cooling apparatus, heat pump system using cooling system, cooling method, and method for operating cooling *Takahashi, Y. Hi* Heat pump type hot water supply outdoor apparatus

these papers are competent in technology and scientific research subjects since their studies are well-rounded and compliant with scholarly standards. More importantly, different researchers explore dissimilar HPWH technologies, availing diverse knowledge about the alternatives available at the market. For instance, while Yan et al. (2020) explore the distributed steam generation air-source heat pump, Zhou et al. (2016) investigate the solar-powered expansion heat pump technology. Therefore, the papers contain in-depth information about the features and relative performances of different heat pump systems. Students and other researchers can tap this knowledge in their secondary research studies.

8.4.4.2 Patents

The presence of patents also indicates the superiority of the highlighted sources. The patents mean that the heat pump systems are innovative discoveries that require protection from replication by unauthorized personnel. Therefore, the papers are reliable and top-quality for students and researchers interested in studying HPWH technologies.

The table provided is an important one for various reasons; the first reason is that it indicates the growth in literature per country and institution on the level of studies that have been done on the topic. Research on heat pumps and related technologies is something that has been ongoing for many years now (Al-Juwayhel et al., 1997, p. 34). Companies, academic institutions, governments, and nongovernmental institutions are undertaking major studies related to various forms of heat pump technologies that could greatly change the landscape of how people leverage and use heat pump technologies (Demir et al. 2008, p. 2382).

Since studies and technologies change yearly, it is possible to change the numbers appearing beside the countries and organizations. The main reason for the above possibility is that more organizations are intensifying research on various heat pump technologies, and related concepts (Ji et al. 2005, p. 231). It means that as more research is done, the ranking by organizations and countries change; those that do lag in the number of patents and research drop in the list, while those that increase the number and quality of patents, and research (Noailly and Batrakova 2010, p. 7812). Therefore, it is important to affirm that given the above argument, the numbers can change in the next decade, or even a year.

8.5 Heat Pump Water Heaters Bibliometrics Application

Table 8.5 illustrates the work of this research by selecting the most important affiliations and their concentration of technologies as shown in Table 8.6.

Table 8.5 SNA top analysis application.

				Technology					
Affiliation	**Country**	**Type**	**A/I[a] rank**	**Solar**	**Thermal**	**Refrigerant**	**Liquification**	**HVAC[b]**	**Heat storage**
University of Hull	UK	A	448	✓					
Purdue University	USA	A	109		✓	✓	✓	✓	✓
Federal Institute of Technology	Switzerland	A	8	✓					
Shanghai Jiao Tong University	China	A	70	✓					
Tianjin University	China	A	387			✓			
Tsinghua University	China	A	72		✓		✓		
Korean Institute of Energy Research	Korea	I	634	✓					✓
LG Electronics Inc.	Korea	I	207	✓					
Mitsubishi Electric Corporation	Japan	I	249		✓			✓	✓
Hitachi Ltd	Japan	I	593		✓		✓	✓	

a) A, Academic; I, Industry.
b) HVAC, Heating, ventilation, and air conditioning.

Table 8.6 Keywords analysis indicated by Pajek software from Scopus Data (SNA 2020).

Heat pump systems	Energy efficiency	Waste heat
285.0004	99.0004	79.9986
Cooling	**Heating**	**Heat transfer**
71.9993	75.9994	60.9995
Carbon dioxide	**Cooling systems**	**Refrigerants**
57.9989	66.9997	68.9993
Air conditioning	**Heat storage**	**Performance assessment**
56.9997	49.9989	24.0016

Table 8.6 is vital because it shows the historical contribution of organizations in terms of HPWHs technologies. However, it is vital to note that research on heat pumps and related technologies is something that has been ongoing for many years now (Al-Juwayhel et al. 1997, p. 34). Companies, academic institutions, governments, and nongovernmental institutions are undertaking major studies related to various forms of heat pump technologies that could greatly change the landscape of how people leverage and use heat pump technologies (Demir et al. 2008, p. 2382). Major organizations concentrate on research and development of solar energy and solar technologies, Purdue University as an example, leading the research of HPWHs according to the experts that they have a long history, and a lot of focus has gone into refrigerants and the refrigerants of the future as well. As they are actively engaged in a highly flammable refrigerant hydrocarbons research. As well as a project in Southeast Asia right now where they are developing inverter-based air conditioners for the Southeast Asian marketplace focused on. As a result, Purdue University has the facilities available now and is bringing them online, where they can test a highly flammable refrigerant. It is a surprise that the two most collaboration countries in the world, USA and China, are working together to improve the research and development of HPWHs technologies, which points to significant changes in the future of energy and cost efficiency.

8.6 Overall Affiliation Ranking 2010–2021

Table 8.7 shows important ranking affiliations on this research in the world in the last decade. The research conducts all the organizations who mostly contribute to the HPWHs research. Table 8.7 shows the ranking period from 2010 to 2021

Table 8.7 Overall affiliation ranking 2010–2021.

Affiliation	Country	2010	2011	2012	2013	2014	2015	2016	2017	2018	2019	2020	2021
University of Hull (Round University Ranking, n.d.)	UK	308	336	437	438	546	567	468	555	494	464	489	448
Purdue University	USA	76	69	93	114	101	431	121	126	124	104	98	109
Federal Institute of Technology	Switzerland	9	9	10	10	11	10	23	15	14	7	9	8
Shanghai Jiao Tong University	China	170	173	182	177	193	171	190	110	134	74	81	70
Tianjin University	China	298	326	313	315	402	329	316	389	335	390	332	387
Tsinghua University	China	87	81	81	94	85	102	98	75	68	46	46	27
Korean Institute of Energy Research (Scimago Lab, n.d.)	Korea	558	592	572	547	523	540	572	510	516	625	670	634
LG Electronics Inc.	Korea		171	196	225	194	175	180	201	178	185	207	
Mitsubishi Electric Corporation (LG 2013)	Japan	432	430	440	457	466	475	276	262	279	266	316	249
Hitachi Ltd	Japan	475	493	512	498	495	500	548	553	180	522	615	593

according to the available data. Since studies and technologies change yearly, it is possible to change the numbers appearing beside the countries and organizations (Li, p. 43). The main reason for the above possibility is that more organizations are intensifying research on heat pump technologies and related concepts (Ji et al. 2005, p. 231). It means that as more research is done, the ranking by organizations and countries will change; those that do lag in the number of patents and research drop in the list, while those that increase the number and quality of patents will rise on the list (Noailly and Batrakova 2010, p. 7812). Therefore, it is understandable that over the years, some firms have improved in ranking while others have subsided. For example, Shanghai Jiao Tong University, Federal Institute of Technology, and Tsinghua University are the top 70 schools in the world, but Korean Institute of Energy Research LG Electronics Inc., Mitsubishi Electric Corporation, and Hitachi Ltd. are top old industries, and they are increasing in the patent and innovation research with a fluctuation ranking in the world.

8.7 Co-word Results

This section presents the coword analysis results.

8.7.1 Keywords Related to Use of Heat Pump Water Heater and Measures of Development Degree

The cluster analysis of the author keywords reveal only 11 highly cited articles from 2010 to 2020, demonstrating the slow evolution of the research related to heat pump systems in the academic field. Nevertheless, in 2020, the cumulative number of highly cited articles increased to 124 members. Figure 8.5 is the visualization and the calculation results of metrics analysis of the keywords related to the technologies of the HPWH. Table 8.3 is the results taken from the visualization graph. In terms of the most used heat pump system, the result demonstrates that heat pump systems are the most cited keywords of the HPWH. Most importantly, most of the highly cited studies focused on the application of water heaters. Heat pump systems ranked in the highest degree with 285.004 in the sum of related keywords of heating and cooling system, gas, absorption, cost-effectiveness, cylinder liners, geothermal energy, low-temperature production, waste heat, refrigerants, and air conditioning. We consider these reports instrumental in establishing a heat pump water system concept at the forefront of the studies related to natural source energy use and efficiency studies. This finding is consistent with previous results by Zhang et al. (2020) that show heat pump technologies are increasingly gaining attention and gradually becoming a top area for researchers.

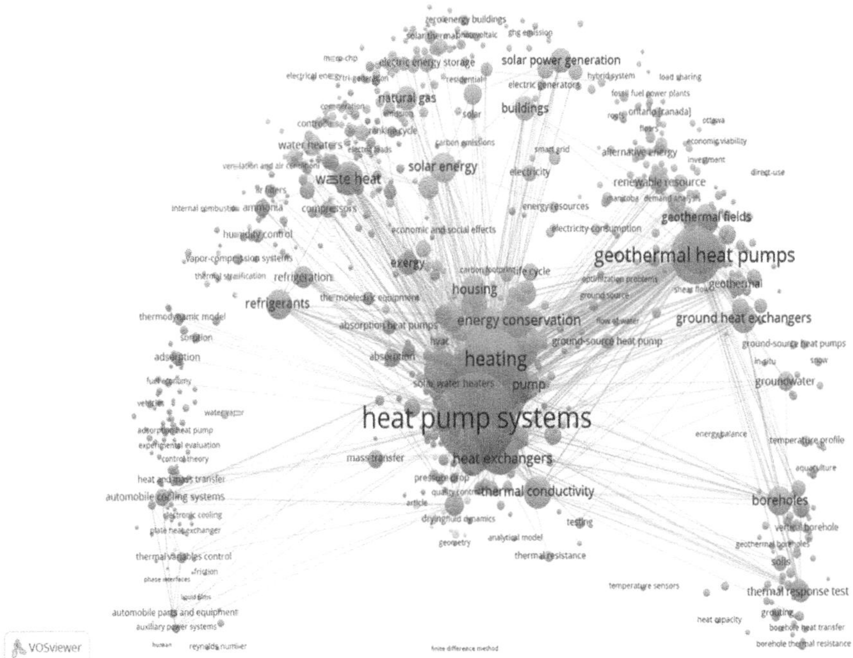

Figure 8.5 Co-occurrence map for author keywords (Jan van Eck and Waltman 2023).

Figure 8.6 is a co-occurrence map for author keywords. It shows the most important 1001 keywords used in the top research related to HPWHs according to the Derwent Innovation index for patent research engines. Heat pump is the most used word in the documents to propose energy efficiency and reduce the cost in the United States, Canada, Germany, United Kingdom, and France based on the data collection to identify the heat pump technologies. Overall, the Figures 8.5 and 8.6 underscore a plethora of ideas and concepts that the expert discussed with researchers during the interview. The first concept that the expert illustrate that is clearly represented in the graphs is the fact that heat pump technologies are complex and involve a litany of technologies that are applicable in different situations and occupations. From Figure 8.6 and Table 8.3, the interconnection of concepts is clearly presented, starting the relationships between financial needs and the different technologies. One of the things that the interviewer discussed is the need to reduce the cost of energy in homes and offices. The goal of developing heat pump technologies is mainly to reduce the cost by leveraging different techniques.

The significant concept in Figure 8.7 is the concept of water absorption in the heat pump technologies. Heat adsorption is an important technology in refrigerant technologies. The idea has also been explored in a range of existing literature.

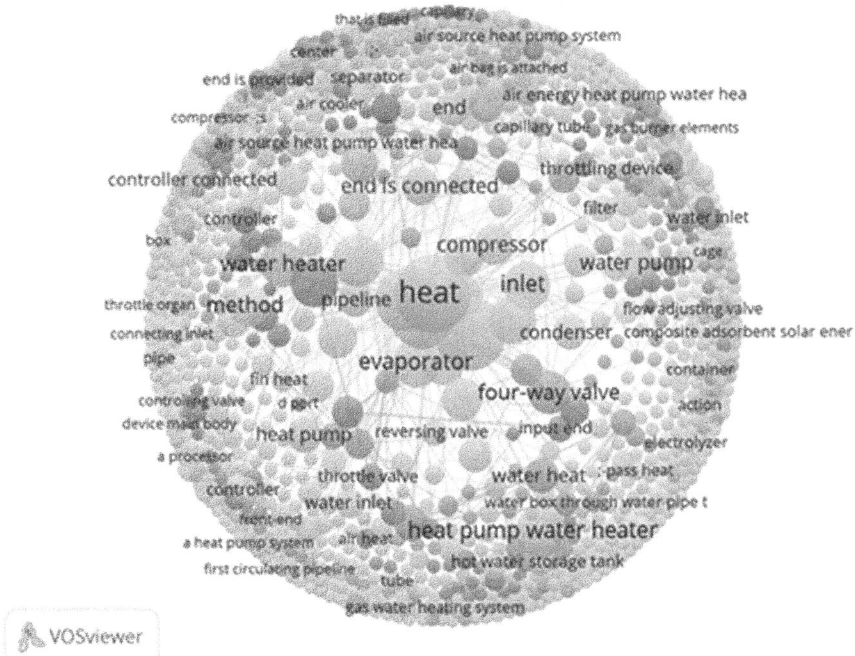

Figure 8.6 Co-occurrence map (Jan van Eck and Waltman 2023).

The technologies have existed for years but have been changing over the years. Traditionally, ships used adsorption technology using seawater and chiller units. As the interviewee said it, studies also stressed that water-driven and heat pump technologies are the best choices when it comes to saving the world from the menace of high cost of fuel energy. Over the years, pollution through emissions has been a major problem for governments and nongovernmental institutions. Governments, just as the interviewee, the three graphs also indicate that the goal of developing heat pumps was mainly to save the world from the menace of pollution that is posing a great danger of climate change and global warming. They somewhat agreed that the most used keywords air conditioning, heat storage, and performance assessment appeared in the Figures 8.5–8.7.

Figure 8.8 shows the selected 200 keywords that demonstrate a clear and strong relationship between the technologies codes keywords in which HPWH is strongly related to heating, liquefaction, ventilating, refrigeration (Table 8.8).

According to the experts, Figures 8.7 and 8.8 gives incredible information about the complexity of heat pump technologies and the types of energies that the heat pumps use. In the highly complex web, one could see different forms of clean energies, including but not limited to solar energy, wind energy, and geothermal heat

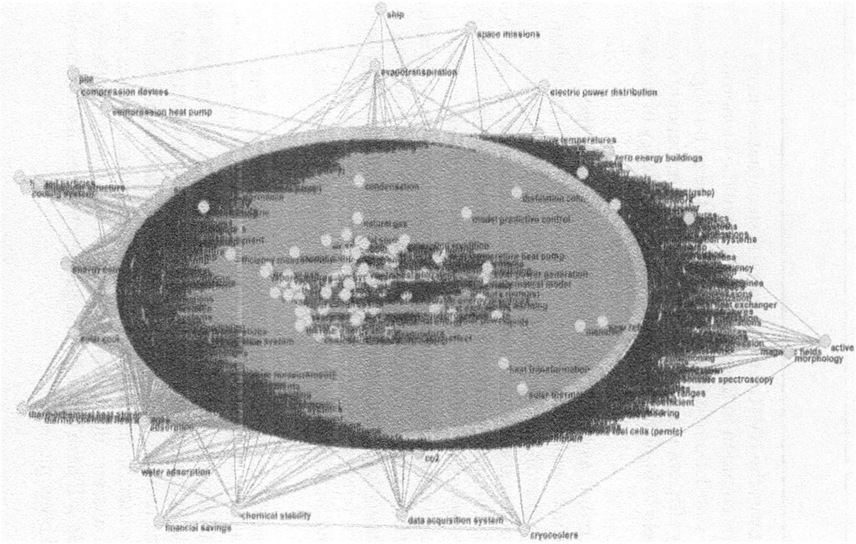

Figure 8.7 Keywords linked to use of HPWH technologies (SNA 2020).

Figure 8.8 Most used heat pump water heater technologies.

Table 8.8 List of heat pump water heater.

Technology	Explanations
The HVAC system	The first HPWH that the interviewee talked about was the HVAC water heating system that is energy efficient for domestic and office space use. HVAC is common in the market because of the numerous applications in houses. Other than water heating, ventilations, and air conditioning, the technologies could help to heat bathing water for people in an energy efficient manner. Consequently, the respondents were confident that the project's use of HVACs could help to improve the state of energy efficiency in homes around the country.
Chemical heat pump	The interviewer also discussed the chemical heat pump, as one the critical technologies that the organization that they work for was developing. The chemical heat pump consists of different parts that help to reduce energy consumption when heating water in homes and office buildings. The major components of the system include the reactor, which is also the generator, absorbers, evaporators, and the condenser. The interviewee was optimistic that the chemical heats pumps will significantly reduce the cost of heating water in homes and offices. The technology used to develop the chemical heat pump uses thermal energy to fulfill its role. The gadgets are also important innovations for various reasons. One is that they have an incredible ability to maintain higher heat storage capacities, as well as high heat reactions. However, the limitation of the CHPs is that they easily lose heat; hence, sometimes cost ineffective. The interviewee was hopeful that placing chemical heat pumps in houses could offer better water heating systems, especially when one needs instant bathing and doing other normal chores that require hot or warm water.
The washer dryer	The third technology that the respondent discussed during the interview was the washer dryer. The machine is a combination of washing and drying technologies. The machine is effective and energy efficient compared to the novel single washers and dryers. Based on the experts, the above machine could improve the rate of energy consumption for people in the houses because its washers and dryers appeal simultaneously. The working of the above machine requires the installer to place the washing machine closer to the heating machine, making it easier for users to access both the hot water and the clothes being washed. People are used to the normal dryers and washers, but combining the two technologies is likely to change the dynamics in the market because people will have cost-effective gadgets that are also economical when it comes to energy utilization in homes and other places.

Whirlpool water heater	Another technology that the experts discussed in the interview was the Whirlpool heater. The company that produces the machine is a specialized maker of washing machines, refrigerators, and a plethora of other energy-dependent gadgets commonly found in homes, restrooms, and offices. The machines are available online in leading shops, including but not limited to Amazon and Alibaba stores. However, the best attribute of the Whirlpool water heater is that it is not as energy intensive as other traditional water heaters in the market today. User can achieve and save a lot in terms of energy cost management. It is of the technologies that the experts speaks highly about in the interview, even though he does give detailed description of the gadgets.
Refrigerant's technologies	Refrigerant's technologies available in the market today include evaporative cooling or swamp coolers, mechanical-compression refrigeration systems used for industrial and commercial refrigeration, absorption refrigeration based on low heat and thermoelectric systems.
Smart thermostat	Smart thermostats comprise of two major components, smart modules and a novel thermostat. The smart module connects to the water boiler and to the home internet connection (WiFi). The role of the thermostat is to communicate changes in temperatures in the water heating system.
Smart appliance	Smart appliances use the internet of things (IoT) and rely heavily on information and communication technologies. They are available for different purposes in the kitchen, offices, living rooms, and water heating technologies. In the recent past, such appliances have been developed to leverage artificial intelligence (AI), machine learning (ML), deep learning (DL), and quantum computer and machine learning (QCML) technologies.
Sensible loads	Sensible load is applicable in cases where one needs to alter mean radiant temperature (MRT).
Gas stream in the hydrogen economy	Gas stream in the hydrogen economy helps to protect the environment from pollution by overcoming polarization and ohmic losses caused by flow of electric current under different operational conditions. The technology is energy efficient and less costly on the environment compared to carbon fuels.
Sensible load into single application	Sensible loads operate by means of sensible cooling and single applications that can achieve the level of performance include but not limited to refrigerant technologies and the evaporator.
Membrane technologies	Membrane technologies are sustainable solutions for health, water, and energy conservation for a cleaner environment. They come with membrane bioreactors, membrane contactors, forward osmosis, microbial fuel cell, and microbial fuel cell.

(Continued)

Table 8.8 (Continued)

Technology	Explanations
Sensible load from latent load with small details in emerging technologies	Sensible load and latent are important technologies used in humidification and dehumidification. Sensible load is applicable in cases where one needs to alter MRT, while latent load is used when the humidity is high.
Regenerative desiccant	It is effective in integrating the heat exchanger and the desiccant element to allow high-efficient humidification and dehumidification without supplying or draining water. The technology is effective in saving on the cost of energy.
Heat treatment technologies	Several heat treatment technologies with better performance metric and cost effectiveness exist. They include case hardening, quenching, normalizing, tempering, strengthening, and precipitation. The three major stages of heat treatment include heating, soaking, and cooling.
New technologies in HPWH	The new heat pump water heater comes with exciting technologies, including variable-speed motor functionalities to minimize electricity usage and create clean air environment for home and office dwellers.

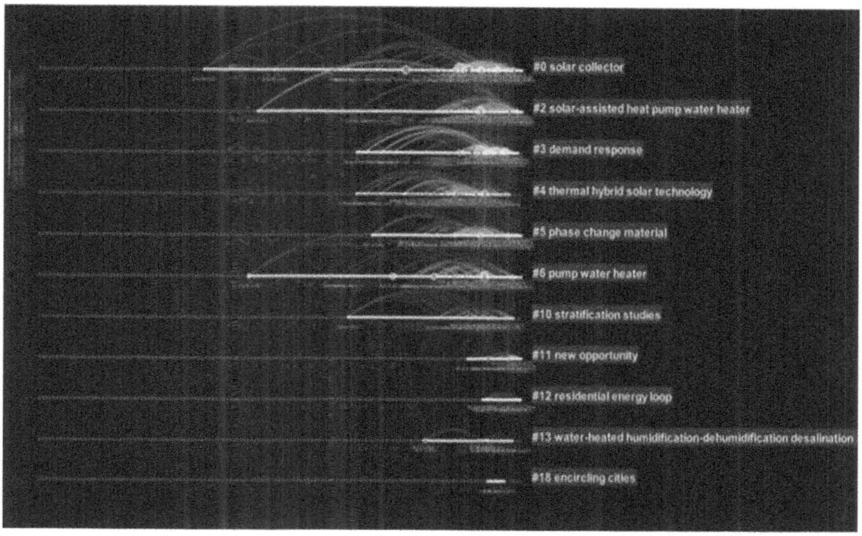

Figure 8.9 An overview cluster analysis of the largest 11 co-citation publications, 2010–2020.

pumps (Li, p. 45). Like the other two graphs, Figures 8.7 and 8.8 also stresses the adsorption, condensation, refrigeration, ambient, and liquefaction heat technologies that are instrumental in driving the heat pump gadgets in an eco-friendly way to ensure environmental safety in the age of global warming and climate change.

Figure 8.9 shows the divided co-citation network into a number of clusters of co-cited references, their light connections between different clusters and strong connections within the same clusters. This timeline visualization of clusters and their labels are plotted horizontally. Each cluster's labels can be seen vertically at the end of the cluster's timeline. While the cluster ID which is at the top of the figure represents the largest clusters, at the bottom of the figure illustrates the newest clusters (Liu et al. 2021). This figure illustrates 11 major clusters whose quality is identified based on the silhouette score that is a consistency or homogeneity indicator. While the oldest and largest hot topic among the top topics is solar collector that has 124 members and a silhouette value of 0.819 – solar collector by LLR, tube by TFIDF, and proposed solar humidification–dehumidification method (HDH) water desalination system and energy system (5.33). Articles on green transition were the two significant citations.

Timeline view of cluster analysis illustrates the cited references and cluster labels. That contributes to recognizing the most applied keywords. While the cited references are generated by the citing articles, cluster labels are derived from abstracts, key words or titles to show trend topics (Liu et al. 2021). Figure 8.10

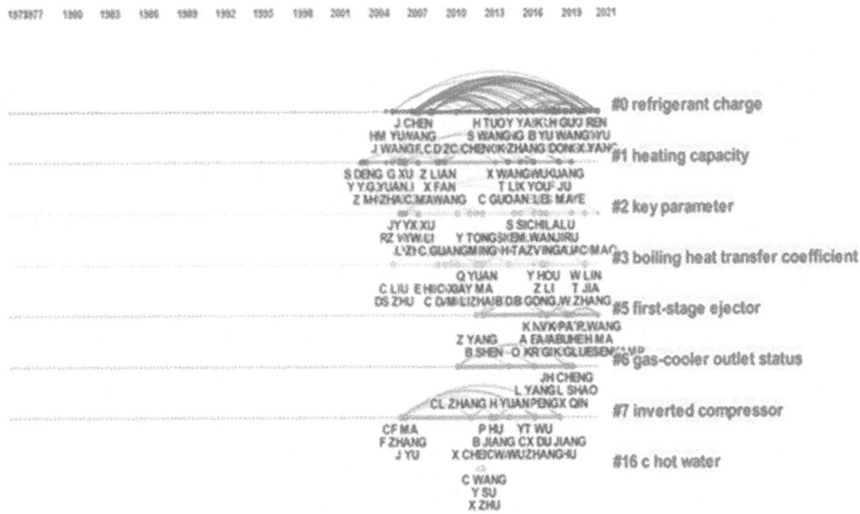

Figure 8.10 Timeline view of cluster analysis of co-cited articles.

illustrates the top trend topics that are refrigerant charge, heating capacity, key parameter, boiling heat transfer coefficient, first-stage ejector, gasa cooler outlet status, inverter compressor and hot water, respectively.

The idea of compression heat pumps as discussed by the interviewer also appears vividly in the graphs. A wide range of research is available on compression heat pump technologies discussing a range of factors related to the technology, as well its advantages to the environment. One of the most interesting facts related to the information given by the interviewer. Heat pump technologies are the devices that could absorb heat at an extremely lower temperature to release it at elevated temperatures. Heat pumps could as well use the ambient heat to improve the rate of energy output more efficiently compared to carbon sources of energy, which is impressive when it comes to saving energy.

Nonetheless, there lacked any disagreements noted, even though the three graphs were more detailed than the information by the interviewee. The disagreement detected is in the detailed examination in the graphs that the interviewee did not fail concretely.

8.8 Interview Results

Northwest Energy Efficiency Alliance (NEEA) and Portland State University (PSU) members interviewed HPWHs experts from Purdue University to explore the emerging technologies of HPWHs that could help and reduce the energy costs in the future and come up with the role that could help improve utilities and find new opportunities such as, reduction in energy use, to maintain the same comfort, such as with a washer or clothes dryer.

Taking advantage of the opportunity with variable-speed equipment, we can drive in such a way that they can crank up the compressor speeds, slow down the evaporator fan speed. And they can go into a very deep dehumidification mode or they can go the other way where we can speed up the evaporator fan and slow the compressor way down and drive that to a fully dry coil because of the nonlinearities that exist inside of the system. They are finding a net gain in efficiency of something on the order of 15% overall annual energy savings.

First, the interviewee was asked to introduce themselves and the organization that they are working in. NEEA is working on projects that are fully funded by a nonprofit organization, the electric and gas utilities in the Northwest states – Oregon, Washington, Idaho, and Montana. They added that the objective of the project is to help utilities with what they call filling the pipeline of new opportunities and identifying the novel technologies of HPWHs.

Second, on the need for energy efficiency, the response indicated that the organization cares for energy consumption by households in the country. However, it is important to note that the interviewer spent most of the time during the interview explaining numerous technologies related to the HPWH that they are trying to develop to improve energy efficiency in residential and office settings.

However, beyond naming the above available HPWH and water heating systems, the experts also talked about important things regarding energy efficiency and technologies to make it easier for users to reduce their energy costs. For example, they said that one way of improving energy consumption using the HPWH would be to use tunable PCMs. They mentioned that the PCMs make it easier to modulate your face temperature; both in heating and cooling to take advantage of the technology. Therefore, a next important step is making the PCMs affordable in commercial applications.

The experts discussed thermal storage in HPWHs in the market. They had an engineer from who developed thermal storage based on the idea of Juul bank and got it patented. This technology was only superficially discussed when talking about cost management and the need to be innovative when developing a HPWH.

Finally, the experts indicated that they were collaborating with numerous faculty members to ensure that the project is complete to save the cost of energy, especially in domestic and commercial apartments. One of the team members has developed a latex paint that is potentially a little cheaper than currently available latex paints. He said that they have demonstrated at noon and full sunlight radiative cooling to the space. The team was at the point with them working on the material side of things; they are generating cooling at sufficient capacity to handle 100% of the loads on residential spaces, probably 75% of commercial space loans.

8.9 Conclusion and Discussion

This chapter provides some background on HPWHs and the most interacting technologies that lead to a better use of the energy efficiency and play an substantial of the cost reduction in the near future. The chapter found that HVAC (heating, ventilation, and air conditioning), thermal, and refrigerant are the most important technologies in the field of HPWHs and are being researched by the two largest contributors countries in the world, USA and China.

Co-authorship analysis, co-word analysis, and interview results sections proved the most highly cited paper and the heat water heaters keywords with the HVAC technologies are Groll, E.A. and X. Ma from Purdue University and the University of Hull – the top experts with the most significant papers and citations. They discussed performance and cost of HPWHs as a main criterion that also need further research for future technologies improvement by integrating artificial intelligence and robot's technology to maintain a better use of HPWHs.

There are many advantages of HPWHs. Investing in a HPWHs technology, which is more energy-efficient water heater technology than alternative electric water-heater systems, can provide benefits to householders and investors for point-of-use operating efficiency, energy saving, and payback period, except higher purchase price. When HPWHs compared with conventional water heater technologies, its energy savings (40%), energy efficiency (>2.0 EF), and lifespan (13–15 years) is always in better conditions.

In the future, domain experts who identified based on SNA in this study can develop some roadmaps, scenarios, or strategic planning for future scientific and technical developments so that investors can forecast new opportunities and threats with HPWHs. Furthermore, one of the disadvantages of HPWH technology is that while market share of HPWHs is low, prices of HPWHs are costly. To solve this problem, experts can help to create business policies so that in the future, HPWHs technologies can produce more devices with a lower price.

References

Al-Juwayhel, F., El-Dessouky, H., and Ettouney, H. (1997). Analysis of single-effect evaporator desalination systems combined with vapor compression heat pumps. *Desalination* 114 (3): 253–275.

Amirirad, A., Kumar, R., Fung, A.S., and Leong, W.H. (2018). Experimental and simulation studies on air source heat pump water heater for year-round applications in Canada. *Energy and Buildings* 165: 141–149.

Ashdown, B.G. (2004). Heat pump water heater technology: Experiences of residential consumers and utilities (No. ORNL/TM-2004/81). Oak Ridge National Lab.(ORNL), Oak Ridge, TN, USA.

Barbosa, F.G. and Schneck, F. (2015). Characteristics of the top-cited papers in species distribution predictive models. *Ecological Modelling* 313: 77–83.

Bornmann, L., Haunschild, R., and Hug, S.E. (2018). Visualizing the context of citations referencing papers published by Eugene Garfield: a new type of keyword co-occurrence analysis. *Scientometrics* 114 (2): 427–437.

Cabeza, L.F., Chàfer, M., and Mata, É. (2020). Comparative analysis of web of science and Scopus on the energy efficiency and climate impact of buildings. *Energies* 13 (2): 409.

Cho, Y., Park, H.J., and Lim, H (2020). Research in user-centered design 2009 to 2018: a systematic keyword network analysis. In: *International Conference on Human-Computer Interaction*, 11–18. Cham: Springer.

Daim, T.U. and Gerdsri, N. (2009). Research and development progress assessment through technological and scientific intelligence. *International Journal of Technology Intelligence and Planning* 5 (4): 341–356.

Daim, T.U., Gomez, F.A., Martin, H., and Sheikh, N. (2013). Technology roadmap development process (TRDP) in the medical electronic device industry. *International Journal of Business Innovation and Research* 7 (2): 228–263.

Demir, H., Mobedi, M., and Ülkü, S. (2008). A review on adsorption heat pump: problems and solutions. *Renewable and Sustainable Energy Reviews* 12 (9): 2381–2403.

Dey, P. and Roy, S. (2016). Social network analysis. In: *Advanced Methods for Complex Network Analysis* (ed. N. Meghanathan), 237–265. IGI Global http://doi:10.4018/978-1-4666-9964-9.ch010.

Henninger, M. (2012). Locating scholarly papers of interest online. *Social Media for Academics: A Practical Guide* 59–83. https://doi.org/10.1016/B978-1-84334-681-4.50004-9.

Hoeschele, M., Springer, D., German, A., et al. (2012). Strategy guideline: proper water heater selection (No. DOE/GO-102012-3602). National Renewable Energy Lab.(NREL). Golden, CO, United States.

Jan van Eck, N. and Waltman, L (2023). VOSviewer Manual. 55.

Jeong, Y., Park, I., and Yoon, B. (2019). Identifying emerging research and business development (R&BD) areas based on topic modeling and visualization with intellectual property right data. *Technological Forecasting and Social Change* 146: 655–672.

Ji, J. et al. (2005). Performance of multi-functional domestic heat-pump system. *Applied Energy* 80 (3): 307–326.

LG (2013). Ranking per brand. http://www.rankingthebrands.com/Brand-detail.aspx?brandID=379 (accessed 24 May 2024).

Liu, J., Li, X., and Wang, S. (2020). What have we learnt from 10 years of Fintech research? A scientometric analysis. *Technological Forecasting and Social Change* 155: 120022.

Liu, H., Li, X., and Wang, S. (2021). A bibliometric analysis of 30 years of platform research: developing the research agenda for platforms, the associated technologies and social impacts. *Technological Forecasting and Social Change* 169: 120827.

McMillan, G.S. and Hamilton Iii, R.D. (2000). Using bibliometrics to measure firm knowledge: an analysis of the US pharmaceutical industry. *Technology Analysis & Strategic Management* 12 (4): 465–475.

Meyer, M. (2002). Tracing knowledge flows in innovation systems. *Scientometrics* 54 (2): 193–212.

Noailly, J. and Batrakova, S. (2010). Stimulating energy-efficient innovations in the Dutch building sector: empirical evidence from patent counts and policy lessons. *Energy Policy* 38 (12): 7803–7817.

Norling, P.M., Herring, J.P., Rosenkrans, W.A. Jr., et al. (2000). Putting competitive technology intelligence to work. *Research-Technology Management* 43 (5): 23–28.

Park, I. and Yoon, B. (2018). Technological opportunity discovery for technological convergence based on the prediction of technology knowledge flow in a citation network. *Journal of Infometrics* 12 (4): 1199–1222.

Roslizar, A., Alghoul, M.A., Bakhtyar, B. et al. (2014). Annual energy usage reduction and cost savings of a school: end-use energy analysis. *The Scientific World Journal* 2014 (Article ID 310539): https://doi.org/10.1155/2014/310539.

Round University Ranking (n.d.). World university rankings. roundranking.com/ranking/world-university-rankings.html#world-2020 (accessed 24 May 2024).

Sachs, H.M., Talbot, J., and Kaufman, N. (2011). *Emerging Hot Water Technologies and Practices for Energy Efficiency as of 2011*. American Council for an Energy-Efficient Economy.

Scopus (2021). Scopus – Search for an author profile. https://www.scopus.com/freelookup/form/author.uri?zone=TopNavBar&origin=NO ORIGIN DEFINED (accessed 24 May 2024).

Setyowati, L. and Heriyanto, H. (2019, December). Research data management in higher educations: knowledge mapping using bibliometric analysis. *Proceeding of International Conference on Documentation and Information* 2: 21–33.

Shen, X., Liu, P., Qiu, Y.L. et al. (2021). Estimation of change in house sales prices in the United States after heat pump adoption. *Nature Energy* 6 (1): 30–37.

Shuying, L., Garces, E., and Tugrul, D. (2019, November 1). Technology forecasting by analogy-based on social network analysis: The case of autonomous vehicles. *ScienceDirect* https://www.sciencedirect.com/science/article/abs/pii/S0040162518315099.

SNA (2020). *ESNAwP: Revised and Expanded Edition for Updated Software*, 3e. Cambridge University Press http://mrvar.fdv.uni-lj.si/pajek/be3.htm.

Sumobrain.com. (2021). Expert search. https://www.sumobrain.com/search.html

Trianni, A., Merigó, J.M., and Bertoldi, P. (2018). Ten years of energy efficiency: a bibliometric analysis. *Energy Efficiency* 11 (8): 1917–1939.

Trout, J. (2021). Are heat pump water heaters worth the cost? https://www.consumeraffairs.com/homeowners/heat-pump-water-heater-value.html (accessed May 3, 2021).

Wang, X., Yang, X., Wang, X. et al. (2020). Evaluating the competitiveness of enterprise's technology based on LDA topic model. *Technology Analysis & Strategic Management* 32 (2): 208–222.

Willem, H., Lin, Y., and Lekov, A. (2017). Review of energy efficiency and system performance of residential heat pump water heaters. *Energy and Buildings* 143: 191–201.

Xu, Z., Ge, Z., Wang, X., and Skare, M. (2021). Bibliometric analysis of technology adoption literature published from 1997 to 2020. *Technological Forecasting and Social Change* 170: 120896.

Yan, H., Hu, B., and Wang, R. (2020). Air-source heat pump for distributed steam generation: a new and sustainable solution to replace coal-fired boilers in China. *Advanced Sustainable Systems* 4 (11): 2000118.

Yoon, B. and Lee, S. (2012). Managing technological knowledge for supporting R&D activities: scientometrics-based approach. *Knowledge Management Research and Practice* 10 (3): 237–251.

Zhang, J., Wang, R.Z., and Wu, J.Y. (2007). System optimization and experimental research on air source heat pump water heater. *Applied Thermal Engineering* 27 (5–6): 1029–1035.

Zhang, Y., Akkurt, N., Yuan, J. et al. (2020). Study on model uncertainty of water source heat pump and impact on decision making. *Energy and Buildings* 216: 109950.

Zhou, J., Zhao, X., Ma, X. et al. (2016). Experimental investigation of a solar driven direct-expansion heat pump system employing the novel PV/micro-channels-evaporator modules. *Applied Energy* 178: 484–495.

9

Technology Intelligence: Burst Analysis for RFID in Hospitals

Haydar Yalçın

Department of Business Administration, Ege University, Izmir 35800, Turkey

9.1 Introduction

Hospital software is designed to ensure the smooth execution of planned processes. These software systems provide high-quality and efficient process information depending on the quality of the data entered. With radio frequency identification (RFID) technology, information about the processes transmitted by the software is automatically compiled in the background without user input, allowing the production of data that can inform process improvement decisions (Frisch 2019). The key to maximizing the benefit after acquiring RFID applications is for hospitals to organize their workflows accordingly and raise awareness among all stakeholders. In this way, revealing all sub-dimensions is critical for the correct management of the process. RFID is a technology used to remotely identify objects, animals, or people with unique identification information (Duroc and Kaddour 2012; Lahiri 2011). This system includes RFID tags or cards that interact with an RFID reader device. An RFID system generally consists of a set of basic components. It is a small device that is attached to objects or people as hardware, also called an RFID tag or card. This tag or card consists of a microchip and antenna. An RFID reader device functions as a device that reads information from RFID tags or cards. This reader usually includes an antenna, an RFID reading module, and a connection unit. The database used as an area where the read RFID information is stored is another component. This database is the area where data is stored that can be used to track objects or people, improve inventory management, or enhance security. When evaluated in the context of these components, it is clear that RFID technologies are quite versatile. RFID technologies are used in inventory management (Saygin 2007), especially in the retail sector, logistics, and anti-theft systems, product tracking in industrial

Future-Oriented Technology Assessment: A Manager's Guide with Case Applications, First Edition.
Edited by Haydar Yalçın and Tugrul U. Daim.

production, patient tracking in hospitals (Chowdhury and Khosla 2007), animal tracking, and many other application areas (Baars et al. 2009; Bottani et al. 2014; Brusey and McFarlane 2009).

9.2 Methodology

There is a deep connection between social network analysis (SNA) and technology management, across many different dimensions. SNA helps businesses strengthen their innovation processes and implement technology management more effectively. Social networks encourage people from different disciplines to come together and generate new ideas. This enables businesses to develop innovation projects and better exploit collaboration opportunities. Technology management includes the processes of planning, implementing, and tracking these innovations (Zabala-Iturriagagoitia 2014). SNA provides important insights at every stage of these processes and allows strategies to be managed more effectively. Additionally, SNA helps businesses optimize customer relationships and marketing strategies (Bayer and Servan-Schreiber 2011). Social media platforms are an important resource for tracking customer feedback, understanding customer behavior, and personalizing marketing campaigns (Castronovo and Huang 2012). At this point, technology management assumes the function of collecting, analyzing, and integrating this data into business strategies. By combining SNR with technology management, businesses can better respond to customer expectations (Lee 2018), optimize marketing strategies, and increase customer loyalty (Ibrahim 2022). Additionally, SNR helps businesses gain a competitive advantage (Singla and Durga 2015). Competitive analysis involves the process of monitoring market trends, analyzing competitors, and tracking changes within the industry. Technology management strategies aim to gain competitive advantage through such analysis. SNR supports performing these analyzes in a data-driven and more effective way (Tseng et al. 2019). Kleinberg's burst analysis focuses specifically on information networks and data flows (Kleinberg 2002; Kumar and Jan 2012; Takahashi et al. 2012). It takes its name from Jon Kleinberg, who is known for his effectiveness in the discovery of these explosions. Thanks to this method, it is possible to detect sudden and intense information flows on a certain subject within a certain period. Kleinberg's burst analysis ranges from identifying instantaneous conversations and newsfeeds about similar products, to keeping the pulse of the market by examining intense conversations and interactions about competing companies over a specific period, to customer feedback, industry trends, and even internal team communication. It is a very effective method.

9.3 Data

Descriptive information about the data analyzed within the scope of the research is as follows: The dataset spans the years 1969–2024, incorporating information from a diverse array of 3373 sources, including journals and books. The average age of the documents is relatively recent at 8.01 years, implying a focus on contemporary information. Each document, on average, has been cited 21.54 times, showcasing a notable level of influence within scholarly circles. Furthermore, the dataset includes an extensive network of knowledge, evidenced by the impressive 317 001 references cited. This comprehensive and well-cited collection reflects a rich and interconnected body of information, contributing to a nuanced understanding of the subject matter over the specified timespan. The data set includes 1019 single-authored scientific documents. This can be interpreted as a significant contribution by individual authors.

However, the pattern of collaboration is also observed to be common, with 1372 scientific documents having multiple authors. On average, each scientific document with multiple authors involves the collaboration of 4.06 people, indicating a strong level of teamwork and co-authorship in the conduct of scientific work. International collaboration, which constitutes 21.82% of joint authorships, is remarkable with this rate. In this respect, the international dimension is indicative of how rich the diversity of ideas and expertise shapes the content of scientific research by adding a global perspective to collaborative efforts within the data set. Overall, it is possible to say that the balance between single-authored and collaborative studies and international co-authorships is important, as it highlights the collaborative and globally informed nature of scientific contributions in this dataset.

9.4 Burst Analysis

Today, the explosion of online data and information flow challenges traditional analysis methods. To overcome this challenge and sort through the information we want to understand, we need advanced analytics methods. In this context, Kleinberg burst analysis stands out as an effective analytical approach used especially to understand and analyze information bursts. Its methodology allows us to examine situations when a topic or word becomes significantly popular at a certain time – that is, experiences a significant explosion. This analysis helps us understand when and how a particular topic or event comes to the fore, looking specifically at data sets from social media, news sites, or other online platforms. Essentially, Kleinberg burst analysis determines when a particular word occurs more quickly than would normally be expected and at an increasing frequency over a period of time (Dernis et al. 2016). These explosions generally reflect periods when events develop rapidly or a meaningful event that attracts the attention of society occurs, allowing us to analyze the causes and consequences of information explosions (Kleinberg 2002). The provided Kleinberg's burst analysis results offer a comprehensive view of the temporal dynamics in research interests and the emergence of significant topics over an extended period, spanning from 1992 to 2023. Each burst, marked by its start and end years, is associated with a specific term or entity, and is quantified by a "strength" value that gauges the intensity of the burst. Notably, these bursts encompass a diverse range of fields, from technology and healthcare to information systems. For instance, bursts related to *RFID systems* and *internet of things (IoT)* point to the rapid evolution and adoption of cutting-edge technologies. On the other hand, terms such as *blood transfusion* and *supply chain management* indicate prolonged periods of heightened interest and research activity in the medical and logistics domains. This analysis serves as a valuable tool for tracking the evolution of research interests, revealing changing trends and helping researchers and analysts stay abreast of shifting foci within different disciplines over the years (Table 9.1).

9.5 Cluster Analysis

The data provided describes a network of 17 clusters, each representing a group of RFID items or data points related to hospitals. These clusters vary in size, and the largest are summarized here, illustrated in Figure 9.1. Cluster 0, the largest cluster with 246 elements, is associated with the "chipless RFID tag" and has a high Silhouette score with a mean year of 2012, indicating well-defined coherence within the cluster. Cluster 1, consisting of 179 items, is located on the "supply chain" and has a Silhouette score of 0.783 with an average 2010 year. Cluster 2, consisting of 159 items, relates to "patient safety" and has a Silhouette score of 0.783, with a mean date of 2005. These clusters cover a range of topics and are characterized by their size, completeness, labels, and years associated with the items within them. This type of cluster analysis is valuable for organizing and understanding complex data sets, facilitating research, and identifying patterns within the data. Continuing the interpretation of the clusters, Cluster 3 stands out with its high Silhouette score of 0.796 and is labeled "RFID system" with 147 items and an average date of 2011. Although Cluster 4 is smaller at 96 items, it is labeled "RFID system" with 147 items. It has a high Silhouette score of 0.823 and focuses on the "limiting protocol" with an average year of 2014. Cluster 5, consisting of 77 items, relates to "indoor localization" and has a Silhouette score of 0.81 with an average year of 2016. Cluster 6 is interesting with its exceptionally

Table 9.1 Burst analysis.

Begin	End	Strength	Year	Entity
1992	2013	6.4965	1969	Issue
1992	2008	4.072	1969	Events
1997	2008	5.8336	1969	Errors
1998	2007	19.5642	1969	Blood transfusion
1999	2007	8.7639	1969	Animal identification
2000	2004	21.5308	1969	Patient identification
2000	2006	8.7309	1969	Medicine
2000	2012	6.8317	1969	Adverse events
2002	2007	10.9745	1969	Prevalence
2003	2008	29.144	1969	Ubiquitous computing
2003	2008	15.9745	1969	Adverse drug events
2003	2008	9.2121	1969	Circuits
2004	2012	42.4406	1969	Radio frequency identification
2004	2009	9.7986	1969	Systems
2004	2008	9.4338	1969	Medical errors
2004	2011	5.4489	1969	Automatic identification
2004	2011	5.4489	1969	Inductive coupling
2004	2008	4.2704	1969	Experience
2005	2009	47.0652	1969	Supply chain management
2005	2011	42.4804	1969	Radio frequency identification (RFID)
2005	2017	6.6926	1969	Passive RFID
2006	2012	17.3374	1969	Supply chain
2006	2016	16.6514	1969	UHF
2006	2012	13.0905	1969	Collision resolution
2006	2011	12.4848	1969	Passive tag
2006	2011	12.4814	1969	Sensor network
2006	2010	7.6585	1969	Radio-frequency identification (RFID)
2006	2013	6.7987	1969	RFID reader
2007	2015	16.8027	1969	Inaccuracy
2007	2012	15.9548	1969	Pervasive computing
2007	2011	14.9688	1969	Information technology
2007	2011	14.5416	1969	Information systems

(*continued*)

Table 9.1 (Continued)

Begin	End	Strength	Year	Entity
2007	2012	12.1727	1969	Inventory control
2007	2012	12.1727	1969	Case study
2007	2014	11.4402	1969	Technology adoption
2007	2012	10.9128	1969	Business value
2007	2014	5.5156	1969	Loop antenna
2007	2013	5.0809	1969	Mobile RFID
2007	2011	4.0964	1969	Mobile robot
2008	2016	24.792	1969	RF identification (RFID)
2008	2013	11.7019	1969	Tag antenna
2008	2013	7.6535	1969	Low power
2008	2012	5.4548	1969	RFID systems
2008	2012	5.3378	1969	Power
2008	2012	4.4881	1969	Impacts
2008	2012	4.1879	1969	Competitive advantage
2009	2013	13.971	1969	Active RFID
2009	2015	7.317	1969	Operations
2009	2015	7.0723	1969	Benefits
2009	2013	4.9297	1969	Wireless sensor network
2009	2014	4.7779	1969	Ambient intelligence
2010	2017	6.0476	1969	Slot antenna
2011	2016	9.8361	1969	RFID security
2011	2015	7.1183	1969	Components
2011	2016	6.9144	1969	Support
2011	2015	5.3404	1969	Mutual authentication
2011	2015	5.1646	1969	Sasi
2012	2018	9.0196	1969	Inkjet printing
2012	2018	7.3497	1969	Protocols
2012	2016	6.9591	1969	Authentication protocol
2012	2016	4.6689	1969	Medication errors
2012	2016	4.4296	1969	Circular polarization
2012	2016	4.2443	1969	GPS
2013	2017	6.0282	1969	Tree

Table 9.1 (Continued)

Begin	End	Strength	Year	Entity
2013	2019	4.836	1969	Circuit
2014	2018	6.7856	1969	Coordination
2016	2020	8.7429	1969	Wireless power transfer
2016	2021	6.1003	1969	Quality improvement
2017	2023	5.0652	1969	Phase
2017	2021	4.1746	1969	Cloud
2018	2023	9.3486	1969	Internet of things (IOT)
2018	2023	6.025	1969	Industry 40
2018	2023	4.8508	1969	Diagnosis

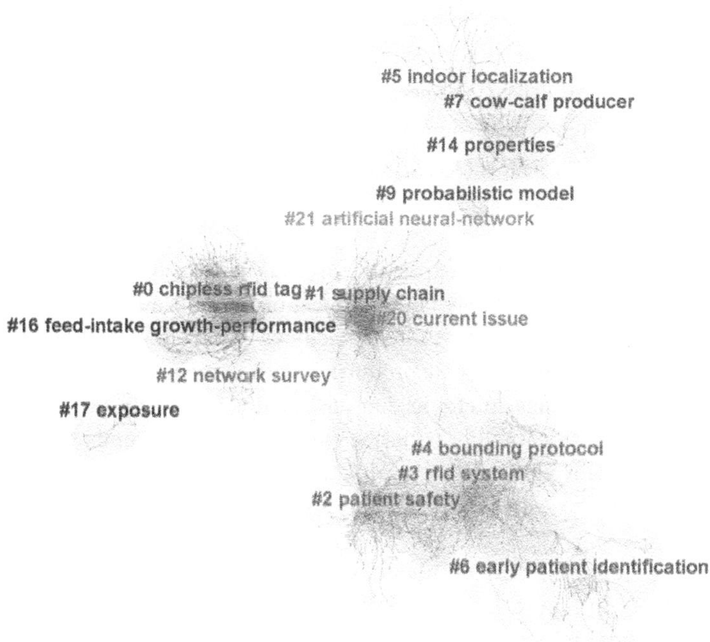

Figure 9.1 Major clusters on RFID in hospitals research.

Table 9.2 Major clusters on RFID in hospitals Research.

Cluster ID	Size	Silhouette	Label (LLR)	Average year
0	246	0.839	Chipless RFID tag (26 443.54, 1.0E-4)	2012
1	179	0.783	Supply chain (29 664.52, 1.0E-4)	2010
2	159	0.783	Patient safety (37 673.44, 1.0E-4)	2005
3	147	0.796	RFID system (27 653.51, 1.0E-4)	2011
4	96	0.823	Bounding protocol (5498.89, 1.0E-4)	2014
5	77	0.81	Indoor localization (4415.62, 1.0E-4)	2016
6	50	0.949	Early patient identification (1318.75, 1.0E-4)	2002
7	19	0.984	Cow-calf producer (3118.74, 1.0E-4)	2004
9	10	0.994	Probabilistic model (26.13, 1.0E-4)	1995
12	8	0.997	Network survey (170.27, 1.0E-4)	1995
14	6	0.996	Properties (26.81, 1.0E-4)	1991
16	6	0.996	Feed-intake growth-performance (25.63, 1.0E-4)	1993
17	6	0.998	Exposure (25.63, 1.0E-4)	1997
20	5	0.998	Current issue (51.58, 1.0E-4)	1993
21	5	0.997	Artificial neural-network (25.63, 1.0E-4)	1994
27	3	0.997	Temperature humidity (47.14, 1.0E-4)	2009
32	3	1	HIV patient (26.13, 1.0E-4)	2003

high Silhouette score of 0.949, which expresses the well-defined harmony of its 50 elements. It is associated with "early patient detection" and its average year is 2002. Cluster 7, although smaller with 19 items, achieves a very high Silhouette score of 0.984 and focuses on "cow-calf producer" with a mean year of 2004. These clusters, which vary in size and topic, demonstrate the versatility and applicability of cluster analysis in revealing patterns, connections, and trends within complex data sets. Such insights can be invaluable to researchers and analysts in various fields, helping to extract information and make informed decisions (Table 9.2).

The term *all degree* represents the number of connections of a node in the network (Burt et al. 2013). In Figure 9.2, the words "SYSTEM" and "DESIGN" indicate connections to other nodes in the network. "PATTERNS" and "BEHAVIOR," refer to different usage patterns or behaviors of these nodes. "PERFORMANCE" and "MODELS" represent the performance and models of the systems. "MANAGEMENT" and "HEALTH," refer to management and health-related issues,

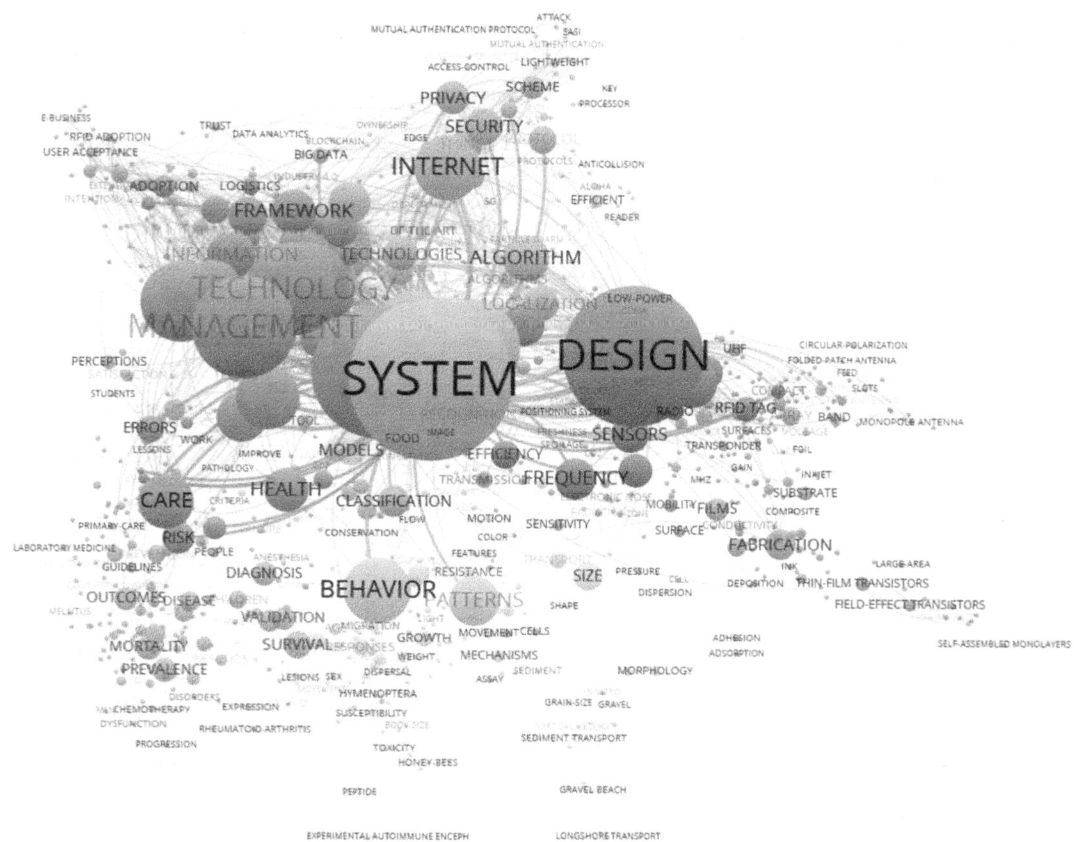

Figure 9.2 Keyword map.

while the terms “TECHNOLOGY” and “IDENTIFICATION” refer to technology and identification processes.

Betweenness centrality determines how important a node is among other nodes in the network (Burt et al. 2013). The terms “SYSTEM” and “IDENTIFICATION” indicate that they play important roles in the foundation of the network. “BEHAVIOR” and “HEALTH,” play a critical role in interacting with other nodes. “MANAGEMENT” and “MODELS,” represent significant research or management functions within the network.

The phrase *high aggregate constraints* indicates that the use of a system or technology is subject to a high level of constraints (Hu and Zhang 2014; Saglietto et al. 2020). These limitations may be related to security or data privacy. “PATTERNS,” “BEHAVIOR,” and “MODELS,” may be subject to these limitations to a high degree of impact.

Low aggregate constraints indicate that the use of a system or technology is subject to fewer constraints. This means that data access and use is freer. “RFID READER,” “RADIATION-PATTERN,” and “FACE RECOGNITION,” appear to have these lower limitations.

In Figure 9.3, LIU, ALEX X (Alex X Liu) stands out as a researcher of particular importance within this network. According to the “all degree” and “betweenness centrality” measurements, it can be said that Alex is effectively connected or in a central position. It should also be noted that its impact on the network may vary, depending on whether it is subject to the constraints specified by “high aggregate constraints” or “low aggregate constraints.” TENTZERIS, MANOS M (Manos M Tentzeris) may be another important actor in the network. It can be said that this actor plays an effective role on the information flow in the network, especially with its high value in terms of betweenness centrality. It should also be considered whether it is subject to “high aggregate constraints” or “low aggregate constraints” constraints. XIAO, BIN (Bin Xiao) stands out as someone who has established important connections within the network. The name can help us further understand his role within the network. HAN, JINSONG (Jinsong Han) stands out as a node that affects the flow of information within the network, while playing an important bridge role between others in the network, according to the “betweenness centrality” measurementCOLELLA, RICCARDO (Riccardo Colella) may have collaborated with others within the network. Besides names, betweenness centrality and other metrics can help us better understand its role in the network.

These data are examined through key metrics used to evaluate the scientific impact and productivity of different universities and research institutions. The h-index measures the number of an institution’s most cited articles, while the total number of citations and number of published articles reflect the overall impact and productivity of each institution’s research. Georgia Institute of Technology

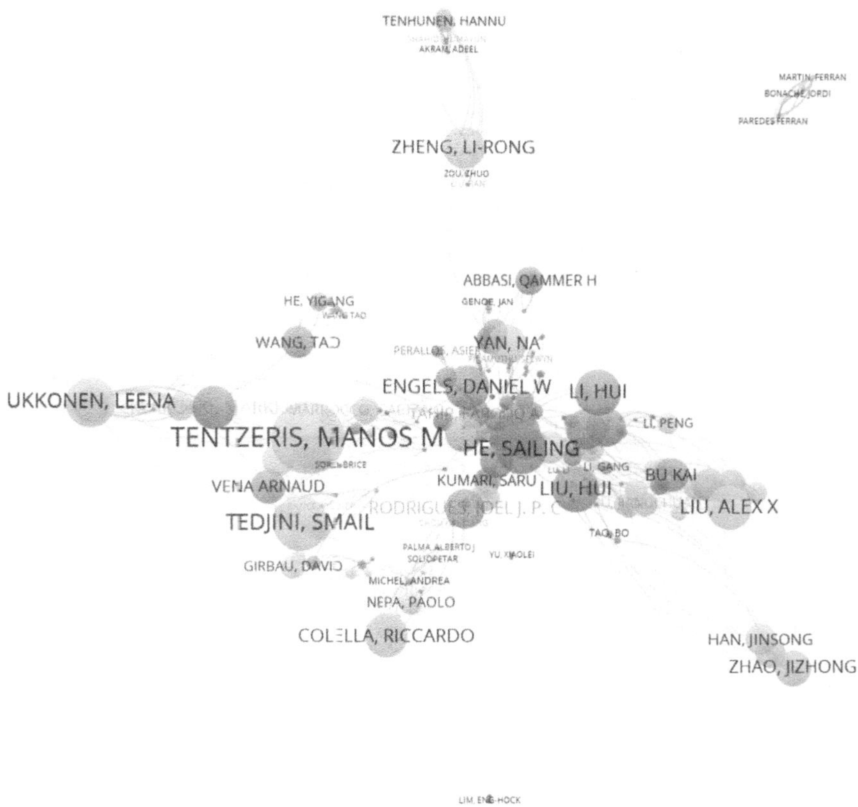

Figure 9.3 Author collaboration network on RFID in hospitals.

stands out with its high h-index as an institution that has received a lot of citations according to the total number of articles. Similarly, Hong Kong Polytechnic University and the University of Hong Kong also have impressive research outputs with high h-indexes and total citation counts. These institutions demonstrate that their researchers are making significant contributions internationally.

Likewise, institutions such as Tampere University of Technology, University of Roma Tor Vergata, and University of Pisa have also created significant impacts in relatively smaller research communities with their h-indexes and total citation numbers. This indicates the existence of studies that are highly cited compared to the number of articles. Finally, world-renowned institutions such as the Massachusetts Institute of Technology (MIT) have made a significant impact in the scientific world with their high h-indexes and total citation counts. These data highlight the quality of research and world-class scientific contributions of these institutions. These metrics provide an important tool for assessing each institution's scientific success and shaping research strategies. Each institution demonstrates that it has research communities of varying sizes and specializations that are capable of contributing effectively to their field (Table 9.3).

Oregon Health & Science University (104) stands out as the institution with the highest number of collaborations in this field. This is important as it shows that the university cooperates effectively and takes a leadership role in RFID technology research. Taipei Medical University (103) also has a high number of collaborations. It is possible to say that the Taiwan-based university plays an

Table 9.3 Institution productivity on RFID in hospital research.

h-index	Unit	Citation sum within h-core	All citations	All articles
52	Georgia Inst Technol	6361	8568	176
49	Hong Kong Polytech Univ	5567	8707	234
42	Univ Hong Kong	4915	6050	113
36	Tampere Univ Technol	2663	3865	127
35	Univ Roma Tor Vergata	3559	4358	110
35	Univ Pisa	2083	3026	124
33	Univ Washington	4018	5340	149
32	MIT	4479	5068	81
31	Hong Kong Univ Sci & Technol	4114	4611	75
31	Chinese Acad Sci	7822	8662	155
31	Tsinghua Univ	1965	2987	136
30	Univ Florida	1994	3132	139
30	Monash Univ	2450	3195	118
28	Michigan State Univ	3533	4320	98
28	Univ Calif Los Angeles	2530	3039	84
28	Univ Elect Sci & Technol China	1804	2649	147
27	Xi An Jiao Tong Univ	1357	1937	109
27	Univ Michigan	2015	2473	92
27	Nanyang Technol Univ	2365	2943	84
27	Zhejiang Univ	1781	2474	112
26	Katholieke Univ Leuven	2062	2467	64
26	Oregon Hlth & Sci Univ	1058	1613	70
25	Concordia Univ	1148	1359	46
25	Univ Adelaide	1420	1937	69
25	Natl Taiwan Univ Sci & Technol	1521	1967	94

important role in RFID research and contributes to collaborations at the global level, while the University of Washington (87) ranks third with an impressive number of collaborations in RFID research, ranking the university in this field. Duke University (82) ranks fourth in RFID technology research and Copenhagen University Hospital (79) ranks fifth in RFID research. The fact that these institutions are in the rankings shows that hospital institutions also have a significant interaction on RFID technology. It is valuable (Figure 9.4).

When we look at the cooperation numbers of the top 15 countries, we see that RFID technology is taking shape in a global effort. The United States of America (USA) is far ahead in terms of the number of cooperation both within itself and with other countries. This highlights the United States' leading position in the development of RFID technology and its interaction on a global level. The People's Republic of China ranks second, especially for its cooperation with the United States, indicating China's growth and international interaction in the field of RFID technology. South Korea, Germany, and Taiwan are in the third, fourth, and fifth places, respectively. These countries make significant contributions to the global development of RFID technology with their cooperation within themselves and their interactions with the USA. Canada, Australia, France, Italy, and the United Kingdom, among other countries, also have significant collaboration numbers. These countries' active role in RFID reflects a global trend toward widespread use of the technology in hospitals and the healthcare sector (Figure 9.5).

The data provided reveals the research impact and productivity of various countries, evaluated against a variety of key metrics. The h-index, a measure of

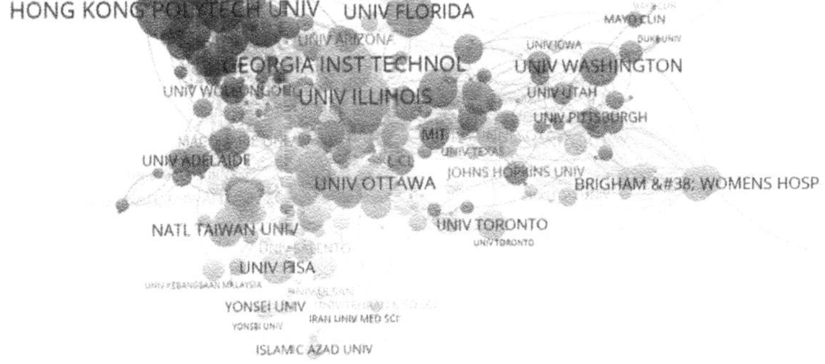

Figure 9.4 Institution collaboration on RFID in hospitals.

Figure 9.5 Country collaboration on RFID in hospitals.

the number of highly cited articles, total citations, and total number of published articles, provides insight into the research output and impact of researchers in each country. The United States has the highest h-index at 182, indicating a significant number of highly cited articles. US researchers have an impressive 76 947 citations in h-core and 230 160 total citations; This demonstrates their significant global research impact in various fields. China, on the other hand, follows closely with an h-index of 130, showing increasing research impact. Chinese researchers have collected 38 037 citations in h-core and 119 741 total citations, highlighting

Table 9.4 Country productivity on RFID in hospital research.

h-index	Unit	Citation sum within h-core	All citations	All articles
182	USA	76 947	230 160	7719
130	Peoples R China	38 037	119 741	6060
99	UK	18 875	45 165	1555
92	Italy	40 758	61 775	1387
83	South Korea	12 499	32 751	1876
83	France	19 705	34 307	1273
73	Germany	14 013	28 100	1323
72	Taiwan	9901	27 819	1608
71	Australia	23 847	40 920	1179
69	Spain	9296	23 727	1133
66	Canada	8558	22 799	1109
57	Netherlands	6997	11 828	449
54	India	5997	13 937	1206
52	Belgium	5287	9012	293
49	Finland	5195	10 185	423
49	Japan	6523	12 012	813
49	Switzerland	5363	8717	317
47	Singapore	5135	8008	255
47	Sweden	5714	9558	355
40	Portugal	2865	4439	185
40	Turkey	3153	6719	526
39	Malaysia	3861	7425	528
39	Greece	4647	6041	171
35	Austria	3215	5380	318
31	Saudi Arabia	1757	3513	343

their significant contributions to the academic community. The UK maintains a strong presence in research with an h-index of 99, although the h-index is slightly lower than the first two countries, given that researchers in the UK have accumulated 18 875 citations in the h-core and 45 165 total citations based on the number of published papers They appear to have made significant contributions. Italy also has a prominent position in the research with its h-index of 92. It reflects the influential research contributions of Italian researchers, with 40 758 citations in h-core and 61 775 total citations. South Korea and France share an h-index of 83, indicating comparable research impact. South Korean researchers accumulate 12 499 citations in the h-core and 32 751 total citations, while French researchers have 19 705 citations in the h-core and 34 307 total citations, both showing significant research impact. With an h-index of 73, Germany maintains a strong research presence with 14 013 citations in the h-core and 28 100 total citations. Taiwan demonstrates a solid research impact, collecting 9901 citations in the h-core and 27 819 total citations, with an h-index of 72. With an h-index of 71, Australia exhibits a significant research impact, with its researchers collecting 23 847 citations in the h-core and 40 920 total citations. With an h-index of 69, Spain maintains a respectable research presence with 9296 citations in the h-core and 23 727 total citations (Table 9.4).

9.6 Conclusion

According to the results of the study, revealing concepts that are likely to emerge suddenly can be used as a very useful analysis method in determining new emerging technologies, revealing important points, and determining sub-technology domains. In this way, this tool stands out as a useful decision-making instrument for managers in technology and engineering.

References

Baars, H., Gille, D., and Strüker, J. (2009). Evaluation of RFID applications for logistics: a framework for identifying, forecasting and assessing benefits. *European Journal of Information Systems* 18 (6): 578–591.

Bayer, J. and Servan-Schreiber, E. (2011). Gaining competitive advantage through the analysis of customers' social networks. *Journal of Direct, Data and Digital Marketing Practice* 13: 106–118.

Bottani, E., Manfredi, M., Vignali, G., and Volpi, A. (2014). Life cycle assessment of RFID implementation in the fresh food supply chain. *International Journal of RF Technologies* 6 (1): 51–71.

Brusey, J. and McFarlane, D.C. (2009). Effective RFID-based object tracking for manufacturing. *International Journal of Computer Integrated Manufacturing* 22 (7): 638–647.

Burt, R.S., Kilduff, M., and Tasselli, S. (2013). Social network analysis: foundations and Frontiers on advantage. In: *Annual Review of Psychology*, vol. 64 (ed. S.T. Fiske), 527–547. Annual Reviews. https://doi.org/10.1146/annurev-psych-113011-143828.

Castronovo, C. and Huang, L. (2012). Social media in an alternative marketing communication model. *Journal of Marketing Development and Competitiveness* 6 (1): 117–134.

Chowdhury, B. and Khosla, R. (2007). RFID-based hospital real-time patient management system. In: *6th IEEE/ACIS International Conference on Computer and Information Science* ICIS.

Dernis, H., Squicciarini, M., and de Pinho, R. (2016). Detecting the emergence of technologies and the evolution and co-development trajectories in science (DETECTS): a 'burst' analysis-based approach. *The Journal of Technology Transfer* 41: 930–960.

Duroc, Y. and Kaddour, D. (2012). RFID potential impacts and future evolution for green projects. *Energy Procedia* 18: 91–98.

Frisch, P.H. (2019). RFID in today's intelligent hospital enhancing patient care and optimizing hospital operations. In: *2019 IEEE International Conference on REFID Technology and Application (RFID-TA)*. IEEE.

Hu, R.J. and Zhang, G.Y. (2014). Structural holes in directed fuzzy social networks. *Journal of Applied Mathematics* 8: Article 452063. https://doi.org/10.1155/2014/452063.

Ibrahim, B. (2022). Social media marketing activities and brand loyalty: a meta-analysis examination. *Journal of Promotion Management* 28 (1): 60–90.

Kleinberg, J. (2002). Bursty and hierarchical structure in streams. In: *Proceedings of the Eighth ACM SIGKDD International Conference on Knowledge Discovery and Data Mining*.

Kumar, S. and Jan, J.M. (2012). Discovering knowledge landscapes: an epistemic analysis of business and management field in Malaysia. *Procedia-Social and Behavioral Sciences* 65: 1027–1032.

Lahiri, S. (2011). *RFID Sourcebook*. IBM press.

Lee, I. (2018). Social media analytics for enterprises: typology, methods, and processes. *Business Horizons* 61 (2): 199–210.

Saglietto, L., Cézanne, C., and David, D. (2020). Research on structural holes: an assessment on measurement issues [Article]. *Journal of Economic Surveys* 34 (3): 572–593. https://doi.org/10.1111/joes.12371.

Saygin, C. (2007). Adaptive inventory management using RFID data. *The International Journal of Advanced Manufacturing Technology* 32: 1045–1051.

Singla, M. and Durga, A. (2015). How social media gives you competitive advantage. *Indian Journal of Science and Technology* 8: 90.

Takahashi, Y., Utsuro, T., Yoshioka, M. et al. (2012). Applying a burst model to detect bursty topics in a topic model. In: *Proceedings on Advances in Natural Language Processing: 8th International Conference on NLP* (22–24 October 2012). Kanazawa, Japan: JapTAL.

Tseng, M.-L., Wu, K.-J., Lim, M.K., and Wong, W.-P. (2019). Data-driven sustainable supply chain management performance: a hierarchical structure assessment under uncertainties. *Journal of Cleaner Production* 227: 760–771.

Zabala-Iturriagagoitia, J.M. (2014). Innovation management tools: implementing technology watch as a routine for adaptation. *Technology Analysis & Strategic Management* 26 (9): 1073–1089.

10

Technology Roadmapping: Data Science Roadmapping of Networked Organizations' Strategic Planning for Artificial Intelligence

Kerem Kayabay[1], Atilla Kılınç[2], Mert Onuralp Gökalp[3], Ebru Gökalp[4], and Tugrul U. Daim[5]

[1] *HLRS, University of Stuttgart, Stuttgart 70569, Germany*
[2] *Energy Institute, Istanbul Technical University, Istanbul 34469, Turkiye*
[3] *Data Analytics Center, Tüpras, Ankara 06510, Turkiye*
[4] *Computer Engineering Department, Hacettepe University, Ankara 06800, Turkiye*
[5] *Mark O. Hatfield Cybersecurity and Cyber Defense Policy Center, Department of Engineering and Technology Management, Portland State University, Portland, OR 97201, USA*

10.1 Introduction

The technological developments in data science and artificial intelligence (AI) fields call for significant organizational change in many data-disadvantaged and data-underachiever industries. Besides the many technical challenges, an organization must overcome strategic, data-related, and organizational challenges to become data-driven and stay competitive (Kayabay et al. 2022a). Many industries and governments have published strategy and policy papers examining the need for sectoral, national, and global coordination and orientation. AI systems already discriminate based on socioeconomic and national factors, inflicting damage (Zhang et al. 2022). Society needs well-coordinated actions implemented by transparent and agile processes focusing on continuous evaluation and improvement to address the challenges in the rapidly evolving data science and AI landscape.

The term *data science roadmapping (DSR),* coined by Kayabay et al. (2022a), is an emerging field incorporating data science, AI, and technology management into a comprehensive strategic planning framework. It integrates data layers into rigorously validated technology roadmapping (TRM) (Phaal et al. 2004a), considering the data life cycle and data-related processes. The DSR provides a clear technology strategy combining AI and infrastructure with business objectives and organizational resources, allowing organizations to design and implement diverse technologies (Kayabay et al. 2022a). Therefore, the DSR

Future-Oriented Technology Assessment: A Manager's Guide with Case Applications, First Edition.
Edited by Haydar Yalçın and Tugrul U. Daim.

approach presents a novel strategic decision-making tool for managing data assets and related technological and organizational resources in for-profit and nonprofit organizations, distinguishing itself from TRM. We foresee it becoming indispensable for strategic planning beyond the organizational level, bringing together chip designers, hardware manufacturers, funding mechanisms, and regulators to communicate requirements and adjust knowledge flows.

Nevertheless, significant knowledge gaps hinder the mass adoption and successful operationalization of the DSR because roadmapping applications require customization to unique objectives and goals (Phaal et al. 2004b; Pearson et al. 2020). The literature needs studies showing how to customize DSR beyond the organizational level, incorporating experiment-based industrial best practices since the industry drives data science and AI advances (McKinsey 2022). This chapter addresses the literature gap by utilizing customized DSR to develop a AI roadmap for networked organizations. The primary objective is to respond to the demands of the ever-changing AI environment by effectively implementing roadmapping and optimizing its outcomes. The networked companies in the case study operate in different sectors but belong to the same parent company. They want to use DSR as a coordination and planning tool to build a joint data center to stay competitive in the large-scale AI era by integrating data and knowledge flows and achieving more computational power. In the short term, we aim to investigate if the DSR can help these organizations agree on pathways forward by establishing communication channels, building trust, understanding the gaps concerning data quality, and experimenting with large models. The following sections will analyze the literature for the DSR, followed by a case study for developing a technology roadmap of networked organizations for AI; after that, the result of the case study will be discussed and analyzed.

10.2 Literature Review

10.2.1 Data Science Roadmapping

In literature, A *technology roadmap* is generally accepted as a time-based chart that displays the linkages and discontinuities among products, business strategy, service developments, and market opportunities (Phaal et al. 2004a). TRM or roadmapping is a widespread technology management approach for examining and conveying commercial-technical linkages as they evolve. Roadmapping helps organizations evaluate disruptive technology, broaden planning horizons, share information, and reach agreement (Phaal et al. 2004a). TRM is a core module for strategically planning organizational aims, improving decision-making processes, mitigating operational risks, and developing collaborations across organizations (Phaal et al. 2006).

Using the TRM model in strategically planning AI, machine learning (ML), and IoT applications has resulted in studies addressing the methodology and practical application (Han and Geum 2020; Han and Jeon 2023; Kayabay et al. 2022a,b; Nazarenko et al. 2021; Simonse et al. 2015). Accordingly, the DSR, building on TRM, has been developed as a strategic planning tool for transforming into data-driven organizations (Kayabay et al. 2022a,b). By considering effect, practicality, and risk, while TRM helps organizations prioritize technical alternatives, the DSR offers a defined technological plan that combines AI strategy and infrastructure with business goals and organizational resources, enabling enterprises to make educated choices about inventing and adopting diverse technologies (Kayabay et al. 2022a).

Therefore, the DSR approach presents a novel strategic decision-making tool for managing data assets and related technological and organizational resources in for-profit and nonprofit organizations, distinguishing itself from TRM. The DSR model may indicate how AI applications relate to future objectives, how well the technical infrastructure processes and stores data, and how structural and regulatory barriers prevent data-driven insights in business processes (Kayabay et al. 2022a). Still, it is essential to understand the roadmapping applications, methods, and tools for sustainable DSR initiatives, investigated in the next section.

10.2.1.1 Technology Roadmapping Applications, Methods, and Tools

DSR builds on TRM, a powerful technique since it is flexible and scalable. TRM is adaptable to multiple purposes, including strategic planning, process planning, and business reconfiguration (Phaal et al. 2004b). Studies in the literature have used roadmapping to deal with strategic capability development (Kerr et al. 2019), hardware development (Pearson et al. 2020), policymaking (Ho and O'Sullivan 2017; Yasunaga et al. 2009), and research and development (Freitas et al. 2019; Ho and O'Sullivan 2020). While roadmapping can help an organization with technology management, it can also help a network of organizations to facilitate coordination and collaboration (Petrick 2013). As government divisions, universities, research organizations, suppliers, manufacturers, and end-users collaborate on strategic planning, cooperation, and networking, they may expand and influence the sectoral level (Nimmo 2013).

Different roadmap development methods are expert-based, computer-based, or hybrid (Pora et al. 2020). While the expert-based approaches rely on qualitative techniques, such as bringing together diverse participants in workshops (Phaal et al. 2004a; Phaal et al. 2007; Phaal et al. 2013), computer-based approaches build on quantitative data analysis (Geum et al. 2015). The hybrid methods leverage expert-based and quantitative approaches, for example, supporting practitioner knowledge in roadmapping workshops with quantitative evidence (Pora et al. 2020) or connecting expert-based approaches (e.g. Delphi) with quantitative methods (Daim et al. 2018). Indeed, one strength of TRM is incorporating tools such as decision models (Daim et al. 2018), patent analysis (Zhang et al. 2021), checklists (Kerr and Phaal 2019), process models (Kayabay et al. 2022a), scenarios (Hansen et al. 2016; Pearson et al. 2020), data mining (Geum et al. 2015), and maturity models (Chutivongse and Gerdsri 2019). Furthermore, roadmapping can be an integrative toolkit platform to which other managerial tools connect (Kerr et al. 2017). However, the connecting tools should adhere to certain principles (Kerr et al. 2013). For example, monolithic tools that do not complement each other can re-create a large amount of data and knowledge (Kerr et al. 2017).

The two proven workshop-based reference processes for rapid initiation of roadmapping are T-Plan and S-Plan (Park et al. 2020; Phaal et al. 2013). While the T-Plan (Phaal et al. 2004a) brings together medium-sized groups to focus on market, product, and technology issues in four half-day workshops, S-Plan (Phaal et al. 2007) is a one-day large-scale workshop to explore, identify, and prioritize opportunities. Kerr et al. (2019) have implemented the S-Plan reference process for the LEGO Group, inserting pre-workshop and post-workshop modules. Pre-workshop encourages preparation to gain time in workshops, while post-workshop allows for post-processing of all the data using what-if and trade-off analysis and synthesis. The proposed approach has enabled the integration of respective tools in pre- and post-workshop modules (Kerr et al. 2017), and inspired by this approach, Kayabay et al. (2022a) have leveraged pre- and post-workshops for building data-integrated roadmaps. The early phase of DSR enables data professionals to connect with stakeholders, locate data resources, comprehend data, and execute preliminary analyzes. This approach improves time allocation between sessions and considers new business opportunities.

10.2.1.2 Toward Sustainable DSR

Implementing comprehensive and sustainable DSR in nonprofit and for-profit organizations requires a multidisciplinary approach, including the collaboration of technology experts, data scientists, policymakers, managers, and industry experts. TRM is a proven technology and strategy planning method, while organizationwide roadmapping deployments are unusual and have mixed results. (Hirose et al. 2022). Updating and revising TRM and translating outcomes from off-site strategy workshops into actions for existing operations are complex (Daim et al. 2018; de Oliveira et al. 2022). Pora et al. (2020) suggest a semi-automated and data-driven decision support system to reflect changes to the roadmap and assess whether the roadmap needs slight adjustments or significant revisions.

Roadmapping experts have established organized and agile engineering-based solutions (de Oliveira et al. 2022). Accordingly, TRM has become a product of multi-stage processes (Pearson et al. 2020). A first iteration might take a short time to show roadmapping's benefits (Kerr and Phaal 2019), and subsequent iterations may require more time, perspectives, and expertise (O'Sullivan and Phaal 2021). For example, the S-Plan reference approach can be a starting point for developing a roadmap before a second T-Plan application when managers are confident about where to focus innovation efforts (Phaal et al. 2007; Phaal et al. 2013). A DSR application in a large oil and gas company follows a similar approach to plan a small number of projects with limited stakeholders (Kayabay et al. 2022b). At the final review meeting, the roadmap participants argued that they would work with one or two business units for the next iteration instead of incorporating everyone because they preferred to see the broader perspective by adding all the projects.

In addition to iterative and agile roadmapping studies, roadmap development processes can leverage experimentation-based approaches (Al-Ali and Phaal 2019). In the DSR context, Kayabay et al. (2022a) argue in favor of executing much of the beginning data science stages (e.g. data discovery, data preparation) pre-workshop and building minimum viable data products during data workshops. Experimentation may uncover possibilities, evaluate commercial value, and establish a high-level data management strategy during roadmapping. However, the industry leads the most important developments in the data science and AI fields (McKinsey 2022), so most of the best practices regarding experimentation-based approaches exist in industrial sources, not scientific technology management literature. Therefore, one of the primary goals of this study is to compile them with the technology management literature in a case study to guide future research and practice.

10.2.2 Strategic Planning for Data-driven Scenarios

Big data, data science, and AI relationship requires cross-organizational process and knowledge alignment. Not only for-profit and nonprofit organizations' governmental and industrial divisions need coordination to manage data and related technologies as strategic assets. This section primarily focuses on how integrated TRM (Heim and Heuss 2017) will emerge at the organizational, supply chain, sectoral, national, and multinational levels to bridge gaps in the interconnected technological, political, and sociological perspectives through the relationship between big data, data science, and AI.

10.2.2.1 Big Data, Data Science, and AI

Big data refers to extraordinarily massive and complicated data collections produced by developing digital technologies such as smart devices, the internet of things (IoT), business processes, social media, and blockchain that overwhelm traditional data processing methods (Power 2014). Due to large volume, variety, and high flow speed, data storage, and timely data processing only become achievable with distributed high-performance hardware and software technologies (Davenport 2014). Distributed technologies and increased processing power have made big data management feasible, enhancing interest in data science and AI. Big data, data science, and AI are complementary trends with individual priorities and purposes that share statistical methods, data-processing steps, and modeling algorithms. Data science applies ideas, techniques, and methods to find patterns in raw data (Gökalp et al. 2021), providing meaningful insights, knowledge, information, and actions critical in diverse nonprofit and for-profit organizations to drive innovation and enhance decision-making (Sarker 2021). Meanwhile, AI attempts to reproduce human cognitive processes by machines intelligently perceiving their environments and taking actions that maximize chances of success (Mikalef and Gupta 2021; The US Council of Economic Advisers 2022). AI may evolve, adapt, and think like humans with image, speech, translation, generation, and auto-driving. AI methods and technologies include neural networks, natural language processing, ML, deep learning rule-based expert systems, and robotic process automation (Benbya et al. 2020).

AI has undergone several eras marked by significant advancements and changes in research focus, technological capabilities, and industrial applications (Sevilla et al. 2022). In 1950, Alan Turing established the Turing Test to evaluate whether a machine may act intelligently as a person (Turing 1950). However, the early neural networks (Wang and Raj 2017) needed computational power and data to achieve significant results. ML-based AI approaches gained significant popularity after 2010 with the emergence of distributed storage and computing frameworks like Apache Hadoop and Apache Spark, among others.

10.2.2.2 The Need for Data-integrated Strategic Planning

AI adoption has plateaued since organizations realized the organizational change required to integrate data-driven approaches in business processes (McKinsey 2022). Kayabay et al. (2022a) organize the challenges of creating cross-functional and strategic value from data into data-related, organizational, technological, and strategic (DOTS) themes, discussing the necessity of implementing practices that can combine these perspectives. Indeed, high-performer AI organizations link AI strategy to business outcomes, engage nontechnical employees in creating AI applications, and standardize processes and protocols (McKinsey 2022). In other words, the organizations witnessing the highest returns from AI follow strategy, data, technology, and talent best practices. The following section discusses the need for strategic planning at the organizational level, considering the DOTS perspectives.

10.2.2.3 Data Science and AI Challenges at the Organizational Level

Over the last decade, software-centric development has driven AI research by focusing on new algorithms. However, the Machine Learning Operations (MLOps) technologies have shifted the attention to data-centric AI research and high-quality datasets (Ng 2021). The MLOps technologies furthermore handle continuous delivery mechanisms since the production models are usually temporary due to the data skews (Google 2020). Therefore, the recommendation is that organizations develop longer-term plans for data management considering the quality and quantity of the data, deployment, and maintenance (Ng 2021).

Alongside the MLOps technologies, many distributed technologies have emerged to process and store big data (Gökalp et al. 2019). Massive technical debts are typical for ML systems (Sculley et al. 2015) due to the fast-evolving and many open-source tools and platforms and these distributed technologies' selection, configuration, and integration (Gokalp et al. 2016; Nazliel et al. 2022). Strategic planning is necessary to minimize the technical debt using technology assessment and forecasting for distributed technologies considering data-related assets, the data science life cycle, technological and industrial trends, and scenarios (Kayabay et al. 2022a).

In addition to the technical challenges, addressing the organizational and strategic issues associated with data-disadvantaged and underperforming sectors (Davenport 2014) is essential for long-term success in data-driven decision-making. The success factors include the data-driven culture promoting openness, trust, lean principles, and effective data management policies (Kayabay et al. 2022a). Business units still need to understand the greater scope and how their objectives, resources, and data services integrate (Kayabay et al. 2022a,b). DSR can support such scenarios by communicating the relationships between strategy, data, and technology perspectives while eliminating information asymmetries and creating a shared understanding in the organization.

10.2.2.4 Data Science and AI Challenges Beyond the Organizational Level

We witness the escalating necessity for coordination and shared orientation for data and AI management at the supply chain, sectoral, research, national, and multinational levels (Peretz-Andersson et al. 2021; Schmitz et al. 2020). The computational power to build the frontier models exists mainly in the private sector actors in a few countries (Langston 2023; Lee and Sengupta 2023; Strier et al. 2023). Consequently, a significant disparity emerges between the industry and academia concerning the implementation of large-scale AI (McKinsey 2022). The private sector actors consistently feel competitive pressure to push AI models into production as quickly as possible. They train and deploy ML models using proprietary practices and private datasets without extensive testing and auditing

(Bommasani et al. 2021). However, in the end, these models have substantial societal and environmental impacts.

Many nations have recognized the significance of the matter and responded by publishing national data and AI strategies (Galindo et al. 2021; HM Government 2021; International Data Spaces Association 2020). In the roadmapping literature, developing the roadmap is more valuable than the result, the strategy document (Phaal et al. 2004a). Therefore, we still need well-structured processes that bring together chip designers, hardware manufacturers, data and software ecosystems, funding mechanisms, policymakers, and lawmakers to communicate the requirements, eliminate information asymmetries, and adjust knowledge flows. In summary, technical debt is one of many issues that will trouble data and AI professionals, absent well-coordinated actions that link the political, societal, and technological perspectives. AI systems are already causing harm, such as facial recognition systems that discriminate based on race and clinical health tools that bias socioeconomic and racial factors (Zhang et al. 2022). The absence of a collective vision and the inability to integrate forecasting into data analysis while considering technical, political, and social factors may lead to increased unemployment, inequality, and political power imbalance (Acemoglu and Restrepo 2020). Since TRM is valuable for coordination and shared orientation at network, sectoral, and national levels (Moehrle et al. 2013), DSR can be an essential tool for data-centric AI.

10.2.2.5 Opportunities for Using DSR

Notably, the data-disadvantaged and underachiever-type organizations have the tremendous potential to exploit data to enhance their decision-making processes and remain ahead of the curve. Kayabay et al. (2022a,b) emphasized the necessity of linking the data layers with the interconnected organizational, technological, and strategic perspectives in such organizations, customizing the DSR accordingly. The DSR can also help technologists, managers, and executives in many more scenarios to resolve the challenges at and beyond the organizational level. For example, a research project spanning multiple organizations can benefit from roadmapping beyond the organizational level to integrate data layers in product-technology roadmaps, for example, to plan the development of next-generation materials (Agrawal and Choudhary 2019).

The European Union (EU) has been trying to counter the data-driven business models of hyper-scale cloud platforms with constantly renewed regulations for competition and innovation (Gawer and Srnicek 2021). Furthermore, the EU recognizes data among most strategic resources and aims to establish data spaces for data exchange among firms, research centers, and other entities (International Data Spaces Association 2020). Still, many strategy and policy documents describing implementations skip the processes to align many perspectives reliably (OECD 2021). The DSR can provide structured processes to unite multiple stakeholders to realize ambitious goals, bearing critical roles in the EU's economic and social frameworks. Policy researchers might examine innovation activities, challenges, and enablers' interdependencies and impact through a TRM integrating data layers (Park et al. 2020). However, the literature needs studies showing the advantages of integrating the data layers into the roadmapping architecture and identifying the strategic tools that support the DSR process beyond the organizational level.

10.2.3 Related Studies

Organizations' various business contexts, objectives, and resource constraints influence roadmap form and implementation. As a result, adjustments to roadmap architecture and roadmapping process are frequently mandatory to customize the roadmap to the organization's unique objectives and goals (Pearson et al. 2020; Phaal et al. 2004b). A few studies investigate custom roadmapping that considers data layers part of the roadmap architecture, although data has become an indispensable asset for AI. Han and Geum (Han and Geum 2020) introduced data-integrated smart service roadmap typologies. In the proposed roadmap architectures, the data layer can be a supporter, mediator, or value generator depending on how data works in the smart service, considering the interactions between data, service, and customer. Han and Jeon (2023) incorporated the data management perspective in the know-how layers for platforms to achieve sustainable growth by strengthening the network effects. Kayabay et al. demonstrated DSR applications in a research organization (Kayabay et al. 2022a), and an oil and gas company (Kayabay et al. 2022b), considering data layers as part of the roadmap architecture and customizing the process accordingly. We could find only one ML roadmapping study at the network level (Uchihira 2021), which does not consider data layers as strategic planning elements. However, many networked companies can benefit from strategically planning and aligning data assets. The literature still needs case studies beyond the organizational level to evaluate and align AI systems with interconnected data-related, political, technological, and societal viewpoints. The following section describes the case study application to alleviate this gap.

10.3 A Case Study of Networked Organizations' Strategic Planning for AI

The pseudo case study concerns the coordination of networked companies, which belong to the same parent company. The companies operate in the finance, energy, consumer durables, and automotive sectors. They want to stay competitive in the data-centric and large-scale AI era by building a joint data center and data lakehouse to share data, build higher-performing and state-of-the-art models with integrated data sources, and build foundational models (Figure 10.1). The companies will achieve higher global efficiency by combining data-based knowledge flows and creating horizontal alignment across the networked organizations. We selected this case study because of one author's experience in an identical initiative.

Considering the three tenets put forward by Phaal et al. (2006), the case study configures the hybrid DSR approach. Even though the recommended process models for DSR are applicable at the organizational level, they can still help beyond that with minor modifications. We follow the recommended process models by Kayabay et al. (2022a), modifying them as necessary throughout the process presented in the subsequent headings to transfer it beyond the organizational level toward generalization through the case study application. We adhere to the principles (Kerr et al. 2017; Kerr et al. 2013) for developing industrially relevant toolkits to achieve data science and AI strategic planning goals.

10.3.1 Kickoff and Planning

The geographically distributed networked companies want to use roadmapping as a coordination and planning tool to build a joint data center. The company representatives meet face-to-face to kick off the roadmapping initiative. However, the responsible teams will carry out the initiative online, sometimes meeting face-to-face to ensure trust. The roadmapping teams will follow the recommended visual process models as a guide throughout the initiative (Kayabay et al. 2022a). The scope includes all four layers, but the organizational data science maturity might be a resource rather than a strategic goal since it would support the data quality ingested into the data center, directly affecting the quality of the ML models. Keeping the scope as small as possible is essential, so assessing models regarding upcoming regulations can wait until this roadmapping initiative is complete (See Section 10.5).

As the industry shifts to large-scale AI, firms need to collaborate to build frontier models with the computing capability for competition. Complementary data assets provide combining forces with a competitive edge. Horizontal integration of data and knowledge flows requires trust and indicates varied data quality. Thus, in the short term, companies should establish communication channels, build trust, discover, and prioritize data sources, close data quality gaps between organizations, compare technologies considering the landscape, understand models that can benefit from integrated data sources, experiment with large models, and agree on pathways forward. Mid-term, they intend to evaluate models based on anticipated legislation and situations. They intend to collaborate long-term to lead the large-scale, data-centric AI era.

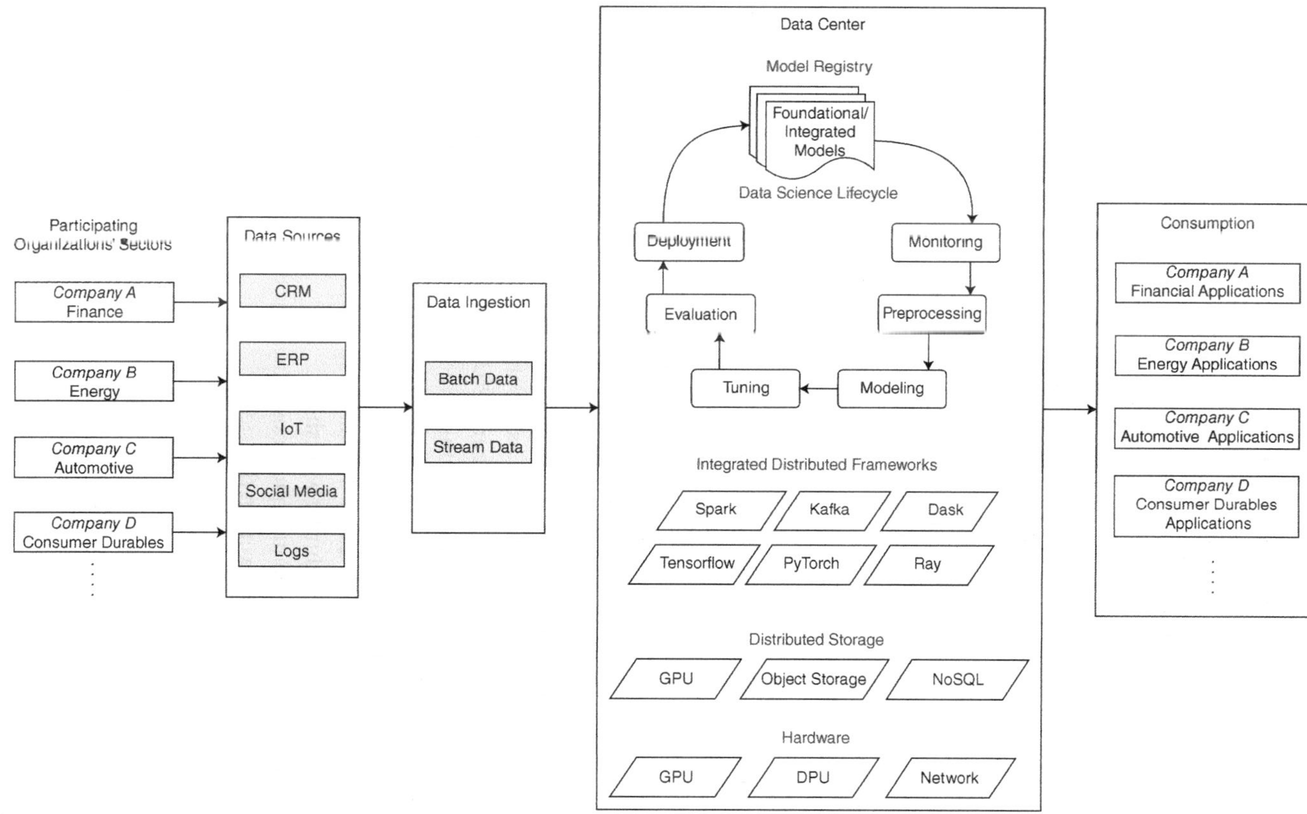

Figure 10.1 Joint data center architecture.

The planning activity prepares the roadmap landscape, considering the outputs of the kickoff activity, followed by four significant activities (Kayabay et al. 2022a): discussing the internal and external data sources, determining stakeholders, selecting stakeholders and teams, and deciding the tools to support roadmapping. While layer mapping can help prepare the roadmap landscape, decision-making approaches can guide team formation to optimize individual differences across working groups (La Torre et al. 2021). Accordingly, the planning team starts with a simple roadmap architecture, tailoring the data science roadmap (Kayabay et al. 2022a). While the market layer comprises only the AI and data science trends and data-driven applications, the value layer uses model cards and data cards to link the resources to the market layer (Gebru et al. 2021; Pascual 2023b; Pushkarna and Zaldivar 2023). After preparing the roadmap architecture (Figure 10.2), the facilitators select the tools that support the remaining process based on the outcomes of this chapter. We cover the tools in the subsequent headings.

10.3.2 Pre-workshop

The pre-workshop aims to pre-populate the roadmap landscape to increase the efficiency of participants' valuable time spent in workshops. Data-driven approaches in the scientific literature suit this task by identifying promising trends and technologies and mapping them to the respective layers on the roadmap. For example, Mülhort and Grottke (2022) developed a data mining model to identify emerging technologies using scientific articles and patents. They can schematically categorize the technologies into four quadrants based on growth (declining vs. increasing) and emergence (emerging vs. popular), recommending actions (i.e. evaluate, establish, explore, and exit) depending on the technology category. Kim and Geum (2021) use network analysis to detect meaningful trends and apply opportunity findings to predict the links between the topics. Furthermore, decision-making approaches can enable prioritization of the improvement areas, pointing to candidate roadmapping components during this stage (Zarrin and Daim 2019). The roadmap can incorporate such findings to be validated by the experts during workshops.

We extend the fundamental questions (Kayabay et al. 2022a) for the pre-workshop team to seek answers in this step:

1. Can the internal and external data assets discussed in the previous planning step provide business value if integrated or use more computational power?
2. Are there unforeseen opportunities created by the additional computational power and integrated data assets?
3. What are the data collection and governance maturities across participating organizations? The team can see if model cards already accompany the models in some organizations.

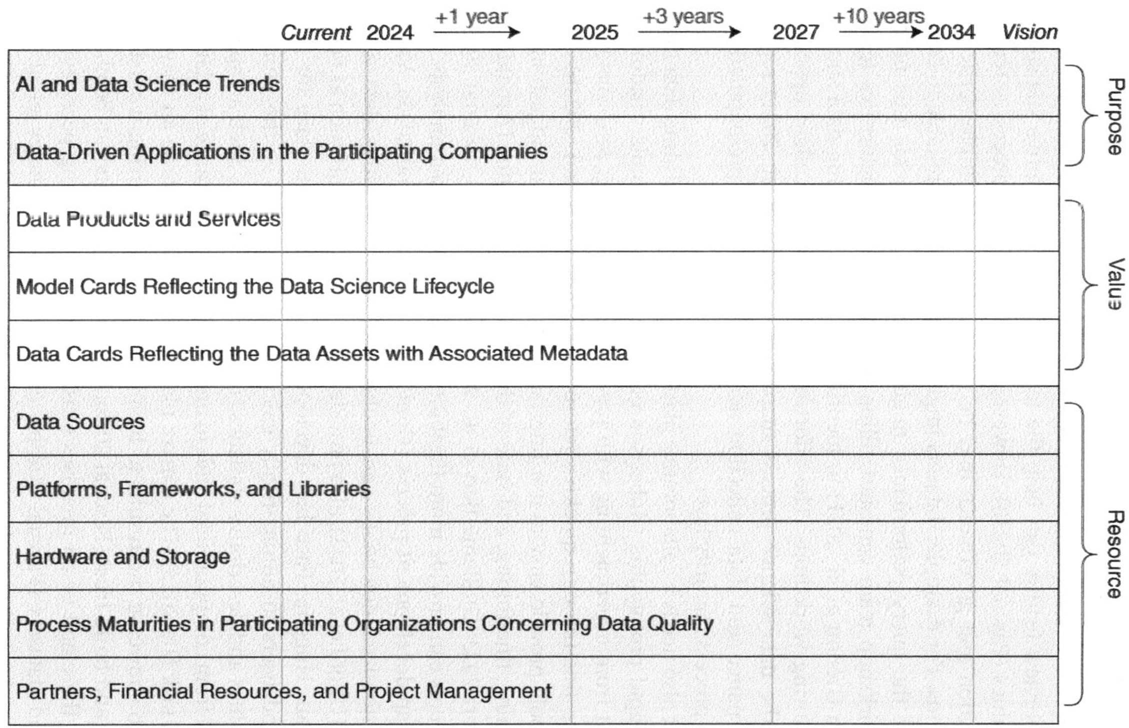

Figure 10.2 Example roadmap architecture to use beyond the organizational level based on (Adapted from [1]).

The model cards improve model discoverability (Pascual 2023b) and help understand the business problems, the linked data assets, and the performance bottlenecks such as lack of computational capacity. On the other hand, data cards (Gebru et al. 2021; Pascual 2023b; Pushkarna and Zaldivar 2023) create an excellent opportunity to discover and understand the existing datasets, enabling metadata insertion. A simple and intuitive data explorer interface with a partial preview feature can help the team calculate descriptive statistics and missing values (Pascual 2023b). However, the data science processes regarding data collection and governance may be different across multiple participating organizations. For example, datasets might need to be more reliable to consider quickly, or the query performance might need help. To tackle this problem, the pre-workshop team can start filling out the data characteristics template (Blackburn-Grenon et al. 2021) to identify the most promising data sources to consider during workshops. The effort also provides an excellent opportunity to assess the data science process maturities (Gökalp et al. 2021) regarding data governance and data collection across organizations, pre-populating the roadmap's process maturity layer.

The DSR also needs a collaborative platform supporting real-time data integration and rapid prototyping (Robertson et al. 2021). For example, roadmapping can benefit from experimenting with pre-trained models (e.g. zero- or few-shot prompting, fine-tuning) quickly before workshops (Pascual 2023b). While no-code tools provide easy-to-use interfaces for real-time data ingestion, automatic training, fine-tuning, and state-of-the-art model evaluation (Arnaldo 2021; Gonfalonieri 2023; Guillame-Bert et al. 2023; Kulkarni 2023; Pascual 2023a,b), low-code tools can provide more flexibility, such as event-based analysis, during the roadmap development process (Gonfalonieri 2023; Pascual 2023a; Vaisburd 2023). Customized notebooks can enable simple and intuitive data explorer interfaces (Ufford et al. 2023), also provided by well-known tools such as PowerBI and Tableau. Hugging Face has developed an interactive interface for dataset creators and users to automatically calculate metrics, such as the number of missing values and descriptive statistics (Luccioni et al. 2023). Such an approach can benefit from computational tools to understand the importance of a data point for an ML task (Jia 2023). The correct platform should also have an easy-to-use interface to showcase the results to business stakeholders during workshops to get buy-in and build trust (Pascual 2023a,b). Such instruments can be platforms hosted on the cloud or on-premises (Pascual 2023b), enabling peer reviews of models and datasets (Pascual 2023b), experimentation (Tingley et al. 2023), and safe notebook sharing between participants (Ufford et al. 2023). However, the pre-workshop team has to consider the computational power required for the next roadmapping steps with the data management regulations and requirements.

10.3.3 Workshops

The DSR process involves four workshops: Strategy, data, technology, and organization, corresponding to the DSR architecture. Since the bottom layers in the roadmap architecture are slightly different beyond the organizational level, we will rename the organization workshop as a consortium workshop. Each workshop can involve iterative large-group discussions and small-group deep dives. The workshop teams can choose the most suitable tool depending on the group size or apply the same tool differently. For example, while validating the outputs of data-driven approaches, Nazarenko et al. (2021) use inputs from internal and external experts during roadmapping development and validation through surveys, in-depth interviews, and expert panels. Zhang et al. (2016) use separate interviews and a two-round workshop for the modification of historical results and forecasting trends.

In addition to validating and modifying results, experts produce valuable data using expert-based approaches. Blackburn-Grenon et al. (2021) use the dot-voting method to select the most critical pain points and the Ishikawa tool to facilitate root cause analysis. The authors also use Business Process Modeling Notation (BPMN) to develop and identify high-level business processes, which might be more suitable for pre-workshops in the DSR. In another roadmapping study concerning ML applications (Uchihira 2021), experts use matching tables based on the outputs of pre-workshop activity to extract noteworthy market trends, issues, needs, technology seeds, and policies. Skulimowski (2019) combines cloud-based Delphi, Political-Economic-Social-Technological (PEST) analysis, and Strength-Weaknesses-Opportunities-Threats (SWOT) analysis, applying them in multiple stages to construct and filter scenarios and determine the relevance of risk factors hindering technology development. Dabab et al. (2018) employ Quality Function Deployment (QFD) to reposition qualitative expert judgments into quantitative parameters.

On the other hand, the decision-making and experiment-based approaches primarily support small-group deep dives. The decision-making approaches use expert inputs and structure the process of ranking options and determining the best alternative for complex problems. For example, Barham and Daim (2020) use Hierarchical Decision Modeling (HDM) to rank the factors influencing big data projects to ensure readiness to start such a project. Nazliel et al. (2022) combine Analytic Hierarchy Process (AHP) and Delphi to select technologies to evaluate data science tools and choose the most appropriate one for project needs. The scientific literature seldom uses experiment-based approaches. Among our findings, Kayabay et al. (2022a) experiment with open-source tools (e.g., Huginn) to collect data from agreed data sources while developing the roadmap, and Robertson et al. (2021) leverage the cloud to implement technologies promoting rapid experimentation and collaboration to enable partnerships with the private sector, academia, and global allies, as required by US government's national AI strategy. In the gray literature, we find numerous approaches to support small group deep dives during workshops, discussed in the pre-workshop heading. However, the roadmapping teams need considerable time and effort to clarify all the points. The teams can agree on open issues published with the roadmap to increase roadmapping agility (Kayabay et al. 2022a).

Therefore, the participating organizations gather together in four consecutive strategy-data-technology-consortium workshops. In the strategy workshop, they perform a SWOT analysis using data-driven, maturity-based, and experimentation results, list and prioritize the operational pain points using dot-voting, and use the Ishikawa tool to analyze the root causes. The data workshop links the data products and services to the strategy layers, building proof-of-concept models accompanied by model cards. They also forecast new models and data cards based on available data assets and sources. The technology workshop bridges the gaps based on the forecasts in model and data layers. They must select the best technologies but need more information regarding differentiating functional and nonfunctional requirements. Therefore, they put high-level landmarks to bridge the gaps and create an open issue regarding clarification of the landmarks (Section 10.5). Lastly, they discuss the pathways to bridge the gaps for processes concerning data quality according to dataset requirements and identify the partners that can help with the upper layers in the consortium workshop.

10.3.4 Post-workshop and Final Review

The workshops produce lots of data to construct a proper roadmap in the post-workshop stage. The post-workshop team synthesizes roadmap drafts based on the available data and iterates using stakeholder feedback. A notable gap in the literature is that the time required from stakeholders can be a substantial bottleneck (Uchihira 2021), magnified by multiple companies participating in the process. A potential solution is identifying and filtering the paths between layers in the roadmap architecture (Kayabay et al. 2022b). Filtering allows isolated stories to be confirmed post-workshop with dispersed stakeholders with limited time. Published roadmaps with open problems boost roadmap creation agility.

It takes time for participating experts to find availability in their busy schedules. However, the post-workshop team constructs a limited number of narratives by connecting the linked roadmap items and prioritizing them according to criticality and coherence. The review committee gathers to agree on publishing the roadmap and reflect on the process, considering the actual outcomes, success, and impact. Standard questions help facilitate the reflection activity (Kayabay et al. 2022a). The answers will be critical for the next roadmapping iteration.

The roadmapping team pitches the refined narratives in the final review meeting, asking the review committee to reflect on the initiative's success. They are delighted to see the broader scope, connections, and how to bridge the gaps; however, they worry about execution and keeping the roadmap alive.

10.3.5 Keeping Roadmapping Alive to Implement Data and AI Governance

The review committee must determine the people responsible for monitoring the progress on operationalizing the roadmap and resolving agreed open issues by the next iteration. Moreover, tight integration between the system should keep the data and model cards on the collaboration platform up-to-date. Then, data-driven approaches can use such data sources to determine whether to update and review the roadmap (Pora et al. 2022). Until then, the time between roadmapping iterations enables more sophisticated and time-consuming approaches. The examples include using decision-making approaches to select the technologies that best meet the requirements based on differentiating features (Nazliel et al. 2022) and applying maturity models (Fukas 2022; Gökalp et al. 2021) thoroughly.

Meta argues making AI more explainable requires cross-industry and cross-disciplinary dialogue between companies, regulators, and academics using simple, easy, and digestible tools (Green et al. 2023). Therefore, a collaboration platform enables version control for datasets, ML models, and accompanying data and model cards (Pascual 2023b). Model and data cards can benefit from standards assuring readability across disciplines (Solaiman 2023). During roadmapping, the dataset creators can follow a standard checklist for identified datasets as part of the pre-workshop to answer questions related to motivation for creating the dataset, data composition, data collection process, and so on (Gebru et al. 2021; Pushkarna and Zaldivar 2023) A standard checklist for model cards can target answering questions such as the problem definition, the model's performance metrics, the model's inputs and outputs, and circumstances for the model to perform well and poorly (Richards et al. 2020).

Organization for Economic Co-operation and Development (OECD) AI principles and the EU's ongoing regulatory proposals aim to engrain responsibility into developing, releasing, and using AI systems to reduce inequalities, such as socioeconomic and gender (Ferrandis 2022). However, applying the principles and regulations is challenging, considering each technical and operational reality of businesses (Primmer and Andrade 2023). Responsible AI licenses can complement tools like model cards, sending use restrictions down the value chain (Ferrandis 2022) and addressing the challenge. Singapore launched a voluntary testing framework introducing technical and process checks to test the AI system against the developer's claims and internationally accepted principles

(Kin 2023). AI classification frameworks can establish a common language between stakeholders, providing foundations for standards, benchmarking, auditing, certification, regulations, and impact assessments (Curtis and Miailhe 2023). OECD also provides a framework to compare the tools for implementing trustworthy AI systems (OECD 2021), which the facilitator team can benefit from while selecting the tools supporting DSR during the planning stage.

The work on the joint data center is still ongoing. Nevertheless, the collaboration platform enables the monitoring team to inspect the roadmap's new data and model cards. They check the data and model cards against the standards and evaluate the model cards against the responsible AI licenses on the collaboration platform. The automated system can create an alarm to update the roadmap before the date approved by the review committee. It would be a minor inconvenience since the team gets ready for the next iteration by considering the visual roadmapping process models against reflections from the previous iteration. If they finish evaluating the data science maturities in participating businesses, they will have much more data to consider in the next iteration.

10.4 Discussion

This chapter attempts to satisfy the most prominent research gap in academic literature that undertakes roadmapping in the data science and AI context, which has implications for research and practice. From the former perspective, more case studies and applications are necessary to improve the generalizability of the proposed solutions. However, accessing the required environments and conducting action research-based roadmapping studies are challenging, and practitioners' time spent in workshop-based roadmapping initiatives is precious. Furthermore, the data science landscape is changing fast. Therefore, we must identify the gaps and methods to enhance new studies' contribution potential, agility, and generalizability. Although the practice has mainly driven the development of roadmapping (Park et al. 2020; Phaal and Muller 2009), a limited number of studies give details on the adaptations and any improvements based on feedback from the actual implementation (Kerr et al. 2019). So, such elements improve the contribution potential of new studies. Visual process models, such as BPMN, can help get feedback during the design process, decide and apply agreed revisions, and catch all the comments and modifications (Kayabay et al. 2022a), strengthening the customizability and generalizability of the research.

From the practical perspective, the AI alignment topic (Leike et al. 2023) attracts the industry's attention toward AI systems aligned with human values. The industry undertakes the issue as a technical problem, missing the socio-technical standpoint. Meanwhile, policymakers focus on developing individual tools and publishing roadmap documents, neglecting how these tools come together in iterative, continuous, and possibly transparent processes. The regulators must set a human-centric and trust-oriented vision and execute well-coordinated actions, bringing together multiple stakeholders to evaluate and align AI systems with interconnected data-related, political, technological, and societal viewpoints. One of the most critical components is the culture that promotes experimentation and learning, democratizing creative contributions by opening up data-driven suggestions regardless of seniority and expertise (Tingley et al. 2023).

Technology management research tools may help practitioners implement such approaches. First, DSR, which customizes TRM, may be a transparent platform that democratizes innovative, data-driven, and well-aligned contributions regardless of seniority and experience. This method may imitate Netflix's learning and experimenting culture (Tingley et al. 2023) and ensure alignment. The DSR uses experimentation-based methods to benefit from developing industrial practices. It may easily include model and data cards as landmarks in the roadmap design since emerging best practices recommend producing model cards as early as the business understanding and problem-framing phases of the data science workflow and upgrading them. Model cards with responsible AI licenses and classification frameworks may be used to assess AI systems between workshops. Policymakers may evaluate AI systems and their implementers using maturity-based techniques beyond the corporate level.

10.5 Conclusion

Strategic planning in the dynamic and constantly changing AI and data science landscape has numerous challenges and uncertainties. Integrating data layers into agile roadmapping is crucial for businesses to overcome challenges and stay competitive by effectively planning data and ML assets. The DSR is an emerging strategic and agile planning framework incorporating data science, AI, and technology management. We envision it as a platform that democratizes and aligns data-driven contributions from everyone, regardless of seniority and experience, ensuring innovation and strategic thinking are ingrained within an organization, steering it toward success and prosperity. The results of this study indicate it is also critical to integrate data layers into roadmapping beyond the organizational level in many fields. The time has come to set a shared vision and design processes that continuously evaluate and align AI systems, considering human values and incorporating forecasting into connected data, technological, political, and societal angles. DSR can be the central hub that can unleash the power of existing instruments and best practices by intelligently combining them in an industrially relevant and strategic toolkit.

This study has several limitations. First, even though the pseudo case study of the networked companies is a first baby step for DSR beyond the organizational level, more studies should incorporate forecasting into data layers at the sectoral, national, and multinational levels, considering the societal and political perspectives. Second, the data science and AI landscape is changing very fast, and we could find even more strategic tools from the industry, especially from the generative AI domain, which has experienced numerous developments. Since AI and data science technologies change rapidly, causing massive technical debts, future research should develop dedicated instruments that facilitate technology evaluation by enabling faster comparison of the functional and nonfunctional requirements and considering the data science life cycle end-to-end and execution on the heterogeneous resources. Furthermore, the industry should consider the application of standard approaches to assess the organizations implementing AI systems and not just the AI systems, for example, encouraging continuous improvement in the small- to medium-sized enterprises (SMEs) regarding transparency, reliability, and sustainability to democratize AI system development.

References

Acemoglu, D. and Restrepo, P. (2020). The wrong kind of AI? Artificial intelligence and the future of labour demand. *Cambridge Journal of Regions, Economy and Society* 13 (1): 25–35. https://doi.org/10.1093/cjres/rsz022.

Agrawal, A. and Choudhary, A. (2019). Deep materials informatics: applications of deep learning in materials science. *MRS Communications* 9 (3): 779–792. https://doi.org/10.1557/mrc.2019.73.

Al-Ali, A.G. and Phaal, R. (2019). Design sprints for roadmapping an agile digital transformation. In: *2019 IEEE International Conference on Engineering, Technology and Innovation (ICE/ITMC)*, 1–6. IEEE. https://doi.org/10.1109/ICE.2019.8792597.

Arnaldo, M. (2021). Intro to Rapidminer: A no-code development platform for data mining (with case study). *Analytics Vidhya* (October 4), https://www.analyticsvidhya.com/blog/2021/10/intro-to-rapidminer-a-no-code-development-platform-for-data-mining-with-case-study/ (accessed 11 May 2024).

Barham, H. and Daim, T. (2020). The use of readiness assessment for big data projects. *Sustainable Cities and Society* 60: https://doi.org/10.1016/j.scs.2020.102233.

Benbya, H., Davenport, T.H., and Pachidi, S. (2020). Artificial intelligence in organizations: current state and future opportunities. *MIS Quarterly Executive* 19 (4).

Blackburn-Grenon, F., Abran, A., Rioux, M., and Wong, T. (2021). A team-based workshop to capture organizational knowledge for Identifying AI proof-of-value projects. *IEEE Engineering Management Review* 49 (2): 181–195. https://doi.org/10.1109/EMR.2021.3063688.

Bommasani, R. et al. (2021). On the opportunities of risks of foundation models. *Arxis* (August 16), http://arxiv.org/abs/2108.07258 (accessed 11 May 2024).

Chutivongse, N. and Gerdsri, N. (2019). Creating an innovative organization: analytical approach to develop a strategic roadmap guiding organizational development. *Journal of Modelling in Management* 15 (1): 50–88. https://doi.org/10.1108/JM2-05-2018-0067.

Curtis, S. and Miailhe, N. (2023). Three takeaways from classifying AI systems designed to fight the COVID-19 pandemic. OECD.AI The AI Wonk. https://oecd.ai/en/wonk/three-takeaways-classifying-ai-systems-designed-to-fight-covid-19

Dabab, M., Craven, R., Barham, H., and Gibson, E. (2018). Exploratory strategic roadmapping framework for big privacy issues. In: *Portland International Conference on Management of Engineering and Technology: Managing Technological Entrepreneurship: The Engine for Economic Growth, Proceedings.* PICMET 2018 https://doi.org/10.23919/PICMET.2018.8481834.

Daim, T.U., Yoon, B.-S., Lindenberg, J. et al. (2018). Strategic roadmapping of robotics technologies for the power industry: a multicriteria technology assessment. *Technological Forecasting and Social Change* 131: 49–66. https://doi.org/10.1016/j.techfore.2017.06.006.

Davenport, T.H. (2014). *Big Data at Work: Dispelling the Myths, Uncovering the Opportunities.* Boston Massachusetts: Harvard Business Review Press.

Ferrandis, C.M. (2022). Responsible AI licenses: a practical tool for implementing the OECD principles for trustworthy AI. OECD.AI (October 21). https://oecd.ai/en/wonk/rails-licenses-trustworthy-ai (accessed 30 July 2023).

Freitas, J.S. et al. (2019). Adapting the roadmapping approach to science-intensive organizations: lessons from a drug development program for neglected diseases. *Journal of Engineering and Technology Management* 52: 3–15. https://doi.org/10.1016/j.jengtecman.2018.01.006.

Fukas, P. (2022). The management of artificial intelligence: developing a framework based on the artificial intelligence maturity principle. *CEUR Workshop Proceedings* 3139: 19–27. https://www.scopus.com/inward/record.uri?eid=2-s2.0-85130763237&partnerID=40&md5=0470ee8a2332eeaf51d2d7d78d596809.

Galindo. L., Perset, K. and Sheeka, F. (2021). An overview of national AI strategies and policies. https://doi.org/10.1787/c05140d9-en (accessed June 24, 2023).

Gawer, A. and Srnicek, N. (2021). Online platforms: Economic and societal effects, European Parliamentary Research Services. https://data.europa.eu/doi/10.2861/844602 (accessed July 6, 2023).

Gebru, T. et al. (2021). Datasheets for datasets. arXiv, (December 01). https://doi.org/10.48550/arXiv.1803.09010.

Geum, Y., Lee, H., Lee, Y., and Park, Y. (2015). Development of data-driven technology roadmap considering dependency: an ARM-based technology roadmapping. *Technological Forecasting and Social Change* 91: 264–279. https://doi.org/10.1016/j.techfore.2014.03.003.

Gokalp, M.O., Kayabay, K., Akyol, M.A. et al. (2016). Big data for Industry 4.0: A conceptual framework. *2016 International Conference on Computational Science and Computational Intelligence (CSCI)* 431–434. https://doi.org/10.1109/CSCI.2016.0088.

Gökalp, M.O., Kayabay, K., Zaki, M. et al. (2019). Open-source big data analytics architecture for businesses. *1st International Informatics and Software Engineering Conference (UBMYK)*, 1–6. https://doi.org/10.1109/UBMYK48245.2019.8965572.

Gökalp, M.O., Gökalp, E., Kayabay, K. et al. (2021). Data-driven manufacturing: an assessment model for data science maturity. *Journal of Manufacturing Systems* 60: 527–546. https://doi.org/10.1016/j.jmsy.2021.07.011.

Gonfalonieri, A. (2023). No-code/low-code AI: new business models and future of data scientists. *Medium* 30: https://towardsdatascience.com/no-code-low-code-ai-new-business-models-and-future-of-data-scientists-a536beb8d9e3.

Google (2020). MLOps: continuous delivery and automation pipelines in machine learning. https://cloud.google.com/solutions/machine-learning/mlops-continuous-delivery-and-automation-pipelines-in-machine-learning (accessed October 26, 2020).

Green, N., Procope, C., Cheema, A., and Adediji, A. (2023). System cards, a new resource for understanding how AI systems work. https://ai.facebook.com/blog/system-cards-a-new-resource-for-understanding-how-ai-systems-work/ (accessed July 30, 2023).

Guillame-Bert, M. et al. (2023). Introducing simple ML for sheets: a no-code machine learning add-on for Google Sheets. https://blog.tensorflow.org/2022/12/introducing-simple-ml-for-sheets.html (accessed July 30, 2023).

Han, M. and Geum, Y. (2020). Roadmapping for data: concept and typology of data-integrated smart-service roadmaps. *IEEE Transactions on Engineering Management* 1–13. https://doi.org/10.1109/TEM.2020.3013295.

Han, M. and Jeon, J. (2023). Roadmap incorporating data management perspective for platform business model innovation. *Sustainability* 15 (4) Art. no. 4: https://doi.org/10.3390/su15043151.

Hansen, C., Daim, T., Ernst, H., and Herstatt, C. (2016). The future of rail automation: a scenario-based technology roadmap for the rail automation market. *Technological Forecasting and Social Change* 110: 196–212. https://doi.org/10.1016/j.techfore.2015.12.017.

Heim, U. and Heuss, R. (2017). Building an integrated technology road map to drive successful innovation. McKinsey. https://www.mckinsey.com/capabilities/operations/our-insights/building-an-integrated-technology-road-map-to-drive-successful-innovation (accessed June 3, 2023).

Hirose, Y., Phaal, R., Farrukh, C. et al. (2022). Sustaining organizational roadmapping implementation–lessons learned from subsea 7. *Research-Technology Management* 65 (3): 50–57. https://doi.org/10.1080/08956308.2022.2048555.

HM Government (2021). National AI strategy. https://www.gov.uk/government/publications/national-ai-strategy

Ho, J.-Y. and O'Sullivan, E. (2017). Strategic standardisation of smart systems: a roadmapping process in support of innovation. *Technological Forecasting and Social Change* 115: 301–312. https://doi.org/10.1016/j.techfore.2016.04.014.

Ho, J.-Y. and O'Sullivan, E. (2020). Toward integrated innovation roadmapping: lessons from multiple functional roadmaps beyond technology R D. *IEEE Transactions on Engineering Management* 1–13. https://doi.org/10.1109/TEM.2020.3029783.

International Data Spaces Association (2020). Implementing the European strategy on data role of the international data spaces. https://internationaldataspaces.org/wp-content/uploads/IDSA-Position-Paper-implementing-european-data-strategy-role-of-IDS.pdf (accessed July 6, 2023).

Jia, R. (2023). What is my data worth? The Berkeley Artificial Intelligence Research Blog. http://bair.berkeley.edu/blog/2019/12/16/data-worth/ (accessed July 30, 2023).

Kayabay, K., Gökalp, M.O., Gökalp, E. et al. (2022a). Data science roadmapping: an architectural framework for facilitating transformation towards a data-driven organization. *Technological Forecasting and Social Change* 174: 121264. https://doi.org/10.1016/j.techfore.2021.121264.

Kayabay, K., Gökalp, M.O., Gökalp, E. et al. (2022b). Customizing and deploying data science roadmapping: a case study of an oil and gas company. In: *2022 IEEE International Conference on Technology Management, Operations and Decisions (ICTMOD)*, 1–6. IEEE https://doi.org/10.1109/ICTMOD55867.2022.10082858.

Kerr, C. and Phaal, R. (2019). Defining the scope of a roadmapping initiative: a checklist-based template for organizational stakeholders. In: *2019 Portland International Conference on Management of Engineering and Technology (PICMET)*, 1–10. https://doi.org/10.23919/PICMET.2019.8893851.

Kerr, C., Farrukh, C., Phaal, R., and Probert, D. (2013). Key principles for developing industrially relevant strategic technology management toolkits. *Technological Forecasting and Social Change* 80 (6): 1050–1070. https://doi.org/10.1016/j.techfore.2012.09.006.

Kerr., C., Phaal, R., and Thams, K. (2017). Roadmapping as a Platform for developing management toolkits: a collaborative design approach with the LEGO. In: *2017 Portland International Conference on Management of Engineering and Technology (PICMET)*, 1–11. https://doi.org/10.23919/PICMET.2017.8125436.

Kerr, C., Phaal, R., and Thams, K. (2019). Customising and deploying roadmapping in an organisational setting: The LEGO Group experience. *Journal of Engineering and Technology Management* 52: 48–60. https://doi.org/10.1016/j.jengtecman.2017.10.003.

Kim, J. and Geum, Y. (2021). How to develop data-driven technology roadmaps: the integration of topic modeling and link prediction. *Technological Forecasting and Social Change* 171: https://doi.org/10.1016/j.techfore.2021.120972.

Kin. Y.Z. (2023). Singapore's A.I. verify builds trust through transparency. OECD.AI The AI Wonk. https://oecd.ai/en/wonk/singapore-ai-verify (accessed July 30, 2023).

Kulkarni, A. (2023). Understand text mining using no code tool orange. Analytics Vidhya. https://www.analyticsvidhya.com/blog/2022/05/understand-text-mining-using-no-code-tool-orange/ (accessed July 30, 2023).

La Torre, D., Colapinto, C., Durosini, I., and Triberti, S. (2021). Team formation for human-artificial intelligence collaboration in the workplace: a goal programming model to foster organizational change. *IEEE Transactions on Engineering Management* 1–11. https://doi.org/10.1109/TEM.2021.3077195.

Langston, J. (2023). Microsoft announces new supercomputer, lays out vision for future AI work. https://news.microsoft.com/source/features/ai/openai-azure-supercomputer/ (accessed June 18, 2023).

Lee, K. and Sengupta, S. (2023). Introducing the AI research supercluster – Meta's cutting-edge AI supercomputer for AI research. https://ai.facebook.com/blog/ai-rsc/ (accessed June 18, 2023).

Leike, J., Schulman, J., and Wu, J. (2023). Our approach to alignment research. https://openai.com/blog/our-approach-to-alignment-research

Luccioni, S., Jernite, Y., and Mitchell, M. (2023). Introducing the data measurement tool: an interactive tool for looking at datasets. https://huggingface.co/blog/data-measurements-tool (accessed July 30, 2023).

McKinsey. (2022). The state of AI in 2022 – and a half decade in review. https://www.mckinsey.com/capabilities/quantumblack/our-insights/the-state-of-ai-in-2022-and-a-half-decade-in-review

Mikalef, P. and Gupta, M. (2021). Artificial intelligence capability: conceptualization, measurement calibration, and empirical study on its impact on organizational creativity and firm performance. *Information & Management* 58 (3): 103434. https://doi.org/10.1016/j.im.2021.103434.

Moehrle, M.G., Isenmann, R., and Phaal, R. (2013). Basics of technology roadmapping. In: *Technology Roadmapping for Strategy and Innovation* (ed. M.G. Moehrle, R. Isenmann, and R. Phaal), 1–10. Berlin, Heidelberg: Springer.

Muhlroth, C. and Grottke, M. (2022). Artificial intelligence in innovation: how to spot emerging trends and technologies. *IEEE Transactions on Engineering Management* 69 (2): 493–510. https://doi.org/10.1109/TEM.2020.2989214.

Nazarenko, A., Vishnevskiy, K., Meissner, D., and Daim, T. (2021). Applying digital technologies in technology roadmapping to overcome individual biased assessments. *Technovation* 102364. https://doi.org/10.1016/j.technovation.2021.102364.

Nazliel, K., Kayabay, K., Gokalp M.O. et al. (2022). Data science technology section: development of a decision-making approach. In: *Presented at the 2022 IEEE Technology and Engineering Management Conference: Societal Challenges: Technology, Transitions and Resilience Virtual Conference*, 172–178. TEMSCON EUROPE https://doi.org/10.1109/TEMSCONEUROPE54743.2022.9802054.

Ng, A. (2021). AI doesn't have to be too complicated or expensive for your business. *Harvard Business Review* (July 29), https://hbr.org/2021/07/ai-doesnt-have-to-be-too-complicated-or-expensive-for-your-business (accessed August 19, 2021).

Nimmo, G. (2013). Technology roadmapping on the industry level: experiences from Canada. In: *Technology Roadmapping for Strategy and Innovation* (ed. M.G. Moehrle, R. Isenmann, and R. Phaal), 47–65. Springer.

OECD (2021). *Tools for Trustworthy AI: A Framework to Compare Implementation Tools for Trustworthy AI Systems*. Paris: OECD https://doi.org/10.1787/008232ec-en.

de Oliveira, M.G., Routley, M., and Phaal, R. (2022). The digitalisation of roadmapping workshops. *Journal of Engineering and Technology Management* 65: 101694. https://doi.org/10.1016/j.jengtecman.2022.101694.

O'Sullivan, E. and Phaal, R. (2021). Agile roadmapping: an adaptive approach to technology foresight https://doi.org/10.17863/CAM.63366.

Park, H., Phaal, R., Ho, J.-Y., and O'Sullivan, E. (2020). Twenty years of technology and strategic roadmapping research: a school of thought perspective. *Technological Forecasting and Social Change* 154: 119965. https://doi.org/10.1016/j.techfore.2020.119965.

Pascual, F. (2023a). Getting started with sentiment analysis on Twitter. https://huggingface.co/blog/sentiment-analysis-twitter (accessed July 30, 2023).

Pascual, F. (2023b). Introducing the private hub: a new way to build with machine learning. https://huggingface.co/blog/introducing-private-hub (accessed July 30, 2023).

Pearson, R.J., Costley, A.E., Phaal, R., and Nuttall, W.J. (2020). Technology Roadmapping for mission-led agile hardware development: a case study of a commercial fusion energy start-up. *Technological Forecasting and Social Change* 158: 120064. https://doi.org/10.1016/j.techfore.2020.120064.

Peretz-Andersson, E., Lavesson, N., Bifet, A., and Mikalef, P. (2021). AI transformation in the public sector: ongoing research. In: *2021 Swedish Artificial Intelligence Society Workshop (SAIS)*, 1–4. SAIS https://doi.org/10.1109/SAIS53221.2021.9483960.

Petrick, I.J. (2013). Networked innovation: Using roadmapping to facilitate coordination, collaboration and cooperation. In: *Technology Roadmapping for Strategy and Innovation* (ed. M.G. Moehrle, R. Isenmann, and R. Phaal), 31–46. Springer.

Phaal, R. and Muller, G. (2009). An architectural framework for roadmapping: Towards visual strategy. *Technological Forecasting and Social Change* 76 (1): 39–49. https://doi.org/10.1016/j.techfore.2008.03.018.

Phaal, R., Farrukh, C.J.P., and Probert, D.R. (2004a). Technology roadmapping – A planning framework for evolution and revolution. *Technological Forecasting and Social Change* 71 (1): 5–26. https://doi.org/10.1016/S0040-1625(03)00072-6.

Phaal, R., Farrukh, C., and Probert, D. (2004b). Customizing roadmapping. *Research-Technology Management* 47 (2): 26–37.

Phaal, R., Farrukh, C., and Probert, D. (2006). Technology management tools: generalization, integration and configuration. *International Journal of Innovation and Technology Management* 03 (03): 321–339. https://doi.org/10.1142/S021987700600082X.

Phaal, R., Farrukh, C.J.P., and Probert, D.R. (2007). Strategic roadmapping: a workshop-based approach for identifying and exploring strategic issues and opportunities. *Engineering Management Journal* 19 (1): 3–12. https://doi.org/10.1080/10429247.2007.11431716.

Phaal, R., Farrukh, C., and Probert, D.R. (2013). Fast-start roadmapping workshop approaches. In: *Technology Roadmapping for Strategy and Innovation* (ed. M.G. Moehrle, R. Isenmann, and R. Phaal), 91–106. Springer.

Pora, U., Gerdsri, N., Thawesaengskulthai, N., and Triukose, S. (2020). Data-driven roadmapping (DDRM): approach and case demonstration. *IEEE Transactions on Engineering Management* 1–19. https://doi.org/10.1109/TEM.2020.3005341.

Pora, U., Thawesaengskulthai, N., Gerdsri, N., and Triukose, S. (2022). Data-driven roadmapping (DDRM): approach and case demonstration. *IEEE Transactions on Engineering Management* 69 (1): 209–227.

Power, D.J. (2014). Using 'Big Data' for analytics and decision support. *Journal of Decision Systems* 23 (2): 222–228. https://doi.org/10.1080/12460125.2014.888848.

Primmer, N. and Andrade, N. (2023). "Businesses are applying the OECD AI Principles. How is it going?," OECD.AI The AI Wonk [Online]. https://oecd.ai/en/wonk/businesses-applying-oecd-ai-principles (accessed 30 July 2023).

Pushkarna, M. and Zaldivar, A. (2023). The data cards playbook: a toolkit for transparency in dataset documentation. https://ai.googleblog.com/2022/11/the-data-cards-playbook-toolkit-for.html (accessed June 30, 2023).

Richards, J., Piorkowski, D., Hind, M. et al. (2020). A methodology for creating AI factSheets. arXiv. https://doi.org/10.48550/arXiv.2006.13796.

Robertson, J., Fossaceca, J., and Bennett, K. (2021). A cloud-based computing framework for artificial intelligence innovation in support of multidomain operations. *IEEE Transactions on Engineering Management* https://doi.org/10.1109/TEM.2021.3088382.

Sarker, I.H. (2021). Data science and analytics: an overview from data-driven smart computing, decision-making and applications perspective. *SN Computer Science* 2 (5): 377. https://doi.org/10.1007/s42979-021-00765-8.

Schmitz, P., Brunson, D., Mizumoto, C. et al. (2020). Towards a national best practices resource for research computing and data strategic planning. *Zenodo* https://doi.org/10.5281/ZENODO.4344057.

Sculley, D. et al. (2015). Hidden technical debt in machine learning systems. *Advances in Neural Information Processing Systems* 2503–2511. https://proceedings.neurips.cc/paper_files/paper/2015/file/86df7dcfd896fcaf2674f757a2463eba-Paper.pdf.

Sevilla, J., Heim, L., Ho, A. et al. (2022). Compute trends across three eras of machine learning. *arXiv* (Mar. 9): https://doi.org/10.48550/arXiv.2202.05924.

Simonse, L.W.L., Hultink, E.J., and Buijs, J.A. (2015). Innovation roadmapping: building concepts from practitioners' insights. *Journal of Product Innovation Management* 32 (6): 904–924. https://doi.org/10.1111/jpim.12208.

Skulimowski, A. (2019). Forward-looking activities supporting technological planning of AI-based learning platforms. In: *Presented at the Advances in Web-Based Learning* (ed. M. Herzog, Z. Kubincova, P. Han, and M. Temperini), 274–284. ICWL https://doi.org/10.1007/978-3-030-35758-0_26.

Solaiman, I. (2023). Comments on US National AI Research Resource Interim Report. https://huggingface.co/blog/us-national-ai-research-resource (accessed July 30, 2023).

Strier, K., Clark, J., and Khareghani, S. (2023). Measuring compute capacity: a critical step to capturing AI's full economic potential. https://oecd.ai/en/wonk/ai-compute-capacity (accessed June 24, 2023).

The US Council of Economic Advisers (2022). The impact of artificial intelligence on the future of workforces in the European Union and the United States of America. https://www.whitehouse.gov/cea/written-materials/2022/12/05/the-impact-of-artificial-intelligence/ (accessed June 4, 2023).

Tingley, M. et al. (2023). Netflix: a culture of learning. Medium. https://netflixtechblog.com/netflix-a-culture-of-learning-394bc7d0f94c (accessed July 30, 2023).

Turing, A.M. (1950). Computing machinery and intelligence. *Mind, New Series* 59 (236): 433–460.

Uchihira, N. (2021). Dialogue tool for value creation in digital transformation: roadmapping for machine learning applications. In: *Advances in the Human Side of Service Engineering*, Lecture Notes in Networks and Systems (ed. C. Leitner, W. Ganz, D. Satterfield, and C. Bassano), 518–526. Cham: Springer International Publishing https://doi.org/10.1007/978-3-030-80840-2_60.

Ufford, M., Pacer, M., Seal, M. et al. (2023). Beyond interactive: notebook innovation at Netflix. Medium. https://netflixtechblog.com/notebook-innovation-591ee3221233 (accessed July 30, 2023).

Vaisburd, A. (2023). Low-Code AI in Enterprise: Benefits, Use Cases, Limitations. Medium [Online]. https://towardsdatascience.com/low-code-ai-in-enterprise-benefits-and-use-cases-b9692ee13168 (accessed 30 July 2023).

Wang, H. and Raj, B. (2017). On the origin of deep learning. *arXiv* preprint arXiv:1702.07800.https://doi.org/10.48550/arXiv.1702.07800.

Yasunaga, Y., Watanabe, M., and Korenaga, M. (2009). Application of technology roadmaps to governmental innovation policy for promoting technology convergence. *Technological Forecasting and Social Change* 76 (1): 61–79. https://doi.org/10.1016/j.techfore.2008.06.004.

Zarrin, S. and Daim, T.U. (2019). Strategic technology planning in product-service systems with embedded customer experience requirements. In: *Proceedings of the PICMET 2019 – Portland International Conference on Management of Engineering and Technology: Technology Management in the World of Intelligent Systems*. IEEE https://doi.org/10.23919/PICMET.2019.8893889.

Zhang, Y., Zhang, G., Chen, H. et al. (2016). Topic analysis and forecasting for science, technology and innovation: methodology with a case study focusing on big data research. *Technological Forecasting and Social Change* 105: 179–191. https://doi.org/10.1016/j.techfore.2016.01.015.

Zhang, H., Daim, T., and (Peggy) Zhang, Y. (2021). Integrating patent analysis into technology roadmapping: a latent dirichlet allocation based technology assessment and roadmapping in the field of blockchain. *Technological Forecasting and Social Change* 167: 120729. https://doi.org/10.1016/j.techfore.2021.120729.

Zhang, D. et al. (2022). *The AI Index 2022 Annual Report*. AI Index Steering Committee, Stanford Institute for Human-Centered AI, Stanford University.

11

Technology Roadmapping: Nano Technology in Construction in Saudi Arabia

Ashfaq AlKhalil and Tugrul U. Daim

Mark O. Hatfield Cybersecurity and Cyber Defense Policy Center, Department of Engineering and Technology Management, Portland State University, Portland, OR 97201, USA

11.1 Introduction

Nanotechnology is emerging as one of the technological approaches that will help solve some of the major problems the world is experiencing in the twenty-first century and beyond. In the construction industry, nanotechnology is presenting numerous opportunities for application given that use of nanomaterials is enabling development of smarter architectural designs and environmentally friendly zones. Most importantly, nanotechnology approaches are being recommended in the building construction industry because they have been proven to be effective in reducing carbon emissions. Nanotechnology provides a base for creation of a wide variety of materials by way of altering the atomic and molecular structures of original materials. The new products have different properties and features that make them more preferred when considering aspects such as energy efficiency, reduction of air pollution, water treatment, and modification of conditions within build environments (Sapna and Dinesh 2020). Based on this, nanoarchitecture presents a feasible approach to building construction, as it can meet the required environmental performance standards that will help address climate change.

Scholars have identified many areas where nanomaterial can be applied, particularly in medicine where engineered products are used in making artificial tissues and diagnostic equipment and devices, and in drug delivery systems. This technology has also been proposed for use in areas such as the automobile industry, in water purification, construction, and development of renewable energy and energy saving components (Bayda et al. 2019). In the building and construction

Future-Oriented Technology Assessment: A Manager's Guide with Case Applications, First Edition.
Edited by Haydar Yalçın and Tugrul U. Daim.

industry, scholars continue to identify areas of application of nanomaterials with the view of solving most of the challenges being faced with traditional approaches. A review of past studies shows that the most potential areas of development of nanomaterials for use in the building and construction industry include real-time monitoring technologies, production of paints and surface coatings, and making structural concrete for thermal insulation, among other areas (Khandve 2014). As the world moves into greater adoption of sustainable technologies, it is estimated that nanomaterials will find many areas of application in the construction of green buildings. A more recent study established five classification areas of application of nanostructures in the construction of green buildings: (i) environmental impact control; (ii) indoor quality; (iii) safety and security; (iv) energy generation and storage; and (v) enhancement of material surface (Sapna and Dinesh 2020).

Although there are many proposals for application of nanomaterials in the construction industry, the market share of nanomaterials is still very low. A number of factors are hindering wide adoption of these technologies, the key one being the lack of clarity about the safety of nanomaterials and their likely implication on public health (Boverhofa et al. 2015). The safety concern arises from the fact that at the nanoscale, materials can change their chemical and physical properties in ways that can pose potential health dangers. In addition, at dimensions lower than 50 nm, materials can exhibit unprecedented optical, magnetic, and electrical characteristics (OECD, n.d.). This and other concerns call for a more in-depth analysis of the nanomaterials to identify the directions of application that will meet required safety and technical specifications.

In the past, scholars have primarily focused on proposing use of nanomaterials for making construction materials, such as concrete-based composites and use of nano inclusions into the conventional building materials. However, most of the proposed materials have not progressed beyond the prototype stage despite the successful demonstration of their application potential. There also seems to be a lack of comprehensive studies that demonstrate how the proposed nanocomposites can be seamlessly integrated across all areas of green building designs. This chapter thus seeks to fill this gap by undertaking an inquiry of the current nanotechnology approaches through a comprehensive review of literature and discussion of a technology roadmap that demonstrates how this new technology can be onboarded with the view of making it the mainstream building construction approach within 20 years. The goal here is to critically evaluate past studies, synthesize their findings, and discuss the possible application of nanoarchitecture in the Kingdom of Saudi Arabia. The study also discusses some of the key concerns that are likely to inhibit adoption of nanoarchitecture in green projects.

11.2 Research Methodology

The research methodology in this chapter is a comprehensive literature review from past studies and other scholarly publications, searching past studies from various databases including Google Scholar, IEEE Xplore, and other online databases that publish peer-reviewed journal articles. It also included publications from organizations and professional websites that document information relevant to the topic of study. Searches were done using keywords such as *nanoarchitecture, nanotechnology in architecture, construction nanomaterials, nanotechnology and green buildings,* and *nanotechnology in sustainable buildings*. Overall, around 72 publications were identified to provide relevant information. After a critical analysis of the materials, the information was organized based on the key topics and presented in a logical and comprehensive format while highlighting the key concepts and study findings.

After undertaking this preliminary study, a technology roadmap for the application of nanoarchitecture was also conducted for a 20-year period. Essentially, a technology roadmap presents the action plans that illustrates the initiatives that can be taken in order to transform the current construction paradigms into green architecture where nanotechnology innovations will become widely adopted in the building and construction industry. The roadmap was created based on literature review of the most recent studies that discussed technology trends in the area of green building and sustainable approaches in the building sector. The Quality Function Development (QFD) methodology was used in creating links between the different elements of the technology function. The QFD matrices were modified by assigning weights to key factors based on their relative importance in influencing or causing an effect.

In this study, the technology roadmap considered a period of 20 years, divided into three phases: the current period (0–5 years), medium term (5–10 years), and long-term (10–20 years). The roadmap considered the market drivers for green architecture and attempted to link the products, technology, and the resources that will be needed in the implementation of the plan in the various stages. These components of the roadmap are presented in diagrams that help to visualize how they are interlinked.

The rest of the chapter is structured as follows: Section 11.3 covers the background of nanotechnology and includes an overview of the concept of sustainable building designs and green building. Alongside these concepts, this section also looks at the considerations for application of nanotechnology in sustainable designs. Section 11.4 introduces the concept of nanoarchitecture and its application in making construction materials. Section 11.5 introduces the concept of the building sector and green buildings, and it focuses more on the environmental effects and cost-saving with highlighting the sustainable development issues. Section 11.6 covers the benefits of using nanomaterials in green buildings and the major concerns with using nanotechnology in building designs. Section 11.7 integrates the information gathered in the previous sections and discusses the possibility of application of nanoarchitecture in the Kingdom of Saudi Arabia while emphasizing the need to pursue sustainable building designs across the Kingdom. Section 11.8 provides the technology roadmap that can be adopted while implementing green building projects.

11.3 Background of Nanotechnology

11.3.1 What Is Nano?

Nano (n) is a prefix in a unit of measurement that denotes as 10^{-9} of a unit according to the International System of Units. The nano parameter was derived from Greek term νᾶνος, which is interpreted as dwarf in English. The nano parameter is mostly applied with units of measurements of time (nanosecond, nS), electrical capacitance (nano-farads, nF), and length (nanometer nM). The term *nano* is also commonly used as a prefix of other terms outside the bracket of units of measurement such as nanoscience, nanotechnology, nanomaterial, etc. In this case, these terms are used to denote something relating to the scale of a billionth of a meter or nanometers (Fahmy 2010). For example, *nanoscience* is a branch of science that is concerned with study, characterization, and design of materials at an extremely small scale of 1–100 nm (atomic of molecular level). To put the dimensions into context, it is important to understand that a single strand of human hair is about 60 000 nm thick (Bayda et al. 2019).

11.3.2 Nanotechnology and Nanomaterials

The term *nanotechnology* is defined by the National Nanotechnology Initiative (NNI) as "a science, engineering, and technology conducted at the nanoscale (1–100 nm), where unique phenomena enable novel applications in a wide range of fields, from chemistry, physics, and biology, to medicine, engineering and electronics" (Sapna and Dinesh 2020). It involves manipulation, production, and application of nanostructures. Nanotechnology entails "the production and application of physical, chemical, and biological systems at scales ranging from individual atoms or molecules to submicron dimensions, as well as the integration of the resulting nanostructures into larger systems" (Bhushan 2004).

Nanomaterials are being considered for application in different areas because they can be manipulated to produce new materials with a wide range of properties.

In addition, very small particles of the original material can be used on themselves to create materials with enhanced properties (Bayda et al. 2019). A nanomaterial is a material that has one of its dimensions being 1 nm (equivalent to 10–9 m) (Zhu et al. 2004). In the construction and buildings sector, nanomaterials are being considered because of the properties that present, for example, enhanced strength and durability, ultra-low friction, thermal insulation, better control of adhesion and grip, and light absorption property, among other characteristics (Abdellatif 2010). Nanoscience as a research area can be traced in the fifth century BCE when scientists discovered that an atom is the smallest indivisible component of matter. The commercial application of nanotechnology started in the early 2000s, and its adoption in different fields is fast growing, especially in human health and diagnostics (Altintas 2017).

11.3.3 Role of Nanotechnology

The continued development and application of nanotechnologies in various fields presents great potential to solve some of the problems facing humanity. Use of one-dimensional nanomaterials is a growing research area, and for several decades there have been successful applications in areas such as electronics, chemistry, biology, and in manufacturing and engineering (Poole and Owens 2003). Notably, the silicon integrated-circuit industry is among the leading fields in the application of one-dimensional nanomaterials. Many past and ongoing researches have primarily focused on finding ways to enhance the physical and chemical properties of conventional materials while targeting specific areas of use. It is noted that owing to the large surface area as well as the capability of nanoparticles to self-assemble, researchers are finding new ways of using nanomaterials in developing high-value products. Moreover, two-dimensional nanomaterials have also been developed and found to have superior mechanical and electrical properties, such as the carbon nanotubes (CNTs). Still, nanowires, which are extremely thin threads of less than 50 nm, have found use in developing electrical devices and in the biomedical industry due to their superior electrical and magnetic properties. There is also growing interest in studying and developing three-dimensional nanomaterials such as the C60 fullerenes that are used in architecture to make strong and lightweight and geodesic domes (Niroumand and Zain 2013).

In the building and construction industry, nanotechnology offers great potential of improving the functional properties of built environment, resource and cost savings, and in meeting the goal of sustainable development. The construction industry was identified as one area with huge potential for future application of nanotechnology (Zhu et al. 2004; Gann 2002). It is noted that application in architecture will give superior interactive experience to the occupants, thus enabling people to live more comfortably and in healthier and humane environments

(Niroumand et al. 2013). As it will be more fully described in this chapter, nanotechnology is enabling development of construction materials with superior properties. This material can thus be used in fixing some of the current problems as well as changing the construction processes. For instance, nanotechnology allows development of coating materials that require low maintenance. Other properties include improved flexural strength in cementitious components, higher thermal conductivity, as well as better fire retardant and sound abortion characteristics (Khandve 2014).

11.3.4 Application of Nanotechnology in Sustainable Building Designs

The continued emphasis on sustainable development is creating a growing impetus to adopt sustainable building design and use processes that have minimal environmental impact. It is estimated that the construction industry accounts for 72% of overall electricity consumption in the United States. Construction industry also consumes about 3 billion tons of raw materials, including steel, stones, sand, gravel, and wood. In addition, 17% of water usage goes to buildings. It is also estimated that 25–40% of municipal waste and 30% of carbon dioxide emissions come from buildings (Sustainability Roadmap 2020; Marzouk et al. 2014). Overall, the industry contributes to CO_2 emissions across the entire value chain from production of raw materials, transportation, and construction of buildings and in the operational aspects in the entire life cycle of the building (Fahmy 2010). This clearly shows that buildings do negatively affect the environment. Adoption of sustainable design is identified as a practical solution to curb the effect of building on the environment.

The growing sustainable design paradigm is emphasizing the adoption of resource-efficient approaches, such as energy and water saving technologies and overall reduction of pollution that comes from the construction industry. Other than these economic and utility aspects, sustainable building designs also focus on aspects such as optimizing space, increasing the durability of buildings, and improving the indoor environment, in ways that guarantee comfort to the occupants (Wang and Adeli 2014). The term *green design* or *green architecture* generally refers to architectural design that adhere to the sustainability agenda.

Alongside the increasing demand for green buildings is an increasing focus on how nanomaterials can be incorporated in construction materials to help in meeting the said needs. The discussion pertaining to convergence of nanotechnology with the new concepts of green building is becoming even more critical, as cities continue putting stringent measures to conserve the environment and ensure that the goal of sustainable development is achieved. The application of nanotechnology is yet to become widely adopted in the construction industry. Besides, there are growing concerns about the safety of nanomaterials in aspects such as how such materials can affect human health and other aspects (Khandve 2014). These and other concerns are calling for extensive research on the application of nanomaterials in various fields.

11.4 Nanoarchitecture: Definition and Its Applications

11.4.1 Definition of Nanoarchitecture

Nanoarchitecture refers to the application of nanotechnology in the various stages of development of architectural structures. It deals with working at the atomic or molecular scale in order to transform the way built environments are developed. It entails manipulating matter at the scale of 1–100 nm (Niroumand and Zain 2013; Sanchez and Sobolev 2010). Nanotechnology has brought tremendous changes in the development of materials and processes in the construction industry. The progressive application of nanotechnology in architecture has paved the way for designing and construction of smart and sustainable buildings (Dahlan 2019). Notably, nanotechnology is applied in the development of construction materials and equipment as well as in the forms and design theories. In the building and construction, industry nanotechnology focuses on the aspects of nanomaterial production that offer superior qualities and added capabilities compared with conventional materials (Altintas 2017).

The basis of nanoarchitecture is the development of functional nanomaterials. Previously, nanotechnology emphasized more on altering the mechanical properties of construction materials such as Portland cement through use of nanocement or by incorporating other nanomaterials such as nanosilica, and CNTs (graphene). Nanomaterials come with properties such as higher flexural strength, enhanced flexibility, and finer, thinner, and lighter components. It is expected that CNTs will revolutionize the architecture industry in unprecedented ways in future. Studies have shown that CNTs can be applied as "switchable conduits" due to their superior electrical conductivity. CNTs can also be used in making components with capabilities of generating energy (Sustainability Roadmap 2020). Other properties include better thermal insulation and electrical conductivity, self-healing, and cleaning properties and more durability among other properties (Yasin and Atiyat 2017).

11.4.2 Application of Nanoarchitecture

Nanoarchitectures are poised to become the future of building. As more discoveries of areas of application of nanotechnology continue to emerge, it is expected that nanoarchitecture will be a major disrupter in the entire building and construction industry. Past studies have proposed different areas where nanotechnology can be applied to architecture, as discussed below.

11.4.2.1 Nanocement

Past studies have identified the potential of nanotechnology in concrete-based composites using nano cement or ingredients whose dimensions are in the nanosized range. *Nanocement* is used to refer to Portland cement with particles of nano size. The resultant cement can be used on itself or mixed with other nanomaterials such as Nano-silica, graphene, or other products to give products with enhanced properties such as toughness, flexibility, and strength (Setyowati and Purwanto 2015; Flores et al. 2010; Porro et al. 2005).

A study assessed the performance of casted cubes and prisms that were made using nanocement. The cement was made using ingredients in the microscale such as nano silica (500–100) μm, micro cement sulfate (45–0.7) μm, and micro sand (300–75) μm. The study identified that the composites had higher flexural and compressive strength. The study also identified that using such composites can minimize the volume of materials needed to construct buildings. The result of using this alternative is reduction of the carbon footprint that is attributed to the construction industry (Al-Rifaie et al. 2018). Studies have shown that small size particles fill the spaces that are left when normal concrete is used, which leads to more compact components and higher flexural and compressive strength. For instance, adding nano silica in Portland cement was reported to increase the compression strength by 18% (Fahmy 2010; Sanchez and Sobolev 2010). In another study, ultra-fine amorphous colloidal silica (UFACS) was incorporated in Portland cement together with nano silica, fine particles of hematite (ferric oxide) and water produced concrete structures with compressive strength of 91.2 MPA compared to a maximum of about 40 MPa for Portland cement. The resultant composite structures also had superior hardness density of about 2680 kg/m^3 (Green 2008).

Nanotechnology can also be applied through incorporating other nanomaterials other than cement, in order to improve the stability of the microstructure of concrete-based components. Different past studies have proposed different nanomaterials that can be used in making concrete-based systems. These nanomaterials include use of CNTs to reinforce concrete (Gillani et al. 2017), use of bagasse ash (ash from residue left after sugarcane processing) (Setyowati and Khusnan n.d.) and polypropylene fiber (e.g. Styrofoam) (Gann 2002). A past study proposed a new white cement with photocatalytic characteristics and similar functional properties to conventional cement. The white cement was identified to be suitable for enhancing aesthetics characteristics of concrete-based components and structures while at the same time guaranteeing durability (Cassar et al. 2003). As the world moves toward greater adoption of environmentally friendly approaches, white cement can be a good alternative for Portland cement. The potential of incorporating nanoparticles in white cement presents a good research area.

11.4.2.2 Nanotechnology and Steel

The research on the possibility of making steel to nano-size during the cementite phase have indicated that the resulting steel cables are stronger compared to normal steel. These high-strength cables can be used in construction of high-rise buildings or other structures such as bridges. The high-strength steel cables are ideal for use in areas prone to fatigue, which is one of the causes of structural failure. Structural failure of building components can occur even at stress levels that are lower than yield strength of the constituent materials. This can reduce the useful life of buildings (Sauraav 2012). It has also been identified that use of steel with enhanced strength can help in reducing costs and duration of construction.

Structural failure in construction components occurs when cracks begins to form due to stress raisers. A past study has shown that including copper nanoparticles in steel helps in reducing surface unevenness. This results in minimizing the number of stress raisers and consequently limiting the chances of cracking due to fatigue. It is also suggested that fatigue around welds and around heat-affected zones (HAZ) in steel is a major cause of structural failure. A material known as Sandvik Nanoflex, which is developed using very hard steel nanoparticles has improved tensile strength and young modulus. This property makes the material more suitable for application in high fatigue areas. The material is also resistant to corrosion thus proving for higher durability of structures. MMFX2 steel is another modified steel material that is made using steel nanoparticles. MMFX2 steel was found to be an ideal alternative to stainless steel and in addition, it gives better corrosion resistance. This material is also cheaper than conventional steel (Setyowati and Purwanto 2015).

11.4.2.3 Nano-glass

In most cases, glass is used in the exterior walls of buildings in windows. In recent years, scholars have been looking at how to improve the glass material in order to have better control of heat and light entering into a building. This has emerged as a key aspect of making buildings more sustainability as it has a direct effect on energy consumption within a building. Studies have proposed several approaches to achieve this goal. One of the approaches is using thin film coatings that are spectrally sensitive as window panes. This type of glass has the property of filtering out harmful rays such as infrared rays, which have the effect of heating the interior space.

Another approach is to use glass with improved properties of thermal insulation and at the same time improved adequate light. Another strategy is using photochromic technologies to make glass that is capable of absorbing light when light intensity rises above a specified level. The last strategy is to use glass with a tungsten oxide coating. This material is sensitive to change in applied voltage (OECD n.d.). A recent study identified that using nano glass in buildings can reduce the heating, ventilation, and air-conditioning (HVAC) loads in a building by more than 20%. At the same time, using glass treated with nanotechnology that allows higher levels of light transmission improves energy efficiency by reducing use of artificial lights (Abdin et al. 2018).

11.4.2.4 Nanotechnology and Wood

Application of nanotechnology in wood and wooden composites is another area currently under research. One of the properties of wood is that its thermal conductivity is very low. A number of past studies have explored ways of incorporating nanotechnology in making wood composites. Most of the studies are premised on the idea of adopting sustainable approaches that call for a shift toward renewable resources and recycling of wastes. The most common considerations for recycling of wood include making particle boards and plywood panels. One of the studies explored ways of addressing the problem of formaldehyde emissions from wood composites through application of nanotechnology (Porro et al. 2005). Ordinarily, particleboards and plywood are made using formaldehyde-based resins and urea. When the composites are used in building structures, they release formaldehyde emissions, which are linked to health risk due to their carcinogenic effect. To address this problem, the study proposed incorporating nano additives to modify the formaldehyde resin. The nanoparticles included nanosilica, nano zinc-oxide, and nanoalumina with particle sizes ranging from 15 to 50 nm and at a dosage of 1–3% (Candan and Akbulut 2013).

Another study demonstrated how wood-based fiberboards could be made using wollastonite nanoparticles and agricultural waste in order to improve thermal conductivity of wood composites. Wollastonite is a natural calcium inosilicate ($CaSiO_3$) mineral that also comprises small quantities of manganese, magnesium, and iron. It is mostly used as a fertilizer additive for improving the growth of plants. In the mentioned study, rice straws were used together with nano-wollastonite particles in the levels of 3–6% in making medium-density fiberboards (MDF). The resultant product was identified to have improved physical and mechanical properties.

Adding nano-wollastonite in the wood composites also gave improved hydrophobic properties. Based on the study, addition of nano-wollastonite at the dosage of 6% to a composite with 90% of wood fiber and filing with rice straw particles produced an MDF composite with improved physical and mechanical properties (Tichi et al. 2018).

The possibility of surface treatment of wood using nanomaterials has also been explored in a number of past studies (Papadpoulos and Taghiyari 2019; Goffredo et al. 2017). Surface treatment of wood with nano copper and nano silver was identified to have antimicrobial properties hence contributing to enhancing durability of wood. The addition of zinc and copper nanoparticles in wood coatings has also been reported to provide improved decay resistance and resistance against termites in wood. A different study proposed use of nanocopper and nanosilver in wood coatings because the resultant product was found to give property of low accumulation of heat of the surface and improved durability, while addition of photocatalytic titanium dioxide (TiO_2) in combination with nano-copper and nano-silver gave improved antifungal capabilities in wood (Flores et al. 2010). When wood composites are used in the construction of buildings, they prevent the facades facing receiving direct sunlight from becoming overheated. Generally, nano wood and wood composite are ideal for thermal insulation in buildings. Other than this, the altering of the molecular structure of wood can bring added benefits such as enhanced fire resistance, and improved resistance from degradation by biological agents such as fungi and termites (Taghriyari 2014).

The rising number of studies relating to application of nanotechnology in the making of wood composites is demonstrating the potential for commercial application of engineered wood composites in building and construction. The private sector is leading in investigating how nanomaterials can be applied as raw materials for wood composites or in materials used for surface treatment of wood. This is seen as a viable path based on research that shows how the excess forest biomass can be used in making new, high-value, and commercial wood products (Rains et al. 2014). This move is in line with the sustainable approaches given that finding new ways of using forest excess biomass will be critical in mitigating the problem of deforestation and its environmental consequences. There is therefore the huge potential of applying nanotechnology in creating wood-based composites that give properties that are in tandem with green architectural designs.

11.4.2.5 Nanotechnology and Coating Materials

Nanotechnology can be used in creating coating materials such as paints using nano-sized components. When applied on walls, nano paint gives superior properties such as higher resistance to corrosion and improved thermal insulation. Higher resistance to corrosion comes because the paint is hydrophilic meaning that it repels water and as a result the structures become more durable compared to when normal paint is used (OECD n.d.).

11.5 Building Sector and Sustainable Development Issues

Buildings' design is an essential component in the fabric of cities. However, buildings are the largest contributor of greenhouse gas emissions (Al-Tamimi 2017).

Therefore, governmental initiatives and other efforts have played a significant role in raising public awareness and increasing the adoption of practices to achieve a more efficient and sustainable building design. Also, buildings consume the largest amount of the world's energy use, so high-efficiency designed buildings allow us to save energy and improve energy performance (Xia et al. 2014; Chung 2012).

This has led to addressing issues related to efficiency throughout multiple stages of the building design process, including the early stages (Scheuer et al. 2003), during the construction stage (Hernandez et al. 2008; Carriere et al. 1999; Federspiel et al. 2002) as well as the operation stage (García et al. 2017). It has also fueled the growth of energy efficiency and performance studies across a broad range of industries, from the building sector (Iribarren et al. 2015) especially in the electric power business (Sueyoshi et al. 2017), to single-family dwellings (Sueyoshi et al. 2017), commercial and industrial buildings (Gourlis and Kovacic 2016). The performance of sustainable buildings can be addressed by a comprehensive view or a robust model that covers multidimensional aspects but narrowing down to one aspect is necessary to better assess the efficiency of the building.

11.5.1 Cost-saving Productive Processes and Environmental Remediation

Chung (2012) summarized the building's efficiency using factors into four groups: random factors such as unusual weather conditions; physical characteristics including square footage; incentives faced by building management, including functionality; and differences in how the building is used, such as operating hours. There is overwhelming evidence that suggests the way a building is designed, and the construction material can affect the building efficiency and well-being of its occupants. The simulation method factors construction material and the properties of building as one of the most important factors for building efficiency and related to the energy consumption of the building (Carriere et al. 1999; Federspiel et al. 2002).

The construction industry can reduce waste and save costs through raw material substitution and redesign products effectively. This means choosing the material plays an essential role in cost-saving. While some materials may cost more, it is often the case that environmental impacts results are less that can make the overall building more efficient. We believe that green buildings can help to solve the issues of decoupling economic growth and construction, which impacts climate change and environmental stress. Additionally, the choice of building construction material is one of the most important factors of intervention that provides opportunities to limit the environmental impact and contributes to the achievement of sustainable development goals.

11.5.2 Green Building Concept

Green buildings are a concept that is increasingly becoming famous in the building and construction industry. It entails new building design approaches that focus on saving energy and water, improving the internal environment for the occupants, reducing toxic wastes in the building environment and using building materials that contribute to reduction or elimination of greenhouse gases emissions in the atmosphere. Other than in the interior environment, the green building concept requires landscaping design that include planting of local plant species, having more shade areas (Islam et al. 2015).

Energy efficiency is more emphasized among the main components of green building designs. The idea of enhancing energy efficiency focuses on using high-efficiency windows and orienting them to allow natural light into the interior environment to minimize electricity lighting. The design of the green building also includes walls, ceiling, and floors with improved insulation properties to reduce use of air conditioners. Another strategy for increasing energy efficiency using solar energy to reduce consumption of grid electricity in lighting and water heating and powering appliances. Successful application of these design interventions ensures that the environmental impact of the building is significantly reduced while at the same time (Boverhofa et al. 2015).

11.6 Using Nanomaterials in the Production of Concrete-based Composites

The possibility of using nanomaterials in the production of cement and other concrete composites is well-documented (Marzouk et al. 2014; Demirdoven and Karacar 2015; Mosly 2015). Different studies have proposed the suitability of different nanomaterials for making cement-based composites. Nanotechnology can be applied in making new materials that are engineered to the nanoscale or admixing nanomaterials with the conventional cement in order to improve their technical and mechanical properties such as tensile strength and weight. It is noted in past studies that cement-based materials have poor technical characteristics; for instance, they are brittle, thus requiring reinforcement that are permeable to water and chemicals (Du et al. 2019). When materials are decreased to the nano or atomic scale, they can exhibit different characteristics compared to materials of larger sizes. Specifically, nanomaterials can exhibit different properties such as chemical reactivity, mechanical properties, electronic conductivity, and absorption of light and heat. These changes are primarily due to increased surface area, as size is reduced to the nano size and due to the aspect of increased quantum effects. For example, as the surface area increases due to reduced size of particles, the chemical reactivity also increases (Kumar et al. 2020).

There are majorly two approaches to develop nanomaterials. The first approach is termed as the top-down approach (TDA). This approach entails physically reducing the size of the material to a nanosized through milling. This is purely a mechanical process that is mostly used in making materials such as nanoclay, nano-alumina, and nano silica. The second approach is known as bottom-up approach (BUA). BUA involves changing the molecular structure of materials, meaning that the material is developed through a process of altering the chemical structure of the original material. This second approach requires advanced expertise and can be more costly compared to the TDA approach (Jamshidi et al. 2020).

11.6.1 Carbon Nanotubes (CNTs)

Scholars have proposed different ways through which nanotechnology can be applied in the development of concrete-based composites such as bricks. The most widely quoted material is use of CNTs (Sustainability Roadmap 2020; Sanchez et al. 2009; Makar et al. 2005). Adding CNTs in cement results in composites with improved properties such mechanical strength, porosity, and overall durability of the composite and structures. Specifically, multi-walled carbon nanotubes (MWCNTs) are reported to offer unique properties to the composites because they have a higher aspect ratio (Shah et al. 2009). It is indicted that MWCNTs are exceptionally strong and lightweight, given that the strength to weight ratio is 1.8 g/cm^3 and can reach a maximum strain of 10% that no other material can reach. It is also indicated that CNTs electrical conductivity potential is six times higher than copper (Gann 2002). CNTs have also been found to be sensitive to compressive stress therefore, when incorporated with cement can be used in making self-sensing composites for monitoring traffic flow (Han et al. 2009). The ability to withstand compressive stress as well as good electrical resistance makes CNTs ideal for use in a built environment when specific functional aspects are needed, such as automated traffic flow monitoring.

11.6.2 Polymer Cement

A number of past studies investigated other ingredients that can be added into concrete mortar in order to minimize use of cement. One study proposed using a mixture of plastic waste and sugarcane processing waste (bagasse ash). These ingredients are ground into nanoparticles, mixed with a smaller amount of cement and then used to make the lightweight polymer bricks. The study focused on testing for the technical performance of polymer bricks, including compressive strength, thermal resistance and thermal transfer value, and density. Other aspects that were tested include acoustic abilities (property of absorbing sound). Overall, the study identified that the polymer bricks that were made using the mixture of polymer and bagasse in the nano sizes gives compressive strength that is higher than conventional cement by 10.0–24.57%. It was also identified that the resulting polymer brick gives satisfactory acoustic performance. The value of sound transmission loss (STL) was 53.906 dB, while the absorption coefficient of the new product was 0.26. Lastly, the study indentured that the value of the overall thermal transfer value (OTTV) was lower than that of a building constructed using conventional concrete materials (Marzouk et al. 2014). This study provides proof that materials at nano sizes can have properties that are fundamentally different from the materials in normal sizes, which opens avenues for their application in areas beyond the normal areas of use.

Another study proposed use of Styrofoam waste is glided and mixed with cement. The researchers in this study were interested in finding an innovative way of recycling Styrofoam. This material is commonly used in packaging of electronics and is not decomposable. In the production of nanomaterials for construction, the Styrofoam waste was used in the place of the coarse sand aggregates and mixed with bagasse ash. This mixture was then used to make composite bricks. Testing of the resulting products was done after 7, 14, and 28 days. The study identified that Styrofoam waste can be used in making lightweight polymer bricks. A previous study had identified that Styrofoam can reduce the unit weight of components by 4.01%. It was also identified that the bricks have superior sound absorption qualities compared to concrete (Marzouk et al. 2014).

11.6.3 Advantage of Nanotechnology in the Construction Industry

The discussion in the foregoing emphasizes that nano-based concrete composites have superior mechanical properties such as tensile strength and flexibility and lightweight. Several studies have confirmed this aspect based on laboratory trials of concrete-based components that were produced using nanomaterials (Gann 2002; Dahlan 2019; Yasin and Atiyat 2017). Other aspects relating to composites made using nanomaterials render them more suitable for application in green architectural designs compared to conventional materials as illustrated below.

11.6.3.1 Enhanced Thermal Insulation

People in the building and construction businesses are using different materials to achieve the thermal insulation within the indoor environment. Some of the materials that are traditionally used in this regard include mineral wool, expanded polystyrene, extruded polystyrene, cork, cellulose, and polyurethane. However, there are also a number of state-of-the-art materials being used today, such as vacuum insulation panels (VIPs – the most used), gas-filled panels, aerogels, and a number of phases changing materials (Jelle et al. 2012). A number of studies have also shown great potential of nanomaterials in providing building materials with enhanced thermal insulation properties. The nanomaterials can provide thermal insulation in the range of as low as 4 mW/mK, which matches the lowest possible with VIPs (Marzouk et al. 2014).

In addition to thermal insulation, concrete-based composites made using nanosilica have shown enhanced thermal resistance properties. Studies have identified that nanosilica when used in cement reacts with calcium hydroxide, resulting in formation of more calcium-silicate-hydrate. This makes the matrix of the bonding of cement denser, which results in improved mechanical properties and thermal resistance of the composites (Sikora et al. 2018; Plassard et al. 2004; Bjornstrom et al. 2004). Other studies have shown that incorporating nanosilica in cement results in reduced thermal conductivity (Bjornstrom et al. 2004). Apart from giving superior mechanical properties, studies have also shown that when nanosilica is used as an additive in cement, the hydration process of the motor is reduced significantly (Qing et al. 2007).

These studies provide a basis for proposing use of nanoinclusions in the construction of green buildings. When a material has low thermal conductivity property, it means that heat transfer across the material is low (Porro et al. 2005). Therefore, for a building with walls made of nanocement, the low thermal transfer property of the material makes it possible to maintain the indoor environmental conditions relatively constant for a longer period without requiring air-conditioning. Consequently, this will also bring the advantage of reduced energy consumption hence reduced energy costs.

11.6.3.2 Lightweight Materials with Enhanced Strength

Portland cement is regarded as conventional construction material. One of the major drawbacks of using cement in the conventional form is the weight of structures and low flexural strength. Ordinarily, steel is used to reinforce concrete structure, and as a result, it further increases the weight of the components. Past studies have identified that concrete-based components made of nanocement provide for more compressive strength compared to the normal Portland cement (Gann 2002; Dahlan 2019; Yasin and Atiyat 2017; Li et al. 2004). The mixture of nanocement with polymer materials such as Styrofoam and bagasse ash reduced the amount of cement needed to make construction bricks, while providing enhanced flexural strength. At the same time, the bricks were identified to be lighter in weight by 4% compared to bricks made using Portland cement (Marzouk et al. 2014; Li 2004). While putting other factors constant, the lightweight of construction bricks and enhanced flexural strength are variables that can allow construction of taller buildings given that weight is a major issue when constructing multistory buildings. The structures can be reinforced with steel frames to enhance strength.

The high strength-to-weight factor can be a game changer in the construction of high-rise buildings with sustainable designs. A mixture of nanocement and other noninclusions can also be used to make lightweight beams with improved flexural strength and corrosion resistance. In addition, the beams can improve abrasive resistance, which enhances structural durability (Kohai et al. 2018).

11.6.3.3 Self-cleaning and Photocatalytic Properties

One of the most sought properties of nanomaterials in the construction sector is photocatalysis because it gives the self-cleaning effect. Structures and concrete based-components that are made using nanomaterials such as nanocement have been found to be self-cleaning. The nano-size of materials enables formation of dense matrix that results in very smooth and water-repellent surfaces. In addition, such surfaces exhibit low force of surface attraction. These properties prevent water and dirt particles from attaching to the surfaces. Hydrophobic coatings can be applied on several brisling and construction materials such as concrete, paint, wood, and metal including leather and textiles to give self-cleaning or easy-to-clean surfaces. The mostly applied nanocoating additive that gives photocatalytic property is titanium dioxide (TiO_2). The self-cleaning property leads to enhanced structural durability.

There are many examples of practical application of nanoarchitecture in the world. For example, in Narita International Airport in Tokyo, all buildings are covered with a self-cleaning textile membrane that is also meant to give better comfort for the occupants. Since 2006 when the renovation was done, the airport has reported a significant reduction in cost of maintenance. Apart from the self-cleaning property, nanotechnology is also applied to make surfaces with antimicrobial effects. These properties contribute to enhancing durability of architectural structures (Atwa et al. 2015).

11.6.3.4 Reduction of Air Pollution

Because nanoparticles will reduce the amount of cement used in the construction industry, there is potential for significant reduction in environmental pollution. As already noted, the bundling sector is a major contributor to carbon emissions; therefore, incorporating energy-efficient alternatives such as use of natural lighting in buildings and using nanomaterials with enhanced thermal insulation properties in walls will have a positive effect toward reducing dependence on artificial lighting, heating, and air conditioning. In addition, increased use of green nanomaterials in the construction industry will play a critical role in reducing the amount of energy needed at various stages of construction and in the entire life cycle of the building. For example, using self-compacting nanocement requires no vibrators to compact, and requires up to 80% less time to pour compared to normal cement (Abdellatif 2010).

A. Cost Implications

There is no adequate literature that makes cost comparison between nano-architectural designs and conventional building designs. Application of nanotechnology in architecture can be expensive due to the material and the technologies needed. However, many studies have shown that adopting nano-architectural and sustainable designs can have significant cost reduction implications. For example, use of self-compacting nanocement is said to reduce the time of pouring concrete by 80%. This translates to less labor requirement and hence lower costs (Abdellatif 2010). Other studies have shown that use of natural lighting and thermal insulation contributes to reducing the energy demand in buildings, which translates to less utility bills. Moreover, the self-cleaning property results in less maintenance of build structures and enhanced durability. There is therefore a strong indication that adopting nanoarchitecture can create significant cost reduction.

11.6.4 Drawbacks of Nanotechnology in the Construction Industry

Despite the recent development in the area of nanotechnology in the building and construction industry, there are concerns about possible toxicity of some nanoparticles that can have potential health risks. It is, for instance, indicated that some nanoparticles such as titanium oxide can have health effects comparable to those presented by asbestos. Specifically, it is suggested that inhalation of nano-titanium oxide of about 8.8 mg/m^3 in concentration can lead to lung inflammation. Other studies have indicated that exposure to nanoparticles can potentially lead to damage of the DNA in human cells, which can result in the development of certain types of cancers or cause brain damage. Some of the other nanomaterials that studies continue to find ways used in making concrete composites such as CNTs are already considered to be toxic, for instance, by the US Environmental Protection (Pachero-Torgal and Jalali 2011; Mohajerani et al. 2019). The health effects of nanoparticles can vary depending on the amount and duration of time a person is exposed. There is also the concern that increased application of nanomaterials in the construction industry will lead to generation of nanowastes that may present challenges in disposing and recycling (Kohai et al. 2018).

11.7 Nanoarchitecture Application in Saudi Arabia

The Kingdom of Saudi Arabia is a country located in the Arabian Peninsula. The country is on record as the world's largest producer of petroleum and gas. For this reason, the country does not envision any immediate problem concerning energy meaning that country may be slow in adopting the green building technologies unlike in other developed and developing countries. Despite these phenomena, other factors such as the predominant desert climate in the entire gulf region, which demands heavy reliance on air-conditioning in indoor environments, may create an impetus to adopt technologies that will provide enhanced quality of living and lowering the cost of energy. Another factor that will also contribute to the increased adoption of green building technologies in the kingdom is the need to shift to a non–oil-based economy as outlined in the Saudi Vision 2030, alongside the increasing calls for countries to adopt sustainable practices and minimization of greenhouse emotions (Weeta 2020).

The extreme weather conditions in Saudi Arabia contribute to the increased consumption of energy and water. The special status of Saudi Arabia as the largest producer of oil and gas, and one of the largest consumers contributes to the problem of greenhouse emissions (Dahlan 2019). This means that the country needs to identify measures of reducing its carbon footprint. In the same vein, there is also a growing need to ensure that buildings are more sustainable by enhancing the level of energy and water efficiency. These and other factors are exacerbating the need for green buildings in the country. Alongside the growing demand of green technologies, an increasing number of scholars have demonstrated the possibility of incorporating nanotechnology in the production of building construction raw materials.

The government of the kingdom of Saudi Arabia has made tremendous steps moving toward increasing adoption of green architecture. In this regard, the government launched the Sustainable Houses Forum of Saudi Arabia in May 2018. Also, in line with this step, the ministry of housing has developed a Sustainability Quality and Evaluation System that will be used in the assessment and rating of green building projects across the kingdom. Some of the most promising infrastructural projects where green building technology is expected to be prioritized include in the development of Prince Abdelaziz Bin Mousaed Economic City in Hail, Neom Mega City project as well as the Qiddiya City project (Porro et al. 2005).

The King Abdullah University of Science and Technology (KAUST) is among the world's biggest green building projects with LEED certification. The design of this graduate-level research institution provides a typical example demonstrating some of the design approaches that can be incorporated while utilizing nanomaterials in the implementation of green building projects in other parts of the kingdom. The design of the KAUST campus was primarily focused on delivering a high-efficiency, low-energy building based on the local climatic conditions. For instance, the company design of the compass building was meant to create shading and ventilation between the units. At the same time, the units are oriented such that there is minimal exposure of facades to direct sunlight.

These design components were incorporated in order to improve the internal microclimate within the buildings, thus minimizing the need for air-conditioning. Second, the designs also considered maximizing use of natural lighting from the roof's large clear windows in the social places to minimize energy consumption (Setyowati and Khusnan n.d.).

The example of the KAUST green building project offers renewed hope that can drive increased adoption of construction designs that will meet the needs of the current generation while taking care of the needs of future generations. The proposal for use of nanotechnology in the building materials gives more impetus for continued adoption of green buildings across the Kingdom.

11.8 Technology Road Mapping for Green Architecture

11.8.1 Introduction

Technology roadmapping is a study approach that focuses on identifying a link between technology goals, the resources, strategies of implementing the technology in a manner that is effective in meeting those goals (Phaal et al. 2004). Technology roadmapping has emerged as an alternative analytical methodology that specifically addresses the technological aspects of the business or industry context. This methodology emphases the visualization of the relationship between the market conditions and a given technology. Technology roadmapping is being considered in a wide range of industries, including in the public sector to explore the technology trends and the potential for a particular technology, including development of a plan to guide the adoption and implementation of that technology. Most recently, the United Kingdom government used technology roadmapping approach to develop a plan for adoption of renewable energy in specific areas such as a cost-cutting strategy and in mapping the growth potential (Shim et al. 2019).

For a technology roadmap to be effective, it is supposed to meet three key conditions:

1. It should explain the technological trends as well as the drivers for the outlined trends.
2. It should provide details about the product(s) or service(s) and their application.
3. The technology roadmap should state the resources and other capabilities that are needed in order to drive adoption.

A well-drafted technology roadmap can help in aiding communication among the most critical stakeholders who participate in technological development. In addition, a technology roadmap can facilitate consensus building among the stakeholders. Still, technology roadmapping also provides a guide that can be used to direct technology development within a certain sector or industry (Garcia and Bray 1997).

A technology roadmap is different from a forecast in the sense that it does not give information on whether the future will be good or bad in regard to the development of a given technology. Rather, roadmaps examine the actions that need to be taken in order to reach a specific goal to address a given issue. There are a variety of roadmaps. Some of the common types include project roadmaps, cross industry roadmaps, industry roadmaps, and technology roadmaps. Roadmaps are essential when doing strategic planning. Ordinarily, technology roadmapping takes into consideration a long-term view of the development of a new technology, which in most cases is 10–15 years or more (Garcia and Bray 1997).

The time axis of the technology roadmap comes out as a prominent feature as it outlines the time periods when a given market condition or technology products, events, targets are expected to be achieved. It is indicated that the length of the period in the time axis of a technology roadmap is largely industry specific. For example, the aerospace industry usually considers a planning period of about 20 years. This is unlike in the consumer electronics industry, where periods are approximately 5 years and less. In the building sector, a practical planning period of 20 years is considered to be ideal when developing a technology roadmap, given the length of the time needed to develop legislation (Garcia and Bray 1997).

In this section, a technology roadmap for green building technologies is presented. The roadmap takes into consideration the market drivers that will influence adoption, green products, technology, and resources that will aid will be key in the development of green architecture while taking into consideration the application of nanotechnology within a 20-year period.

11.8.2 Steps in Technology Roadmapping

Figure 11.1 outlines the steps that were followed in creating the technology roadmap for the green building technologies while applying nanotechnology. The roadmapping process involved five major steps. The first step involved identification of the market drivers that will shape the future of green building and nanotechnology concepts. This was followed by a market analysis, which involved evaluation of the market in order to identify the opportunities and customer needs that will inform greater adoption of green building technologies. The third step involved exploration of the product plan through a product features mind map, where a breakdown in the key product features and functionalities as well as the benefits are provided. The next step involved creating a technology plan in which the key green building technologies were evaluated. The last step involved evaluation of the resources that will be needed in order to meet the goal

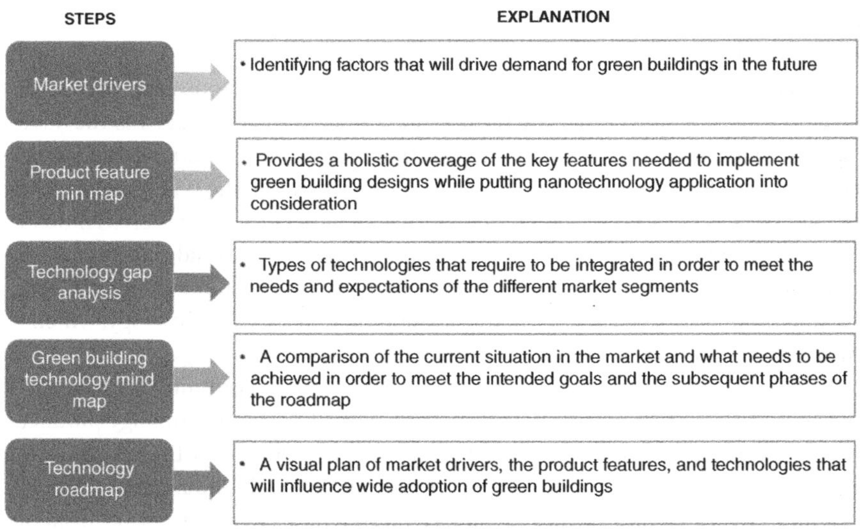

Figure 11.1 Steps in creating the technology roadmap.

of delivering net-zero emission buildings through integration of nanotechnologies within the postulated 20-year period.

11.8.3 Market Drivers

Market drivers entail the forces that are expected to influence the market demand for green buildings and application of nanotechnology in the building sector. This includes the trends in technologies and other factors that will enable the market to grow and become sustainable. The market drivers of the green building are shown in Figure 11.2.

As shown in Figure 11.2, a total of 22 drivers of green architecture were identified to be the most impactful in driving the demand for green buildings. These drivers are put in seven broad categories: (i) Environment; (ii) Government policy; (iii) Economic; (iv) Technical; (v) Culture; (vi) Aesthetics; and (vii) Safety.

The first key driver is the environment-related factors that include LEED platinum rating, which is the highest certification rating of buildings based on the extent to which the building is designed to operate efficiently. The emphasis on net-zero carbon emission is another environmental factor that will drive the adoption of green architectural designs and application of nanotechnology in making building materials. Moreover, the holistic architectural design approach for ensuring that construction adheres to green/sustainable paradigms will also be a key environmental factor that will drive green building adoption. Lastly, the growing impetus to produce sustainable energy such as use of high-efficiency solar panels

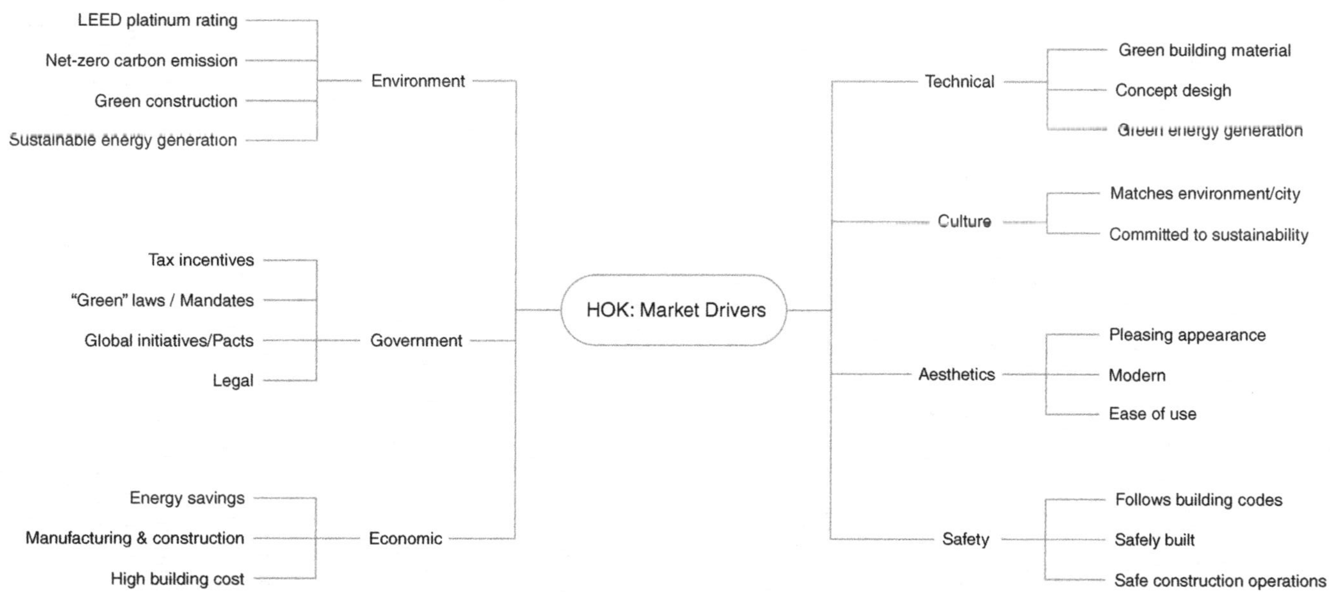

Figure 11.2 Market drivers of green buildings and nanotechnology.

using nanotechnology is projected to influence a greater shift toward green building adoption.

The second key driver is government policy. Key aspects in regard to government policy include tax incentives for building construction projects that incorporate green designs and make use of green materials. Passing and enforcement of "green" laws and mandates for the building and construction industry will also be critical in encouraging adoption. Still, adherence to global initiatives and pacts by the government on matters relating to sustainability will be instrumental in driving demand. Lastly, creation of a legal framework for aiding regulation of the construction sector that discourages unsustainable approaches is also projected to be a key influencing the demand for green buildings. Notably, lack of a regulatory framework for directing the application of nanotechnology in the building sector while addressing the health effects was identified as a major limitation in adoption of nanotechnologies

Economic factors also play a key role in driving market demand for green buildings. The most critical economic factors include the need for cost and energy saving and the overall high cost of building across the entire life cycle of the building. As identified previously, applying nanotechnology in green buildings construction can deliver economic benefits through aspects such as enhanced durability of buildings, reduced need for maintenance of building amenities, and energy conservation, among other benefits (Abdellatif 2010). Another economic factor is the shifting of the manufacturing and construction sectors toward sustainable approaches through increased application of nanotechnologies in constructing sustainable buildings.

The technical factors that are projected to drive the demand of green buildings include availability of green building construction materials, especially nanomaterials that give enhanced properties compared to conventional materials. Other technical factors include concept designs that adhere to green/sustainable standards and the ability to incorporate green energy generation technologies in buildings.

Culture will also play an instrumental part in driving the market demand for green buildings. A culture that aligns with environmentally sustainable cities will be key in influencing adoption. Societal commitment to sustainability and the desire to shift to sustainable approaches will also be an important driver of green building adoption.

The aesthetic component of the buildings also influences demand. Still, the continued emphasis on modern designs and ease of use of building amenities and ecofriendly options, both in indoor and outdoor spaces, is projected to prompt a shift from conventional buildings to sustainable ones. Nanomaterials can enhance the aesthetic appeal of the built environment when incorporated in building materials and when used as coating materials (Cassar et al. 2003).

Lastly, the safety factor will also influence demand for green buildings. Mainly, aspects such as strict adherence to standard building codes, building amenities that are safely built, and safe building operations will contribute immensely to instigating a shift to sustainable building designs. As outlined earlier in this chapter, nanomaterials such as nanocement and CNTs have been identified to provide enhanced strength when applied in building construction (Sauraav 2012). Hence, increased application of nanotechnology in the building sector will influence greater adoption of green buildings.

11.8.3.1 QFD Market Segments and Drivers

Figure 11.3 shows the quality function deployment (QFD) of the market segments and the drivers. The projected market for the green buildings is classified into six key segments including government, mixed use, residential, science and technology, healthcare, and aviation and transport. In the QFD, the rating criteria of the drivers to the market segment was high score with a value of 4, medium score with a value of 2, and low score with a value of 1. The scoring considered the extent to which a given market driver is important to each of the market segments. After assigning the score of all drivers to each market segment, a summation of the scores was done in order to identify the most and least influential drivers.

As shown in Figure 11.2, the legal factors, energy saving, following building codes and need for safely built buildings were identified to be the most important drivers to adoption of green buildings with each scoring 24 points. Tax incentives, green laws and mandates, green manufacturing and construction, and green energy generation were ranked as the second most important driver, with each scoring 22 points. Ease of use had a score of 20, followed by sustainable energy generation, which scored 18 points. The drivers that are ranked fifth in terms of extent to influence demand include LEED platinum rating, global initiative or treaties, design costs, commitment to sustainability, and safe construction operations, where each scored 17 points. The net-zero carbon initiatives, availability of green building materials, and culture that matches sustainability approaches and green cities had a score of 16 points, followed by green concept designs that scored 15 points. The least influential drivers are pleasing designs, with a score of 12 points, and green construction, which scored only 10 points.

In terms of the broad seven categories of the drivers, government policies have a total score of 81 points, followed by safety factors (65 points), economic factors (63 points), environmental factors (61 points), technical factors (53 points), aesthetic factors (46 points), and cultural factors (33 points), in that order. Based on the scores, we can conclude that government policy, safety, and economic factors will be the three most influential drivers of green building adoption.

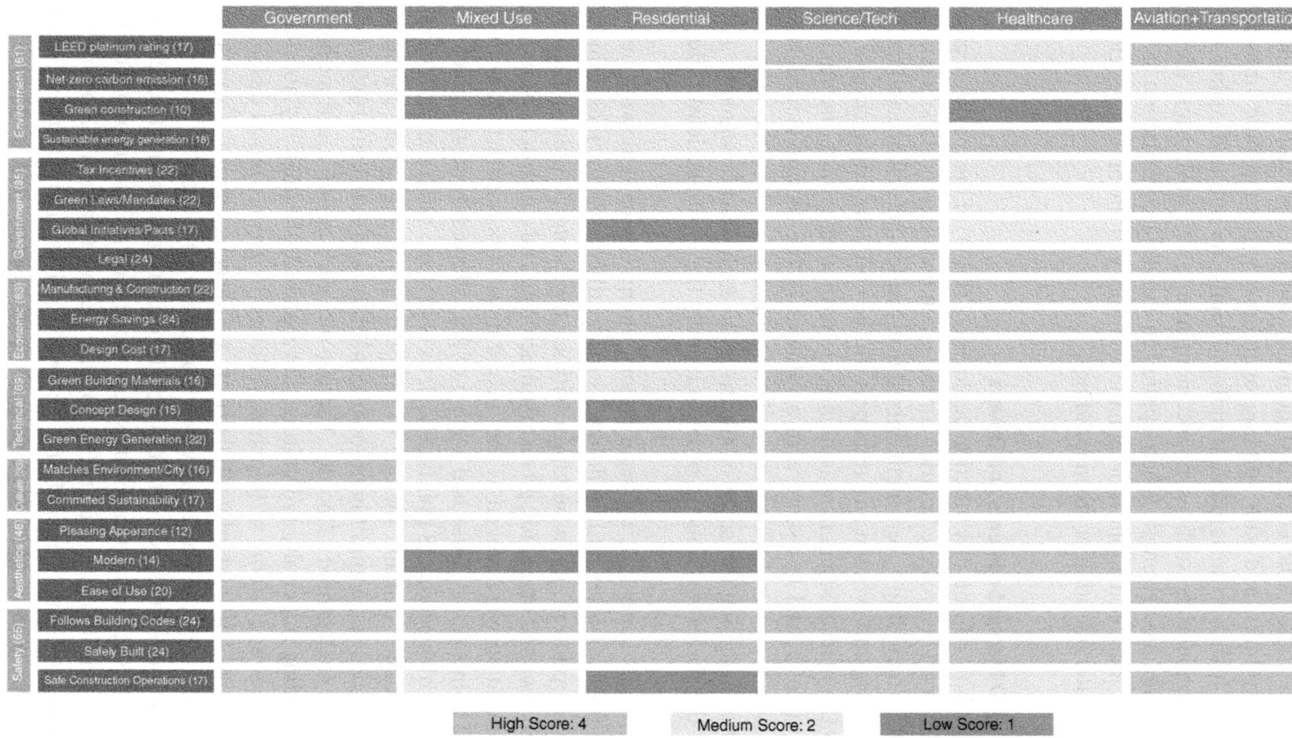

Figure 11.3 QFD market segments and drivers.

11.8.4 Product Feature Mind Map

Figure 11.4 shows the overview of the product feature mind map, which gives holistic coverage of the key features of the product. For the green building, the features are classified into seven categories including safety features, green building design features, resource efficiency features, high-efficiency power generation features, cost-effective features, advertising features, and high-efficiency electronics features.

Safety features of the green building include a backup energy source such as biofuel or wind power to supplement solar-generated power in case of outage, non-toxic building materials, and buildings that are resilient to mechanical or chemical damage and biodegradation. Nanotechnology application in the building sector provides a viable remedy for addressing safety concerns that relate to building construction and operations. The green building design features include optimized design with reduced usage of steel where alternative materials such as CNTs are used instead of steel. The designs also include efficiency status dashboard for monitoring purposes, green roof system design, and operation and maintenance optimization.

Resource efficiency features include daylight-controlled lighting, reduced resources/energy wastage, and water efficiency, while high-efficiency power generation features include sun-tracking solar panels to maximize the amount of solar energy, solar panels with integrated photovoltaic cells that utilize nanoscale components, biofuel generation, geothermal systems, and wind power generation. Green buildings also include cost-effective measures such as reduced operational costs, energy modeling to optimize usage, and reduced construction costs through use of alternative building materials. There are features to advertise the building such as virtual building demos, "design-your-own-building" models, and brand opinions. Lastly, green buildings include high-efficiency electronic systems such as smart lighting and artificial intelligence that allows automated building adjustment, cloud computing operated spaces, and smart heating, ventilation, and air-conditioning (HVAC) systems.

11.8.4.1 QFD of Market Drivers and Product Features

A QFD was conducted that assessed the extent to which the product features relate to the market drivers and to identify the most influential features of the green building that will drive adoption. The QFD of market drivers and product features is shown in Figure 11.5. The QFD applied a scoring criterion in which a high score was assigned 4 points, medium score 2 points and low score 1 point. The market drivers were assigned weights based on the order of their importance, as discussed in Section 11.8.3.1

From Figure 11.4, it is shown that resource efficiency with a score of 49 points is the product feature that is the most important in regard to driving adoption

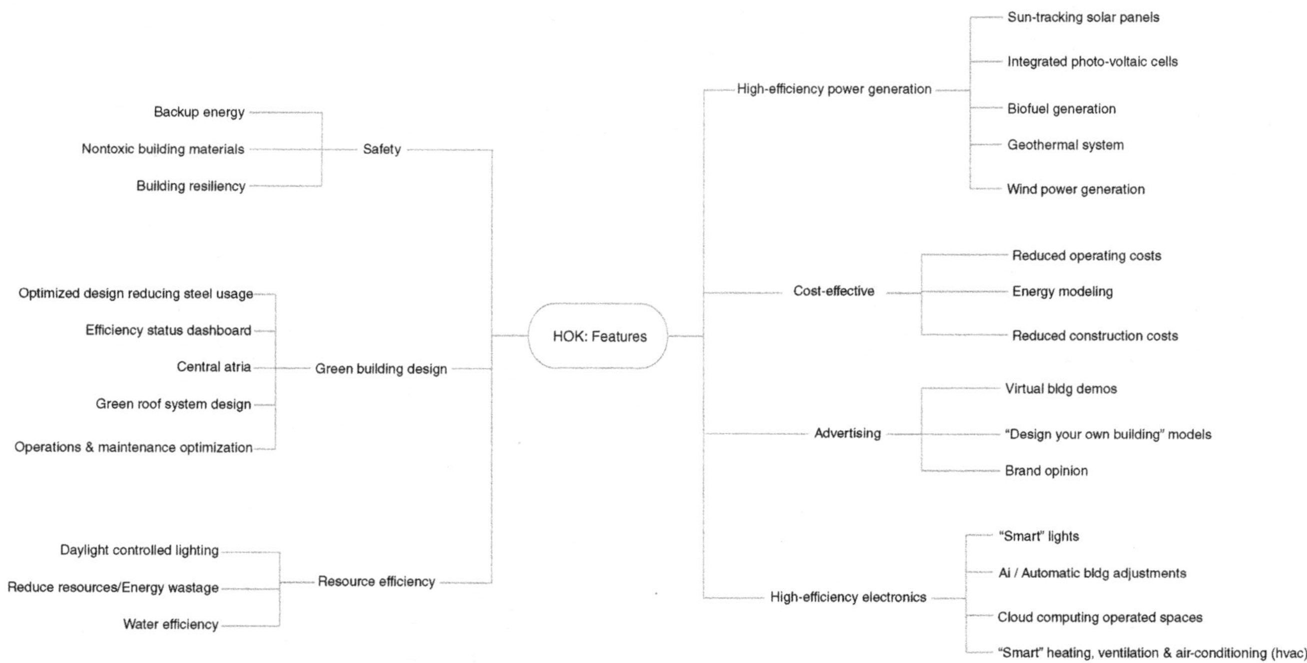

Figure 11.4 Product feature mind map of a green building.

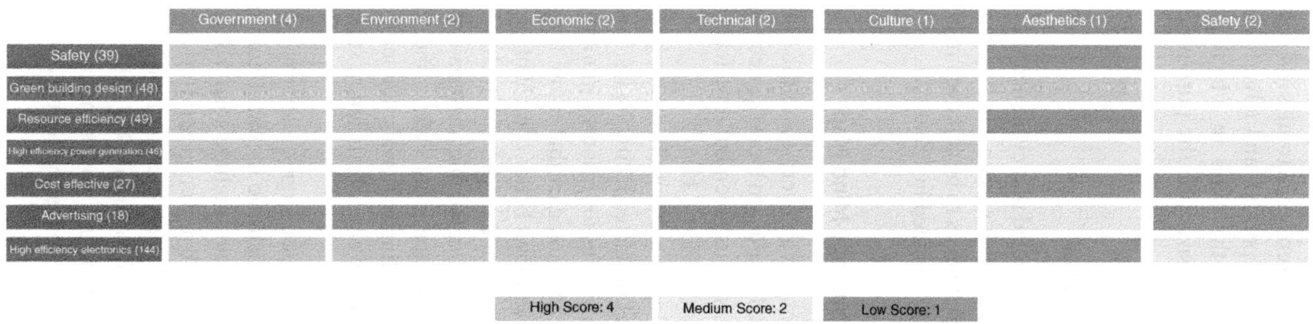

Figure 11.5 QFD of market drivers and product features.

followed by green building design features (48 points), high-efficiency power generation features (46), high-efficiency electronics (44 points), safety features (39 points), cost-effective features (18 points), and advertising features (18 points), respectively. This ranking shows the order in which product features of the green buildings should be emphasized for the purpose of encouraging market adoption.

11.8.5 Technology Gaps Analysis

Technology gap analysis was conducted that evaluated the gaps that must be filled in terms of meeting the requirements for green buildings by comparing the current situation in the market and what needs to be achieved in phase II of the roadmap. The summary of the analysis is provided in Table 11.1.

11.8.6 Green Building Technology Mind Map

The technology mind map provides an overview of the different types of technology that require to be integrated in order to meet the needs and expectations of the different market segments in as far as green building features are concerned. The technology mind map is shown in Figure 11.6.

The green building technologies are classified into five broad categories with each category having a number of related technologies. The first category is smart systems, which includes technologies for resource monitoring, ensuring effective resource usage, user friendly technologies that allow user-defined settings, and electrochromic smart glass that allows better control of energy, depending on the environmental conditions. The second category of technologies is green building technologies, which includes biodegradable materials to minimize pollution, sustainability analysis tools, Occupational Safety and Health Administration (OSHA)-approved recyclable building materials, low-carbon-footprint manufacturing, green insulation to increase energy efficiency, suitable resource sourcing, and water-efficient technologies.

The third category of technologies is green energy generation that includes solar panels that are enhanced to achieve the maximum efficiency level of 46%. Other energy generation technologies include solar windows and roofs, as well as geothermal, biomass, and wind power generation technologies. The fourth category is virtual reality technologies such as holographic virtual reality and haptic suit to maximize user immersion into VR. The last category of technology is energy storage technologies that include power management tools with artificial intelligence, cost-effective energy storage batteries, and thermal energy storage and alternative technologies.

Table 11.1 Technology gaps analysis of the green building product features.

Product features	Current requirement	Phase II requirement	Gaps
Green building design	- Very few net-zero carbons footprint projects.	- All projects are net-zero carbon footprint.	- Need increase in customer acceptance of environmental advantages
Resource efficiency	- Energy generated is supplemented by green energy generation.	- 100% of energy generated is stored or used without any waste or “burning off” of energy and annually meets or exceeds annual use.	- Large-capacity energy storage needs improvement to fully utilize green energy generation.
High-efficiency energy generation	- Current solar up to 23% efficient. - Solar is limited to flat, hard surfaces.	- Solar energy: potentially 46% efficient. - High-efficiency solar, wind, geothermal, biomass energy generation. - Building materials w/integrated PV cells.	- Green energy generation has limited form and low efficiency. - Energy generating building materials are currently unavailable.
Cost-effective	- High costs are associated with cutting-edge systems and materials.	- Increase in government incentives. - Decrease in cost of energy efficient materials and construction. - Decrease in net-zero design costs/simulation.	- Cheap and effective green materials manufacturing. - Accurate and automated design cost and simulation. - Favorable government taxes and incentives for green buildings.
Advertising	Continued project acquisition	- VR/Hologram walkthrough tours of building designs.	- Development of VR/Hologram technology.
High-efficiency electronics	- Some high efficiency electronics are available but not required.	- All electronics derive power from Green Energy Generation. - All electronics communicate effectively with each other.	- Better energy efficiency for lights, heating, cooling, elevators, servers, etc. - “Smart” Buildings/Dwellings

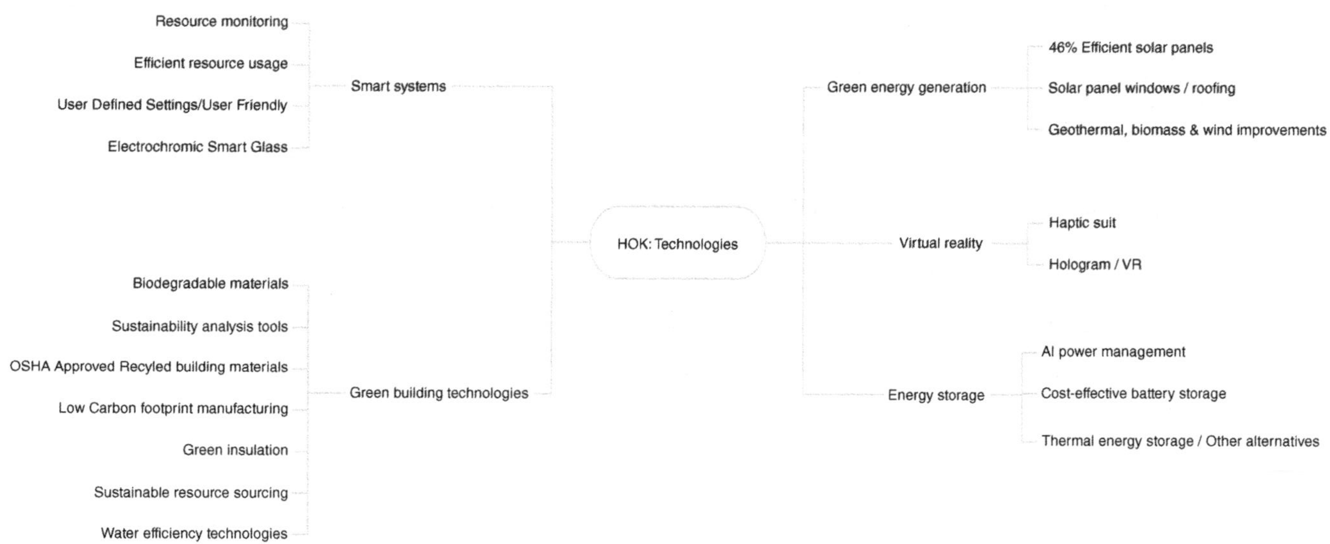

Figure 11.6 Green building technology mind map.

11.8.6.1 QFD of Green Building Technologies and Product Features

A QFD was conducted to assess the green building technologies that are most critical in delivering the key product features. The same scoring criteria was used as with QFDs that were conducted previously. The respective product features are assigned weights in accordance with their importance in driving demand, as discussed in Section 11.8.4.1. Figure 11.7 provides the summary of the QFD results.

It is shown in Figure 11.7 that green building technologies with a score of 75 points play the most important part in meeting the product features that correspond to the key market drivers of green buildings. This is followed by green energy generation and energy storage technologies, which scored 65 points each. Smart systems are ranked third, with a score of 63 points, and virtual reality technologies are the lowest ranked with a score of 30 points. This ranking of technologies shows the technologies that must be most emphasized in order to meet the market needs as far as green buildings are concerned.

11.8.7 Green Building Technology Roadmap

After analyzing the market drivers, the product features and technologies that will influence wide adoption of green buildings, a technology roadmap was developed that placed these factors within a 20-year plan while setting goals to be achieved in the present to medium term (10 years) and in the medium term to long term (10–20 years). The duration covered by the roadmap is divided into two periods: (i) present to 10 years; (ii) 10–20 years. The roadmap also outlines the resources that will be needed in this period in order to achieve the goals. The roadmap is shown in Figure 11.8.

11.8.7.1 Present to Next 10 Years

As shown in the roadmap, in the present to 10 years, emphasis should be put in ensuring that economic and environmental drivers are met. In the next 10 years, the industry should work toward ensuring that all architectural designs and building construction projects are LEED platinum certified. This is the highest level of LEED rating. The building materials and technologies that will fulfill the economic needs should focus on aspects such as enhanced efficiency in aspects such as enhanced energy and water efficiency. Other technologies will include geothermal power generation, integrated PV cells, and wind power generation. Key technologies include generation of biomass and wind energy, solar panel windows and roofing, and sustainable resource sourcing. Application of nano-based building materials will help ensure that the goal of delivering buildings with LEED rating is achieved. In terms of resources, partnering with solar energy companies, and lobbying for favorable laws and incentives will be critical.

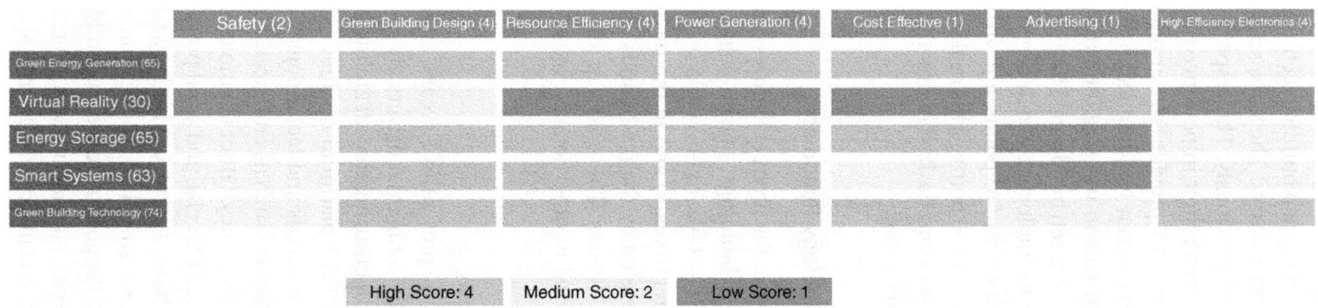

Figure 11.7 QFD of green building technologies and product features.

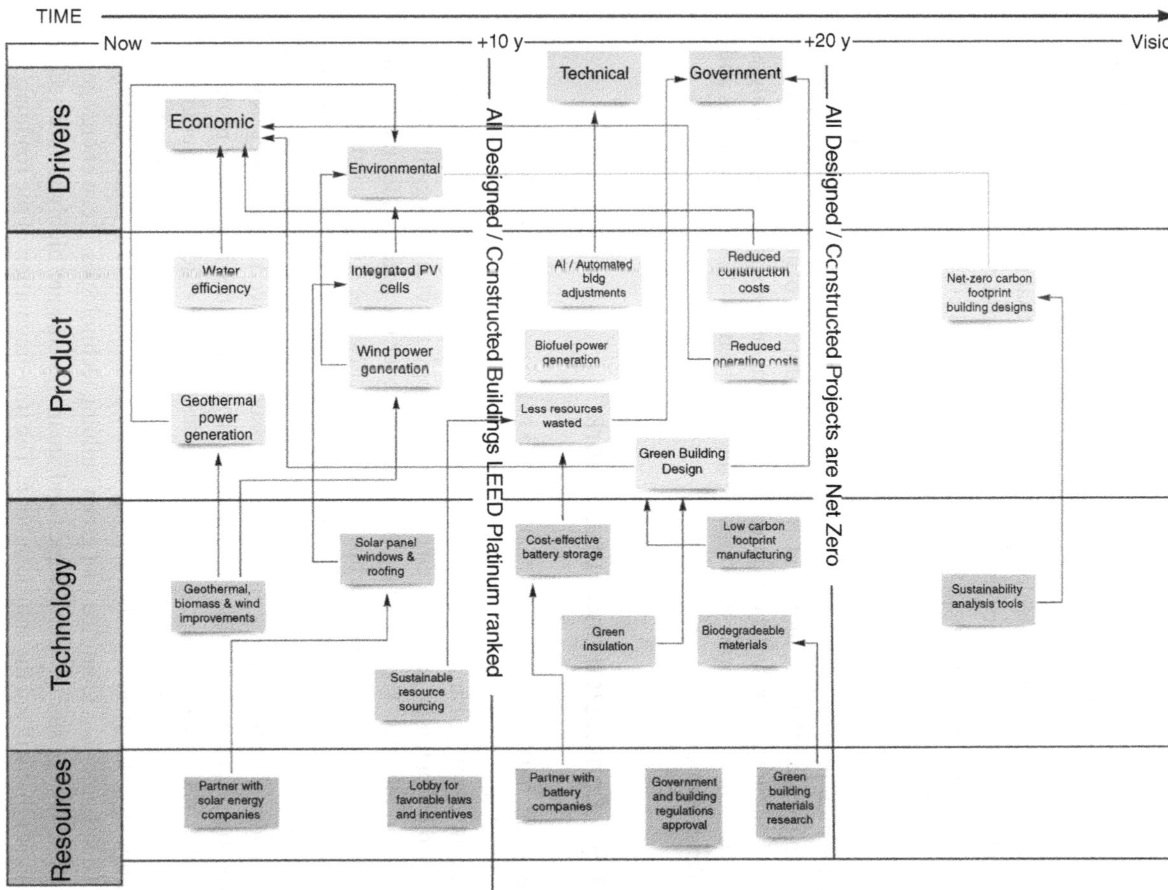

Figure 11.8 Green building technology roadmap.

11.8.7.2 10–20 Years

The long-term goal that needs to be pursued and achieved in the next 10–20 years is to ensure that all designed building and construction projects are net-zero carbon. This will require a greater focus in ensuring that all the technical drivers and government policy drivers are met. The product features that are required to incorporate in green buildings include Ai/augmented building adjustments, biofuel power generation, less wastage of resources, reduced construction and operating costs, and green building designs. The technologies that must be incorporated include cost-effective battery storage, low-carbon-footprint manufacturing, green insulation, and use of biodegradable building materials. In this duration, there will be a need to establish partnerships with battery companies, increase green building materials research, and ensure that governments undertake building approvals based on established green-friendly standards and regulations. Research into the application nanotechnology in making of building materials will be essential in helping to meet the long-term goal. In the future, adoption of net-zero carbon footprint building designs will be required, emphasizing use of sustainability analysis tools to help in monitoring and tracking the effectiveness of the technologies and to direct improvements.

11.9 Conclusion and Future Work

Designing buildings and spaces that respond to the needs of people and the environment through green construction and sustainability have been the focus of both the developed and the developing countries in the last 10 years. This chapter studied the potential growth of emerging technology, nanotechnology, in the construction industry. The aim of the study was to cover the potential for nanotechnology application in green architectural approaches. The study included development of a technology roadmap for green buildings. The technology roadmap shows that the key market drivers of the green architectural trends are broad as demand will be included by environmental issues and government policy as well as economic, technical, culture, aesthetics, and safety factors. In terms of resource features, resource efficiency, green building design, high-efficiency power generation features and high-efficiency electronics, safety features, and cost-effective features we identified to be the most key in driving the demand of green buildings.

This chapter has identified that green building technologies, green energy generation, energy storage technologies and smart systems are highly ranked technologies that will aid the transformation. The implementation of the roadmap in the various stages will require key resources such as partnering with technology companies, lobbying to have supportive laws, partnering with regulatory agencies, and through research and development. The green building technology roadmap provides a clear and concise plan that promises to transform the building and construction industry by applying emerging technologies such as nanotechnology in the development of sustainable solutions.

Using innovative assessment techniques for sustainability analysis of nanotechnology in the construction industry, the research was expanded to integrate technology roadmapping, which is effective in associating the technology development goals, the technologies, and technology products and resources. The potential future research area in the development of green buildings is to undertake an analytical study to create plans for more effective implementation by considering managerial perspectives, financial, environmental, and health matters.

References

Abdellatif, A.A. (2010) Towards sustainable architecture with nanotechnology. In: *Al-Azhar Engineering 11th, International Conference*. Cairo, Egypt: December 21–23, https://doi.org/10.13140/RG.2.1.2930.0247.

Abdin, A.R., El Bakery A.R., and Mohamed, M.A. (2018). The role of nanotechnology in improving the efficiency of energy use with a special reference to glass treated with nontechnology in office buildings. *Ain Shams Engineering Journal* 9: 2671–2682.

Al-Rifaie, W.N., Ahmed, W.K., and Mahdi, O.M. (2018). Green building using nano materials for high performance cement motor composite. In: *2018 5th International Conference on Renewable Energy: Generation and Applications*, 203–206. Al Ain: ICREGA https://doi.org/10.1109/ICREGA.2018.8337578.

Al-Tamimi, N. (2017). A state-of-the-art review of the sustainability and energy efficiency of buildings in Saudi Arabia. *Energy Efficiency* 10 (5): 1129–1141.

Altintas, Z. (2017). *Biosensors and Nanotechnology: Applications in Health Care Diagnistics*. Wiley.

American Hospital Association. (n.d.). Sustainability Roadmap. Building design and construction. http://www.sustainabilityroadmap.org/topics/building.shtml#.Xxbq47PVLIV (accessed July 21, 2020).

Atwa, M., Al-Kattan, A , and Elwan, A. (2015). Towards nano architecture: nanomaterials in architecture – review of functions and applications. *International Journal of Recent Scientific Research* 6 (4): 3551–3564.

Bayda, S., Adeel, M., Tuccinardi, T. et al. (2019 Dec 27). The history of nanoscience and nanotechnology: from chemical-physical applications to nanomedicine. *Molecules* 25 (1): 112. https://doi.org/10.3390/molecules25010112. PMID: 31892180; PMCID: PMC6982820.

Bhushan, B. (2004). *Handbook of Nanotechnology*. Berlin: Springer.

Bjornstrom, J., Martinelli, A., Matic, A. et al. (2004). Accelerating effects of colloidal nano-silica for beneficial calcium-silicate-hydrate formation in cement. *Chemical Physics Letters* 392 (1–3): 242–248.

Boverhofa, D.R., Bramante, C.M., Butala, J.H. et al. (2015). Comparative assessment of nanomaterial definitions and safety evaluation considerations. *Regulatory Toxicology and Pharmacology* 73 (1): 137–150.

Candan, Z. and Akbulut, T. (2013). Developing environmentally friendly wood composites panels by nanotechnology. *BioResources* 8 (3): 3590–3598.

Carriere, M., Schoenau, G.J., and Besant, R.W. (1999). Investigation of some large building energy conservation opportunities using the doe-2 model. *Energy Conversion and Management* 40 (8): 861–872.

Cassar, L., Pepe, C., Tognon, G. et al. (2003). White cement for architectural concrete possessing photocatalytic properties. In: *Proceedings of the 11th international congress on the Chemistry of Cement*. Durban: ICCC.

Chung, W. (2012). Using the fuzzy linear regression method to benchmark the energy efficiency of commercial buildings. *Applied Energy* 95: 45–49.

Dahlan, A.S. (2019). Smart and functional materials based nanomaterials in construction styles in Nano-architecture. *Silicon* 11: 949–1953. https://doi.org/10.1007/s12633-018-0015-x.

Demirdoven, J.B. and Karacar, P. (2015). Green nano-materials with examples of applications. In: *Green symposium*. Mimar Sinan Fine Arts University, Faculty of Architecture Istanbul, Türkiye, 15–17 April.

Du, S., Wu, J., Alshareedah, O., and Shi, X. (2019). Nanotechnology in cement-based materials: a review of durability modeling and advanced characterization. *Nanomaterials (Basel)* 1213: https://doi.org/10.3390/nano9091213.

Fahmy, H.M. (2010). *Nanomaterials & Architecture, "Sustainable Nanoarchitecture"*, 1–100. LAP LAMBERT Academic Publishing.

Federspiel, C., Zhang, Q., and Arens, E. (2002). Model-based benchmarking with application to laboratory buildings. *Energy and Buildings* 34 (3): 203–214.

Flores, I., Sobolev, K., Torres, L.M. et al. (2010). Performance of cement systems with nano-SiO_2 particles produced using sol-gel method. In: *TRB First International Conference in North America on Nanotechnology in Cement and Concrete*. Irvine: TRB May 5–7.

Gann, D. (2002). *A Review of Nanotechnology and its Potential Applications for Construction*. SPRU, University of Sussex.

Garcia, M.L. and Bray, O.H. (1997). Fundamentals of technology Roadmapping. *Sandia National Laboratories* 1–29.

García, Ó., Alonso, R.S., Prieto, J., and Corchado, J.M. (2017). Energy efficiency in public buildings through context-aware social computing. *Sensors* 17 (4): https://doi.org/10.3390/s17040826.

Gillani, A.S., Khitab, A., Ahmad, S. et al. (2017). Improving the mechanical performance of cement composites by carbon nanotubes addition. *Procedia Structural Integrity* 3: 11–17.

Goffredo, G.B., Citterio, B., Biavasco, F. et al. (2017). Nanotechnology on wood: the effect of photovoltaic nanocoatings against aspergillus Niger. *Journal of Cultural Heritage* 27: 125–136.

Gourlis, G. and Kovacic, I. (2016). A study on building performance analysis for energy retrofit of existing industrial facilities. *Applied Energy* 184: 1389–1399.

Green, B.H. (2008). Development of a high-density cementitious rock-matching grout using nano-particles SP-254. In: *Nanotechnology of Concrete: Recent Developments and Future Perspectives* (ed. K. Sobolev and S.P. Shah), 121–131. Farmington Hills: American Concrete Institute.

Han, B., Yu, X., and Kwon, E. (2009). A self-sensing carbon nanotube/cement composite for traffic monitoring. *Nanotechnology* 20: 1–5.

Hernandez, P., Burke, K., and Lewis, J.O. (2008). Development of energy performance benchmarks and building energy ratings for non-domestic buildings: an example for Irish primary schools. *Energy and Buildings* 40 (3): 249–254.

Iribarren, D., Marvuglia, A., Hild, P. et al. (2015). Life cycle assessment and data envelopment analysis approach for the selection of building components according to their environmental impact efficiency: a case study for external walls. *Journal of Cleaner Production* 87: 707–716.

Islam, S., Al Awadh, M.A., and Khan, R.A. (2015). Recent trends in the green building concept. *International Journal of Advanced and Innovative Research* 5 (1): 189–192, 2278–7844.

Jamshidi, A., Kurumisawa, K., White, G. et al. (2020). Use of nanotechnology in concrete pavements. *Smart Nanoconcretes and Cement-Based Materials* 16: 383–401. https://doi.org/10.1016/B978-0-12-817854-6.00016-7.

Jelle, B.P., Gustavsen, A., and Beatens, R. (2012). Innovative high performance thermal building insulation materials – Today's state-of-art and beyond tomorrow. In: *Building Enclosure Science & Technology*. Atlanta, GA: BEST 3–2012 2–4 April.

Khandve, P.V. (2014). Nanotechnology for building materials. *International Journal of Basic and Applied Research Special* 04: 146–151.

Kohai, M., Ismail, O.S., Elnawawy, O., and Ragab, K.S. (2018). Performance of the lightweight concrete with available nano-silica in case of full replacement of coarse aggregate. *International Journal of Scientific and Engineering Research* 9 (1): 223–231.

Kumar, R., Mathur, R, Mishra, A.K. (2020). Opportunities and challenges for use of nanotechnology in cement based materials. NBM & CW. https://www.nbmcw.com/tech-articles/others-article/25054-use-of-nanotechnology-in-cement-based-materials.html (accessed June 6, 2020).

Li, G. (2004). Properties of high-volume fly ash concrete incorporating nano-SiO. *Cement and Concrete Research* 34: 1043–1049.

Li, H., Xiao, H.G., Yuan, J., and Ou, J. (2004). Microstructure of cement mortar with nano-particles. *Composites: Part B* 35: 185–189.

Makar, J.M., Margeson, J., and Luh, J. (2005). Carbon nanotube/cement composites – Early results and potential applications. In: *Proceedings of 3rd International Conference on Construction Materials: Performance, Innovations and Structural Implications*. Vancouver, (August 22–24), 1–10.

Marzouk, M., Metawie, M., Hisham, M. et al. (2014). Modeling sustainable building materials in Saudi Arabia. *Computing in Civil and Building Engineering* 1546–1553.

Mohajerani, A., Burnett, L., Smith, J. et al. (2019). Nanoparticles in construction materials and other applications and implications of nanoparticle use. *Materials (Basel, Switzerland)* 12 (19): 3052.

Mosly, I. (2015). Barriers to diffusion and adaptation of green building in Saudi Arabia. *Journal of Management and Sustainability* 5 (4): 104–114.

Niroumand, H. and Zain, M.F.M. (2013). The role of nanomaterials in nanoarchitecture. *Procedia – Social and Behavioral Sciences* 89: 27–30.

Niroumand, H., Zain, M.F.M., and Jamil, M. (2013). The role of nanotechnology in architecture and built environment. *Procedia - Social and Behavioral Sciences* 89: 10–15.

OECD (n.d.). Opportunities and risks of nanotechnologies report in cooperation with the OECD International Futures Programme. http://www.oecd.org/chemicalsafety/nanosafety/44108334.pdf

Pachero-Torgal, F. and Jalali, S. (2011). Nanotechnology: advantages and drawbacks in the field of construction and building materials. *Construction and Building Materials* 30 (25): 582–590.

Papadpoulos, A.N. and Taghiyari, H.R. (2019). Innovative wood surface treatment based on nanotechnology. *Coatings* 9 (866): 1–14. https://doi.org/10.3390/coatings9120866.

Phaal, R., Farrukh, C.J., and Probert, D.R. (2004). Technology roadmapping – a planning framework for evolution and revolution. *Technology Forecasting and Social Change* 71: 5–26.

Plassard, C., Lesniewska, E., Pochard, I., and Nonat, A.A. (2004). Investigation of the surface structure and elastic properties of calcium silicate hydrates at the nanoscaled. *Ultramicroscopy* 100 (3–4): 331–338.

Poole, C.P. and Owens, F.J. (2003). *Introduction to Nanotechnology*. Hoboken: Wiley.

Porro, A., Dolado, J., Campillo, I. et al. (2005). *Effects of Nanosilica Additions on Cement Pastes (Applications of Nanotechnology in Concrete Design)*. London: Thomas Telford.

Qing, Y., Zenan, Z., Deyu, K., and Rongshen, C. (2007). Influence of nano-SiO2 addition on properties of hardened cement paste as compared with silica fume. *Construction and Building Materials* 21 (3): 539–545.

Rains, M.T., Rudie, A.W., and Wegner, T.H. (2014). The promise of wood-based nanotechnology. *The Consultant* 34–39.

Sanchez, F. and Sobolev, K. (2010). Nanotechnology in concrete – a review. *Construction and Building Materials* 24 (11): 2060–2071.

Sanchez, F., Zhang, L., and Ince, C. (2009). Multi-scale performance and durability of carbon nanofiber/cement composites. In: *Nanotechnology in Construction: Proceedings of the NICOM3 (3rd International Symposium on Nanotechnology in Construction)* (ed. Z. Bittnar, P.J.M. Bartos, J. Nemecek, et al.), 345–350. Prague. Berlin: Springer.

Sapna, R. and Dinesh, K. (2020). Nanotechnology for sustainable future. In: *Handbook of Nanomaterials for Manufacturing Applications* (ed. C.M. Hussain). Elsevier https://doi.org/10.1016/b978-0-12-821381-0.00019-3.

Sauraav (2012). Application of nanotechnology in building materials. *International Journal of Engineering Research and Applications (IJERA)* 2 (5): 1077–1082.

Scheuer, C., Keoleian, G.A., and Reppe, P. (2003). Life cycle energy and environmental performance of a new university building: modeling challenges and design implications. *Energy and Buildings* 35 (10): 1049–1064.

Setyowati, E. and Khusnan, P. (2015). The polymer brick as nano-technology based material to support green building construction. In: *Conference proceedings. The Polymer Brick as Nano-Technology-based Material to Support Green Building Construction*. Singapore.

Setyowati, E. and Purwanto, P. (2015). The polymer brick as Nano-technoogy based material to support green building construction. *Conference Proceedings* https://doi.org/10.5176/2301-394X_ACE15.32.

Shah, S.P., Konsta-Gdoutos, M.S., Metaxa, Z.S., and Mondal, P. (2009). Nanoscale modification of cementitious materials. In: *Nanotechnology in Construction: Proceedings of the NICOM3 (3rd International Symposium on Nanotechnology in Construction)* (ed. Z. Bittnar, P.J.M. Bartos, J. Nemecek, et al.), 125–130. Prague. Berlin.

Shim, H., Kim, T., and Choi, G. (2019). Technology roadmap for eco-friendly building. *Energies* 12: 804.

Sikora, P., Elrahman, M.A., and Stephan, D. (2018). The influence of nanomaterials on the thermal resistance of cement-based composites-a review. *Nanomaterials* 8 (465): 1–33. https://doi.org/10.3390/nano8070465.

Sueyoshi, T., Yuan, Y., and Goto, M. (2017). A literature study for DEA applied to energy and environment. *Energy Economics* 62: 104–124.

Taghriyari, H.R. (2014). Nanotechnology in wood and wood-composite materials. *Journal of Nanomaterials & Molecular Nanotechnology* 3 (1): 1–2.

Tichi, A.H., Bari, E., and Nicholas, D.D. (2018). How nano-wollastonite can change the fundamental properties of wood fibre rice straw composites. *IET Nanobiotechnology* 1–4. https://doi.org/10.1049/iet-nbt.2018.5163.

Wang, N. and Adeli, H. (2014). Sustainable building design. *Journal of Civil Engineering and Management* 20 (1): 1–10.

Weeta. (2020). The Green Buildings in Saudi Arabia and their impact: a closer look. https://www.weetas.com/article/green-buildings-in-saudi-arabia (accessed June 18, 2020).

Xia, J. et al. (2014). Comparison of building energy use data between the United States and China. *Energy and Buildings* 78: 165–175.

Yasin, L.S. and Atiyat, D.I. (2017). The effect of nanotechnology on architecture. *Int'l Journal of Advances in Agricultural & Environmental Engg. (IJAAEE)* 4 (1): 125–129.

Zhu, W., Bartos, P.J., and Porro, A. (2004). Application of nanotechnology in construction. *Materials and Structures* 37: 649–658.

12

Technology Roadmapping: Standards of Healthcare Data Cybersecurity

Alison Nalven, Courtney Wright, Jennifer L. Zeitouni, Nolan Thompson, Sara Ferdousi, and Saumya Saxena

Portland State University, Portland, OR 97201, USA

12.1 Introduction

Maintaining cybersecurity in a continuously evolving digital environment is a challenge for all organizations today and the health care sector is not immune. In 2021, there was a 35% increase in cyberattacks against healthcare systems affecting nearly 45 million people (Landi 2022). Organizations managing health data are particularly more vulnerable to cyberattacks due to the critical information possessed by them, which is valuable on the dark web, be used in ransomware attacks, or otherwise. The need to develop stricter policy standards to protect individuals' healthcare data is more urgent than ever. Another major recent concern is that the digital paper trail from GPS data, texts, or period tracking apps could be used to convict women who get abortion services or those who help women obtain those services in states where abortion is now illegal (Hill 2022; Torchinsky 2022; Koebler & Merlan 2022). While the Health Insurance Portability and Accountability Act (HIPAA) was signed into law in 1996 (Schmidt 2022), it does not actually protect all patients' health data, as it only covers communication between the patient and "covered" entities, such as doctors, hospitals, pharmacies, and insurers (Klosowski 2021). There is a major flaw in HIPAA as it does not protect data on IoT devices or in apps. There is actually no single US law that covers privacy of all types of data, and only three states – California, Virginia, and Colorado – have comprehensive consumer privacy laws (Klosowski 2021). The world has changed in the past 26 years since the creation of HIPAA, and

Future-Oriented Technology Assessment: A Manager's Guide with Case Applications, First Edition.
Edited by Haydar Yalçın and Tugrul U. Daim.

healthcare policy standards need to evolve as well to protect users' health data. As such, this chapter will consider the following three areas through research:

(1) Privacy of women's health data in the post-*Roe* era
(2) Security of medical devices in IoT
(3) Privacy of personal health data from insurance companies

The goal of this chapter is to create a technology policy roadmap that outlines the market drivers, technologies, products, policies, and resources necessary to develop improved standards to safeguard critical healthcare data.

12.2 Methodology

The team adapted a technology roadmap format to help define the issues with cybersecurity policies, along with providing standards that can be used to create policy. The start to the roadmap is the literature review. Next, the team used the gathered information to create a set of market segments to be used, as well as a driver mind map. The drivers were then measured against the market segments in a table, ranking each driver against each market segment with a value of 1, 2, or 4, to determine which drivers were the most important in this roadmap. This creates a driver vs. market segment QFD. After the drivers were ranked, project features were identified using a combination of the literature review and driver rankings. Once these features were fully defined in a mind map, another QFD was generated to determine the value of each product feature. This was done by measuring the features vs. the drivers. Up until this point, this is a fairly normal process of generating a technology roadmap. However, the goal of the team was to create a health data and cybersecurity protection standards roadmap. In addition to researching specific emerging technologies that can help create the desired product features, the team researched policy in order to better understand what policies are currently in place and what policies would be needed to fulfill the goals of the product features. This approach to understanding policy's interaction with technology in emerging markets was employed in another body of research on the growth of wind energy industry in China (Zhou et al. 2011). From the policy research, gaps were then identified, along with a policy mind map to define specific policies to be used. These policies were then measured against product features in another QFD to determine the most valuable policies. Afterward, the team had enough information to create the final roadmap. All of these steps were done collaboratively by the team, discussing and working together to create QFDs and mind maps that were well defined and thought out. The final step was to link all the aspects together in the technology roadmap, providing information on the drivers, product features, technology (policies) and resources available, and how they relate to each other. Conclusions are drawn by using this roadmap to determine which policies are most important for the team to create an effective set of standards for cybersecurity around personal data in healthcare.

12.3 Literature Review

In today's digital world, cybersecurity in healthcare plays a pivotal role in the organization's processes and functions. As technology progresses, so do the threats and crimes associated with it. Many articles and papers were reviewed for this project to understand the challenges faced in healthcare cybersecurity and what type of policies can be used to address these challenges. The most significant findings in the conducted research are briefly discussed here.

12.3.1 Cybersecurity of Healthcare IoT-Based Systems: Regulation and Case-oriented Assessment, 2018

This chapter also discusses the exponentially growing IoT in the field of healthcare and proposes a normative hierarchical model of the international cybersecurity standards. The authors have identified four major issues in networked healthcare and medical devices – random failures, privacy, deliberate disruption, and malware disruption. These issues and the fact that international regulations and standards do not include any formal technique for cybersecurity assessment of IoT healthcare solutions make the implementation of the assessment process one-sided and complicated. That is why the authors have developed an approach that considers multiple aspects of development, regulations, standards, security, risks, etc.

The healthcare system is divided into seven layers, and for each layer the standards and regulations required for security are added resulting in the following hierarchical model (Strielkina et al. 2018).

The roadmapping methodology discussed in this chapter to develop a research roadmap for cybercrime and cyberterrorism. It addresses the fact that cybercrime and cyberterrorism co-evolve with their environment and thus makes the roadmapping task challenging. The authors have developed a four-step methodology based on scenario analysis techniques (Ariu et al. 2015a,b) (Figure 12.1).

12.3.1.1 Step 1: Actual State Scenario

The actual state is described as a scenario and consists of a short summary of the contextual environment, followed by the existing cybercrimes and cyberthreats along with the available defenses.

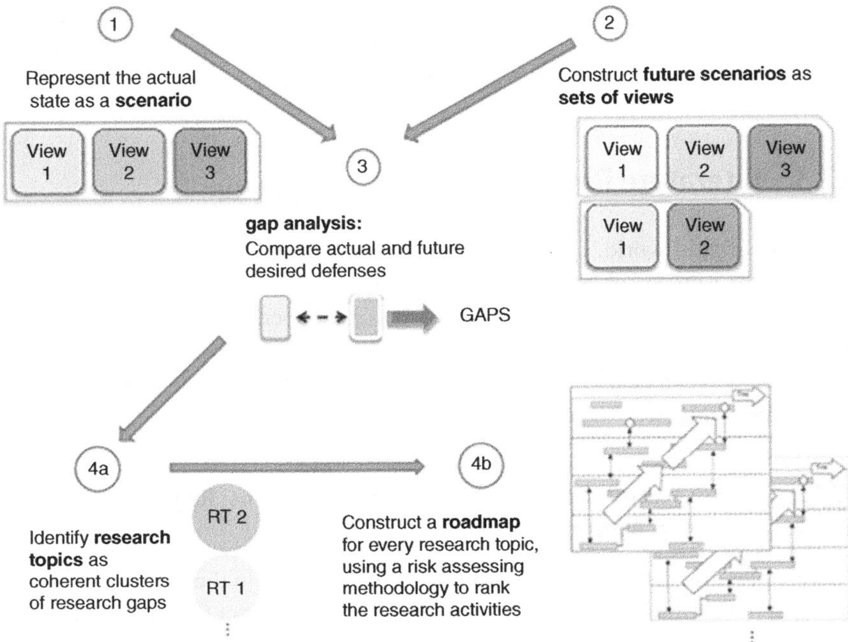

Figure 12.1 Proposed methodology.

12.3.1.2 Step 2: Scenario Building

The goal of this step is to produce a set of possible future scenarios, which should explore a range of potential evolutions of cybercrimes and cyberthreats and of their contextual environment as wide as possible. The step highlights the threats that can emerge, and the corresponding, desirable defenses.

12.3.1.3 Step 3: Gap Analysis

This step identifies the research gaps that arise from the comparison of each of the future views against the actual state views. Research gaps are identified by tracking the changes of the threats from the actual to the future scenarios and comparing the corresponding existing and desired defenses.

12.3.1.4 Step 4: Roadmap Construction

The roadmap addressing the research gaps is built as follows:

- Define a set of broad research topics that are related and can be addressed by suitable research actions.
- Prioritize the research topics using an appropriate risk assessment method.

- Define a vertical roadmap for each identified research topic. This includes first identifying the research actions required to address the identified gaps and then describing the actions in a clear time frame, by considering their interdependencies.

12.3.2 Internet of Things Realizing the Potential of a Trusted Smart World, 2018

This section examines the policy changes for IoT and lists issues to be considered for effective policy. It has divided IoT applications into three broad categories – Industrial, Public Space, and Consumer. It highlights that policies and technologies become more effective if based on the understanding of interdependent social and technical factors. The paper evaluates the following four major frameworks that apply to aspects of IoT. The chapter describes that these frameworks have been developed separately and need to be aligned with each other in their development stages (Taylor et al. 2018).

12.4 Results

12.4.1 Market Segments

As part of this project, the team performed individual research on the topic and identified the market segments. Primary markets deduced through literature review and brainstorming sessions are healthcare institutions, technology industry, consumers, and government.

12.4.1.1 Healthcare Institutions

Every healthcare provider, regardless of size, who electronically transmits health information in connection with certain transactions, is a HIPAA covered entity (Office for Civil Rights, 2013). Healthcare providers include all "providers of services" (e.g. institutional providers such as hospitals) and "providers of medical or health services" (e.g. noninstitutional providers such as physicians, dentists, and other practitioners) as defined by Medicare, and any other person or organization that furnishes, bills, or is paid for healthcare (Office for Civil Rights 2013). Policies on healthcare data protection have a major impact on these institutions. While being HIPAA-covered entities, they must comply with the rules requirements to protect the privacy and security of health information and must provide individuals with certain rights with respect to their health information (Office for Civil Rights 2013); any new laws or policies will affect them too. They are often the data originators, and other entities can access data through them. Once data leaves their control, healthcare providers may or may not be held responsible for the privacy and security of that information.

12.4.1.2 Technology Industry

Big technology companies such as Google, Amazon, Microsoft, and Facebook are all looking for an entry into the healthcare market (Thomason, 2021). There are vast sums of money to be made based on health records. Lobbying efforts by Silicon Valley encouraged the Department of Health and Human Services to draft the Cures Act, which will facilitate data sharing with software companies (Dutcher & Zatkowsky 2021). This in turn will potentially allow these big tech companies a dominant position in the service-based cloud storage of health records.

Health data collected by wearable devices and apps such as Fitbit and Apple Watch are not protected by HIPAA (Dutcher & Zatkowsky 2021; Klosowski 2021). The latest Apple Watches can take echocardiograms and detect falls, but the data set created and the communication systems necessary for transferring this data make it hard to protect privacy (Dutcher & Zatkowsky 2021). Health researchers predict a future with smart homes, which feature technologies like mirrors that detect skin and heart conditions, and mattresses that check vital signs may not be subject to HIPAA regulation if the companies developing the devices are outside the healthcare system (Drees 2020).

12.4.1.3 Consumers

Consumers include patients and anyone who uses healthcare applications, including mental health apps, Flo Health, and wearables such as Fitbit, Apple Watch, or any health-monitoring device, as well as anyone who purchases healthcare products over the internet. Health data protection policies should be created based on consumers interest. In many cases, consumers are not savvy enough to understand that their personal data can be sold once it's outside the healthcare system (Ravindranath 2020) or how many times the data they input through health applications are sold to multiple parties. As a ramification of *Roe v. Wade* overturned by the supreme court, female health information may be used to criminalize consumers. Millions of digital traces from social media accounts and devices that are plugged into algorithms to predict health outcomes, these may be expanded from data mining and could be used for consumer profiling or marketing purposes (Drees 2020).

12.4.1.4 Government

This segment includes both federal, Congress, Department of Health and Human Services Federal Trade Commission, and state-level government. HIPPA, a federal law, was enacted on August 21 1996 (Office for Civil Rights, 2013). However, not all health-related information falls under its purview (Dutcher & Zatkowsky 2021; Klosowski 2021). Due to the lack of comprehensive federal privacy laws regulating health technology companies, some states enacted their own data privacy laws. Currently, three states: California, Colorado, and Virginia have comprehensive consumer privacy laws. Four other states, Massachusetts, New York, North Carolina, and Pennsylvania, have serious comprehensive consumer data privacy proposals in committee (Klosowski 2021).

12.4.2 Market Drivers

Figure 12.2 illustrates the market drivers. Details are provided in the following sections.

12.4.2.1 Technology

Advancement in technology and its omnipresence in current daily life is a major driver for data privacy policies. The rapid pace of health application development and the bureaucratic delay of policy enactment have created a gray area where tech companies can collect, store, and share consumer health data without being accountable. The Federal Trade Commission (FTC) utilizes existing statutory authorities and its power to ensure apps are keeping their promises to consumers around the handling of their sensitive health information; however, there is a need for comprehensive federal privacy legislation to regulate personal health information collected by latest technologies. Such legislation will encourage security measures around health apps and provide consumers of such technologies to be aware of how their PHI is being handled.

12.4.2.2 Safety

Personal health information data breach can cause direct harm to a patient or a consumer of health apps. Cyberattacks on electronic health records and other systems also pose a risk to patient privacy because hackers access to PHI and other sensitive information (Riggi 2022). Particularly, women's fertility health information could be used against them in the states where abortion is banned. HIPAA has provisions for sharing health data safely for research purposes; however, data collected by various apps are not covered by those. In 2021, Flo Health took control of users' sensitive fertility data and shared it with third parties despite expressing privacy claims. This left consumers feeling outraged, victimized, and violated (Reicin 2022). Incidents as such calls for more policies around data safety and privacy. In turn, the tech companies will adopt and change their behavior around consumer PHI.

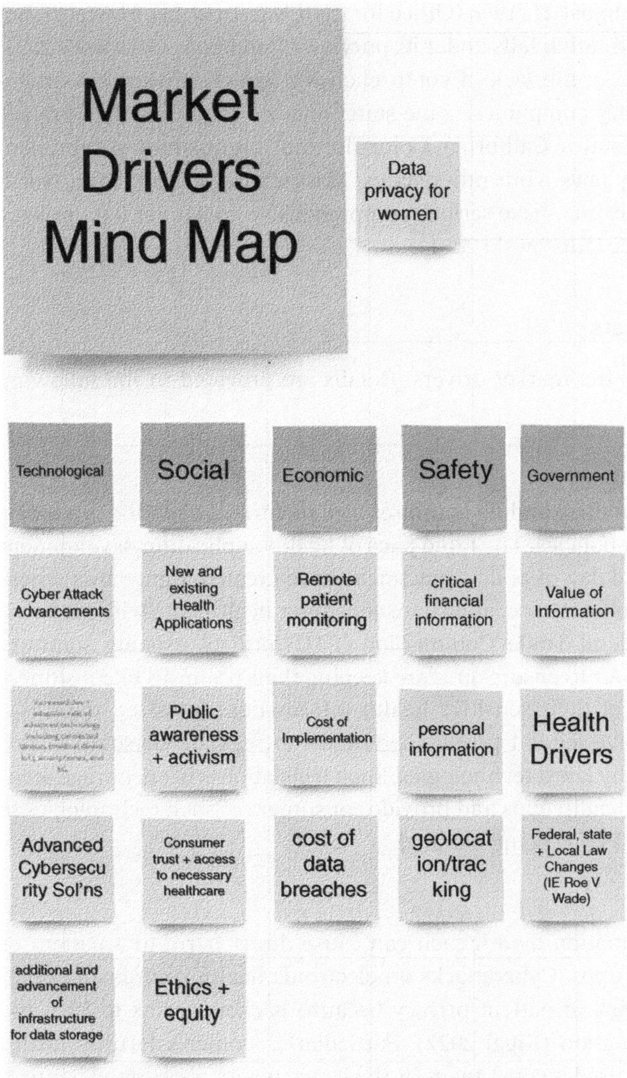

Figure 12.2 Market drivers mind map.

12.4.2.3 Social and Economic

In the age of "surveillance capitalism," data collected through nonstop interactions with digital technologies are being commodified and traded. New digital health technologies and increased awareness about public health due to the COVID-19 pandemic may pave the way for a fundamental shift in healthcare data privacy (Drees 2020). Rise in the demand for virtual care enabled technology companies to proliferate the healthcare market. This has made healthcare a lucrative field. Soaring healthcare costs contribute to the reliance on healthcare applications. Seemingly free health applications often profit through trading in private data of the users. This seismic shift in society's view on data privacy is a major driver for public policies on protecting personal health information.

12.4.2.4 Government

Government plays the most important role in creating public policies to regulate how companies handle consumer data. The Federal Trade Commission (FTC)'s Health Breach Notification Rule ensures that entities who are not covered by HIPAA face accountability when consumers' sensitive health information is compromised (McGuire Woods 2021; HHS 2009).

12.4.3 Market Segments vs. Drivers Mind Map and QFD

The QFD is an imperative part of the technology roadmapping process. With the drivers and segments defined, the team was able to move forward with the comparisons. The mind map for the drivers is shown in Figure 12.3.

Each section of drivers has multiple specific portions that were measured against each of the market segments to create the QFD. When considering what the safety driver was, the team specifically thought of data safety. Is it safe from those who wish to use it maliciously? Is it appropriately stored? Are there laws defining how the data can be used or sold? All of these thoughts come from the safety driver. The team used this to guide the choices in the QFD. When considering personal info in regards to healthcare + data and corporation segments were rated highly, along with geolocation/tracking info along all segments. Consumers of the data do not really care how they use their own information. They want it stored appropriately, but it's much more important to them how corporations use the data they gather from consumers. This is the thought process the team used to rank all the drivers vs. the segments in the QFD (Figure 12.4).

12.4.4 Market Drivers vs. Product Features Mind Map and QFD

After the team ranked the drivers, creating a product features mind map, then comparing the drivers vs. the product features was the next step (Figure 12.5). The team produced the following product features:

- Dynamic Change, meaning the standards can update as technology and needs advance.
- Data Protection, specifying exactly how data can be used, stored, and required levels of protection from malicious attacks.

Figure 12.3 Market segments vs. drivers QFD.

- Effective Federal Policy, meaning it creates a baseline that all states must adhere to, which covers all individuals.
- Data Breach Response, which defines what the company is required to do when a data breach occurs, looking at timeline for a fix, timeline for updated security measures, and restitution requirements for value of data lost.
- Inform Consumers looks at how a company must let a consumer know what their data is being used for, along with the protections they have in place for said data. The mind map and QFD are found on this page (Figure 12.6).

12.4.5 Policy Mind Map

Much of the technology exists to adequately protect consumer data privacy. Public policy will be the key catalyst to expedite the implementation of cybersecurity technology by corporations. Public officials and nonprofit organizations such as

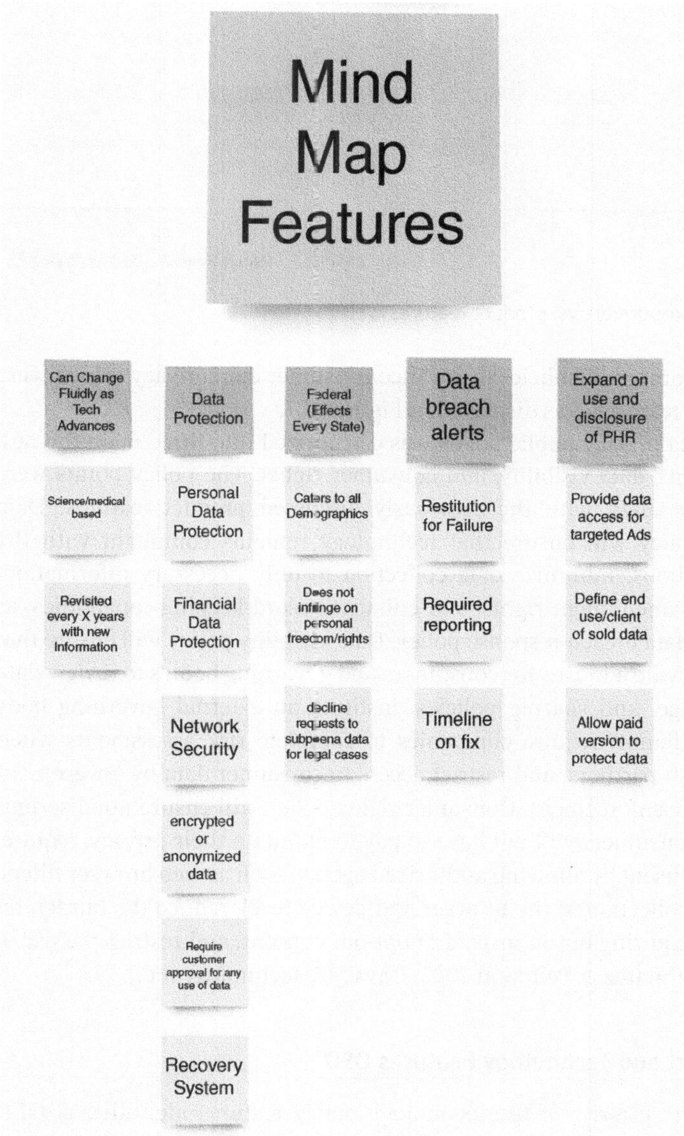

Figure 12.4 Market drivers vs. product features mind map.

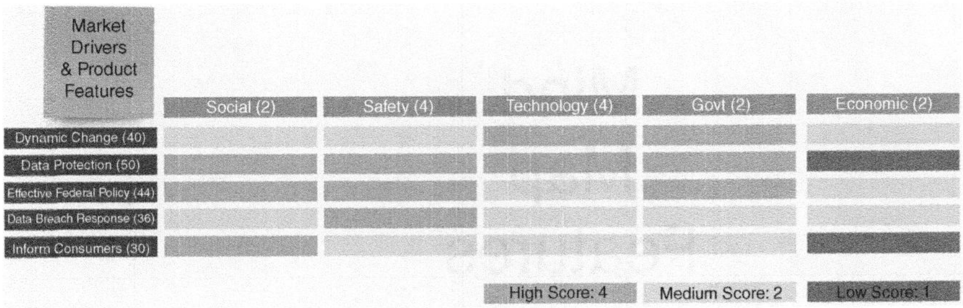

Figure 12.5 Market drivers vs. product features QFD.

the ACLU have proposed policies to advance consumer data privacy, which were reviewed in this research – as displayed in Figure 12.7.

Consumer data privacy public policy was categorized into three main themes: data management, data visibility, and consumer rights. The policy points were then mapped to align with the previously discussed product features. Data management policy will ensure that technology remains compliant with the dynamic regulations, minimize data collection to only necessary information, expand Fourth Amendment rights, and enhance record-keeping capabilities, as well as enforce data breach response policy. Data visibility policy will ensure that corporations provide a means for consumers and governing bodies to review data collection, storage, and sharing policies, institute an external governing body to ensure compliance, require companies to complete risk assessments when sharing data with partners, and restrict access to consumer data by government agencies and law enforcement. Consumer rights policy will ensure nondiscrimination so that consumers will not have to pay to maintain their privacy, require, and prioritize consent by allowing authorized agents, such as web browser filters, to restrict data collection at the browser and device level, reduce the burden on consumers by requiring opt-in instead of opt-out consent, and restrict the use of minors' data (Rahnama & Pentland 2022; Privacy & technology n.d.).

12.4.6 Product and Technology Features QFD

To determine the priority of the technology policy features identified, a QFD was conducted in consideration of the desired product features. The product feature weights were based on those determined with the Market Driver and Product Feature QFD outlined in Section 12.4.4. Data visibility policy was determined to be the most important technology feature, as it significantly impacts data protection, effective federal policy, data breach response, and informing consumers. Consumer rights policy was the second most important technology feature, as it significantly influences data protection and informing consumers.

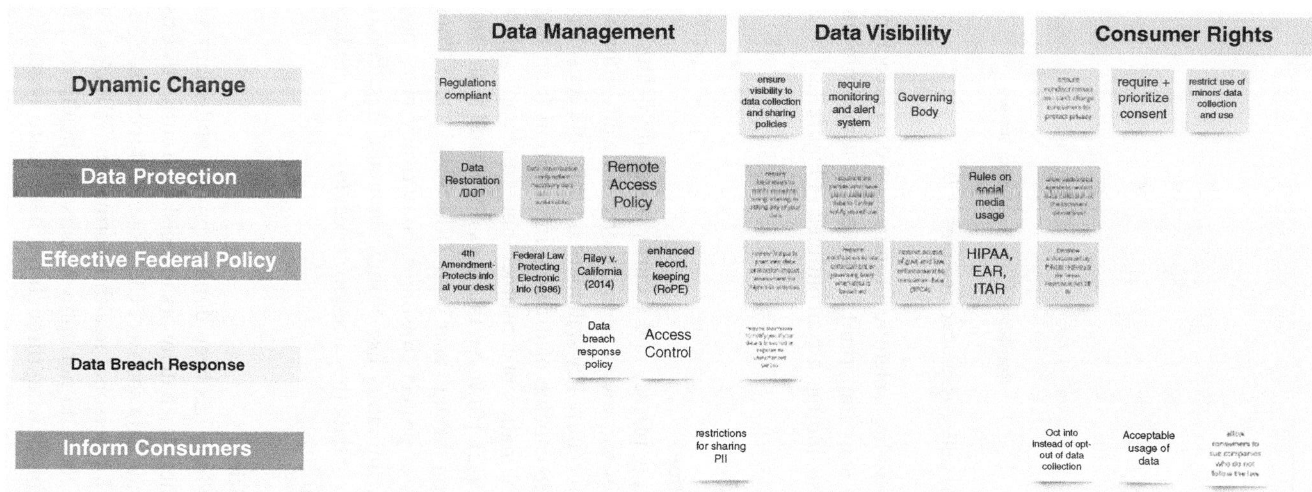

Figure 12.6 Policy mind map.

Product Features	Technology Features: Data management	Data Visibility	Consumer Rights
Dynamic Change (2)	4	4	2
Data Protection (4)	8	16	16
Effective Federal Policy (4)	8	16	8
Data Breach Response (2)	8	8	4
Inform Consumers (2)	4	8	8
Total	32	56	38

High Score: 4 | Medium Score: 2 | Low Score: 1

Figure 12.7 Technology policy vs. product features QFD.

While data management has a significant impact on data breach response, it was determined to be the least important as it has just a medium impact on other product features. These findings are all outlined in Figure 12.8.

12.4.7 GAP Analysis

12.4.7.1 Data Protection

HIPAA rules protect the privacy and security of certain health information and require certain entities to provide notifications of health information breaches (Klosowski 2021). A recent policy statement on this rule considers apps covered by the rule if they are capable of drawing information from multiple sources, such as through a combination of consumer inputs and application programming interfaces (McKeon 2022). However, this does not extend to the smartphone apps that collect information through user input (Klosowski 2021). There is a need for a national framework to protect consumer data privacy, give consumers protections against the discriminatory use of their data, and mandate that companies minimize the amount of data they need to collect to deliver products and services (McKeon 2022; Reicin 2022).

12.4.7.2 Federal Policies

In recent times, very little has been done to protect and secure personal health information (PHI). There is a patchwork of legislation currently existing to protect different types of private data such as FCRA for credit records, FERPA for student records, and HIPAA for personal health information. HIPAA's disclosure rules, which took effect in 2003, do not apply to personal health data in general, just the patient information flowing through the healthcare system (McKeon 2022). Legislators have been trying to enact a comprehensive national privacy standard

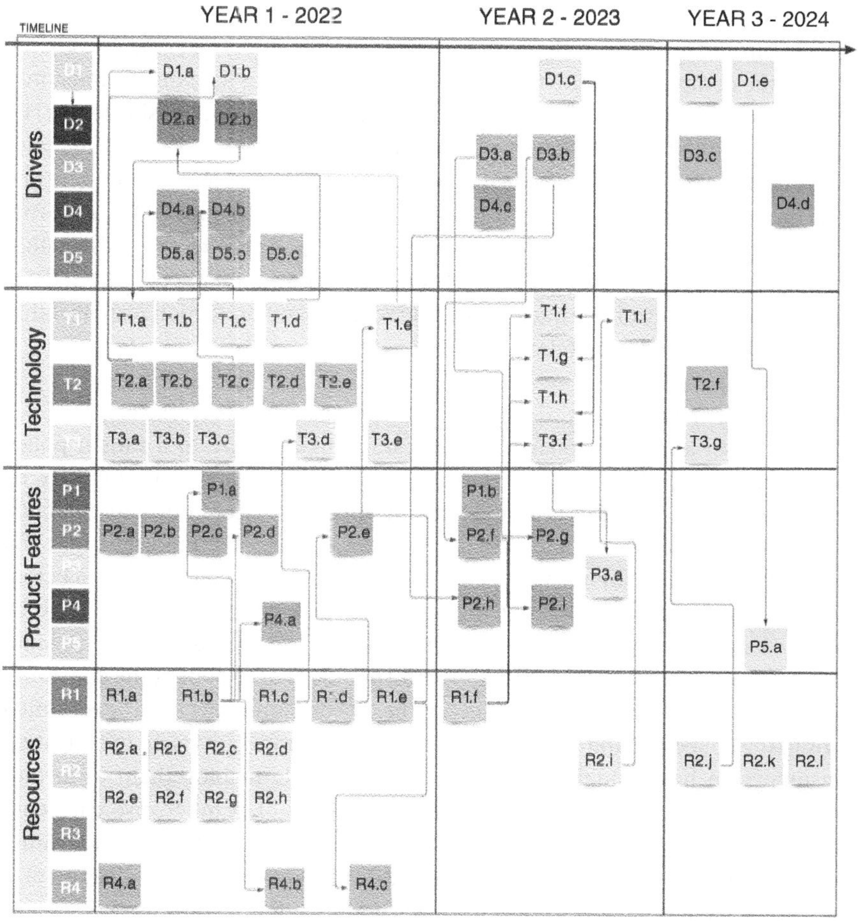

Figure 12.8 Technology roadmap.

for ages; congressional leaders recently released a bipartisan draft bill called the American Data Privacy and Protection Act (ADPPA) (McKeon 2022). There are also a handful of state laws on consumer privacy protection, and this current assortment of laws lead to unnecessary confusion and complexity (McKeon 2022).

12.4.7.3 Dynamic Change

Healthcare technology is advancing at an unprecedented scale. Since the COVID-19 pandemic, the usage of healthcare applications has increased multiple fold. Research from the Organization for the Review of Care and Health Applications found that the COVID-19 pandemic led to a 25% increase in health app downloads, and that, of the 350 000 health apps available on the market, 90 000 were introduced in 2020 alone, an average of 250 a day (Gershman, 2020). While there is a clear need to modernize the healthcare system, the complexity of the undertaking has made a tough go due to consumer personal information privacy and security fear (Dutcher & Zatkowsky 2021). The Cures Act drafted by the Department of Health and Human Services includes provisions to share electronic health information but it does not contain any rules on data privacy (Dutcher & Zatkowsky 2021; Ravindranath 2020). This rule will ease the way for big tech companies to mine personal health information from HIPAA covered organizations. But once the information is out of a HIPAA covered entity's control, the question of data security and protection responsibility remains unanswered (Ravindranath 2020).

12.4.7.4 Inform Consumers

Looking closer at inform consumers the team thinks HIPAA, or a similar policy should be expanded to cover personal health data that may not currently be covered. An example of this is period tracking apps, and other information related to reproductive health. In addition to the HIPAA rules, it's important that if a company collects consumer data, it clearly states the data is being collected before the collection starts. Additionally, the reason for collecting the data needs to be stated. The purpose of data collection could include use to feed an algorithm and improve the application, sale to profit the corporation, or to benefit the consumer in some way. Whatever the use of data may be, the consumer needs to be informed prior to the collection and use of said data. This is an imperative step in educating consumers on how valuable their data may be, while helping them to further protect themselves from malicious entities who might want to use that data inappropriately. An additional objective for the phase II requirement of informing consumers would be to require that the healthcare-covered entity document a mandatory recurring training to all involved personnel, as well as the content of the HIPAA enforcement rule showcasing the penalties of HIPAA violations. In order to have better-informed consumers, authorization, and consent forms must be mandatory in all covered entities, and consumers must know the process to follow in case of any HIPAA violations.

12.4.7.5 Data Breach Response

Data breach response, as discussed earlier, is what a company does when valuable personal data they are storing is released unintentionally to others. An example of an effective and meaningful response was the $5 billion penalty imposed on Facebook by the FTC in 2019 for violating a 2012 order by deceiving users about personal identifiable information (PII) privacy. This was approximately 20 times greater than any data privacy or cybersecurity penalty imposed globally (Federal Trade Commission, 2019). While this response to Facebook's deceptive policies was historically bold, the only current comprehensive federal requirement is that corporations notify affected parties and government agencies about a data breach. While there are many state policies that can impose fines for failing to notify, along with allowing for civil suits, there are no clear ways for the victims of these breaches to regain anything of value from what was lost. The team recommends a federal baseline of restitution for data lost, the value of which can be defined by research within government agencies, and a timeline for when the organization will understand why the breach happened and what they are doing to fix it. Additionally, there needs to be increased or improved monitoring of collected and stored data to understand if a data breach has happened, or if it is currently happening, in order to reduce the amount of damage that occurs from the data breach.

12.4.8 Technology Roadmap

Based on previous information, the roadmap in Figure 12.8 was created. It consists of four main sections, drivers, technology, product features, and resources. These sections as well as the legend for the roadmap are developed in the following paragraphs.

1. Drivers

   The first roadmap section includes five drivers, which are the five sectors that were identified as holding the most impact on the need for health data cybersecurity, the first one being technology (Table 12.1). There are currently a lot of technologies and devices that allow the tracking of health data, including menstruation tracking apps. Big names include Fitbit, Apple Watch, and Garmin. It is also projected that the health data mining by technology giants and technological advancement to include AI and machine learning as well as bodily monitoring technologies will factor into the health data collection.

   The second driver, identified as governmental, was selected due to the recent overturning of the *Roe vs. Wade*, which removed the constitutional right to abortion. The fear of future Supreme Court cases being overturned that will impact health rights and that can be enforced through the collection of health data means that the need for advanced and solid cybersecurity is higher than ever. It is also imperative to tackle social drivers, as consumer needs keep evolving and changing. The expectation in the next couple of years is for the users of applications involving health data to increase, and the demand for virtual care to keep growing as it already has been due to the pandemic. These issues could play a role in the 2024 elections as voters express concern over data security.

   Economically speaking, current healthcare costs and costs of data breaches suggest the need for cybersecurity is already elevated. The

Table 12.1 Technology roadmap: drivers.

Drivers	D1	Technology	D1.a	Health trackers (Fitbit, Apple Watch)
			D1.b	Menstruation tracking apps
			D1.c	Health data mining by tech giants
			D1.d	Advanced use of AI and machine learning
			D1.e	Bodily monitoring technology
	D2	Government	D2.a	*Roe vs Wade* overturned
			D2.b	Concerns other Supreme Court cases will be overturned
	D3	Social	D3.a	Change in social views
			D3.b	Demand for virtual care
			D3.c	Election year
	D4	Economic	D4.a	Healthcare costs
			D4.b	Cost of data breach
			D4.c	Alternative source of revenue for health tech entities
			D4.d	Fines for non-HIPAA entities on health data breach
	D5	Safety	D5.a	Debate if abortion covered in life-threatening cases
			D5.b	Emerging viruses (e.g. Monkeypox)
			D5.c	Ongoing CV-19 pandemic

sometimes-extravagant cost for healthcare should at least include data security, and data breach costs can be detrimental to entities with lower revenues. Projections for economic assistance to solve the issue includes an alternative source of revenue for health tech entities and fines for non-HIPAA entities. From a safety point of view, the current debate over whether abortion should be covered in life-threatening cases, continued cases of COVID-19, as well as the emergence of new viruses such as monkeypox mean that health data privacy is in high demand as healthcare in general is an even greater part of daily lives.

2. Technology
   The second roadmap section is about technology and how it is or should be affecting cybersecurity in health care (Table 12.2). Data management first was one of the first subjects to be tackled. The idea of safe data management currently includes minimizing data collection, minimizing links between data and individual users, collecting and storing data securely, and banning data brokers from selling or transferring health and location data. In the future, the projection is that different processes will be needed for that same level of safe management of health data, including the requirement of enhanced

Table 12.2 Technology roadmap technology.

Technology	T1	Data management	T1.a	Minimize data collection
			T1.b	Minimize links between data and individual users
			T1.c	Collect data securely
			T1.d	Store data securely
			T1.e	Ban data brokers from selling or transferring health and location data
			T1.f	Require enhanced record keeping of data use
			T1.g	Restrict data retention
			T1.h	Require data protection impact assessment for third party activities
			T1.i	Restrict government surveillance
	T2	Data visibility	T2.a	Clearly explain what data is collected and how it is used
			T2.b	Clearly explain how info is shared with others
			T2.c	Ensure data is not used in ways that can harm users
			T2.d	Ensure data is not used in ways that can harm users
			T2.e	Follow stated privacy policy
			T2.f	Publicly release data transparency report
	T3	Consumer rights	T3.a	Allow users to control what data is collected
			T3.b	Allow users to review, correct, and export their own data
			T3.c	Allow users to delete content or terminate their account
			T3.d	Require government to obtain a court order before accessing consumer data
			T3.e	Comply only with valid demands for information
			T3.f	Require opt-in consent to data collection
			T3.g	Allow consumers to sue for data malpractice

record keeping of data use, restricting data retention, requiring data protection impact assessment for third party activities, and restricting governmental surveillance.

Data visibility is the main way data is available for the public and is therefore one the most straightforward ways for data to be used against consumers. Currently the ways to make that data safely visible is by clearly explaining what data is collected and how it is used and how the information is shared with others, and ensuring data is not used in ways that can harm users, notifying users of changes before they go into effect, and following any stated privacy policy. Hopefully in the future the addition of a data transparency report would also contribute to the safe visibility of healthcare data.

Consumer rights are the final subject tackled in the technology section. Giving the consumers power to control the usage of their private data should be a priority. Currently, that power is given to them through allowing them to control what data is collected, to review, correct, and export their own data, and to delete content or terminate their accounts. Other processes include requiring the government to obtain a court order before accessing consumer data, and the right for consumers to only comply with valid demands for information. In the future, requiring opt-in consent to data collection and allowing consumers to sue for data malpractice will also add to that power.

3. Product Features
   Additionally, product features that are essential for the roadmap were identified (Table 12.3). Technologies and processes are always on the path to change, therefore dynamic changes must be considered as part of the product features. Currently there are assessments for algorithm impacts on data, and in the future, the team projects the possibility of having opt out options for authorized agents in web browsers in order to safely keep up with ongoing changes.

   Data protection is currently held out through firewalls, authentication processes, network security, national "do not track" systems and national standards to protect personal reproductive health data. Future protective practices will also include enhanced data inventory (RoPA), data deletion management, compliance monitoring, and data restoration processes. Under product features, future effective policy is expected, which would provide means for consumer opt-in. It will create a framework for a standardized permission requirement for users, and consensual sharing of data. Other future features include penalties for executives that lie to the Federal Trade Commission as a data breach response, as well as the clarification of private and public health records as part of informing consumers.

4. Resources
   Lastly, different types of resources that have an impact on healthcare cybersecurity were identified (Table 12.4). First are the local policies. There are currently many state level policies in effect that strive for the privacy and ethical use of healthcare data and many more proposed and passed and that are to take effect in 2023. Some notable policies include: the California Privacy Rights Act (CPRA), the Virginia Consumer Data Protection Act (VCDPA), and the Colorado Privacy Act (CO) which are all developing privacy compliance

programs (Grant Thornton 2021). There are also acts such as the "Mind Your Own Business Act," the "My Body, My Data Act" and the "Fourth Amendment Is Not for Sale Act" that have the same goal. They would protect citizens' privacy and allow consumers to control how their health data is sold and shared. "The Fourth Amendment is Not for Sale Act" and the Health and Location Data Privacy Act (MA, OR, WA, RI, VT) have similar goals where they stop data brokers from selling any personal information, such as location and health data, to law enforcement and intelligence agencies without legal permission.

Federal policies are a major resource for healthcare data security as well. There are a number of policies that are currently in place that enable safety and security regarding electronic personal data. The "Electronic Communications Privacy Act" or ECPA for example, protects electronic communication while they are being made up to the point when they are stored on computers. The "Health Insurance Portability and Accountability Act" or HIPPA, federally requires national standards creation and application to protect sensitive health information from being shared without the knowledge or consent of their subject. The "Video Privacy Protection Act" forbids the sharing of video records containing identifiable personal information. Another act worth mentioning is the "Children's Online Privacy Protection Act" or COPPA, which requires parental consent for collecting and using any personal information for children

Table 12.3 Technology roadmap: product features.

Product features	P1	Dynamic change	P1.a	Algorithm impact assessments
			P1.b	Authorized agents/web browsers for opting out
	P2	Data protection	P2.a	Firewalls
			P2.b	Authentication process
			P2.c	Network security
			P2.d	Create a national "do not track" system
			P2.e	National standard to protect personal reproductive health data
			P2.f	Enhanced data inventory (RoPA)
			P2.g	Data deletion management
			P2.h	Compliance monitoring
			P2.i	Data restoration process
	P3	Effective policy	P3.a	Provide means for consumer opt-in
	P4	Breach response	P4.a	Penalties for executives that lie to the FTC
	P5	Inform consumers	P5.a	Clarification of private and public health records

Table 12.4 Technology roadmap: resources.

Resources	R1	Local policy	R1.a	State-level policies proposed + passed: CCPA/CPRA – CA, VCDPA – VA, CPA – CO
			R1.b	States pass the Mind Your Own Business Act (2019)
			R1.c	States pass the Fourth Amendment Is Not For Sale Act (2021)
			R1.d	States pass the My Body, My Data Act (2022)
			R1.e	States pass the Health and Location Data Privacy Act (MA, OR, WA, RI, VT)
			R1.f	State-level policies in-effect: CCPA + CPRA (CA), VCDPA (VA), CPA (CO)
	R2	Federal policy	R2.a	Supreme Court overturns *Roe vs Wade*
			R2b	Comply with the Electronic Communications Privacy Act
			R2.c	Comply with the Health Insurance Portability and Accountability Act
			R2.d	Cosponsorship of the Net Neutrality and Broadband Justice Act of 2022
			R2.e	Comply with the Video Privacy Protection Act
			R2.f	Comply with the Children's Online Privacy Protection Act
			R2.g	Determine Value of Data
			R2.h	Women's Health Protection Act (2022)- H.R 3755 Passed the House
			R2.i	Federal expansion of ECPA (1986) to protect consumer data from govt surveillance
			R2.j	Federal expansion of FTCA (1946) to allow consumers to sue for abusive data practices
			R2.k	FTC require businesses to honor GPC
			R2.l	Comprehensive federal data protection policy, similar to CA, VA, and CO
	R3	Funding		
	R4	Organizations	R4.a	ACLU research and legislation to protect consumer data privacy
			R4.b	Hire 175 staff to police the private data market
			R4.c	Empower the FTC to create + enforce laws implementing policy

under 13 years old. Another important act that has passed the house in 2022, especially after the overturn of *Roe vs. Wade* is the "Women's Health Protection Act (2022) – H.R 3755" or WHPA, which would federally protect the right to access abortion care.

However, the need to expand some acts and policies such as the previously mentioned ECPA of 1986 and the "Federal Tort Claims Act" (FTCA) of 1946 – which allows consumers to sue for abusive data practices – is still necessary for the success of those cybersecurity efforts, especially in order to keep up with the development and advancement of technologies and data sharing expansion. By also having the Federal Trade Commission require that businesses honor Global Privacy Control features, and by having a comprehensive federal data protection policy similar to the previously mentioned California, Vancouver, and Colorado acts, there are better opportunities to fulfill cybersecurity needs in the healthcare sector.

Lastly, there are current organizational efforts promoting cybersecurity in the healthcare sector, and those include research and legislation by the American Civil Liberties Union (ACLU) as well as proposed legislation by democratic senators. The ACLU Speech, Privacy, and Technology Project is advocating for legislation to protect consumer data privacy, as well as create a guide for businesses on how to better protect the privacy and free speech rights of individuals. (Privacy & technology n.d.) The Mind Your Own Business Act proposed by US Senator Ron Wyden of Oregon, as well as other state senators, proposes means to better empower the Federal Trade Commission to create and enforce privacy protection, as well as act if laws are breached. If passed, the bill would establish minimum privacy and cybersecurity standards, issue serious fines for first offenders as well as criminal penalties for perjury, and create a system to facilitate consumers opting out of tracking, data sales, and targeted advertising – with provisions to waive fees for low-income consumers who qualify. The bill would also hire 175 staff members to police the unregulated private data market, and require corporations to assess the security, accuracy, and bias of their algorithms. A later draft of the Mind Your Own Business Act incorporates feedback, to provide better funding solutions for the proposal. The bill proposes revenue streams for the FTC through civil suits against companies that violate privacy regulations, then would allow redistribution of the funding to designated "watchdog" non-profits that originally reported the violation. Tax penalties would also be levied on companies that lie about privacy protections (US Senator Ron Wyden of Oregon n.d.).

12.5 Conclusion and Limitations

This paper outlines the team's work to establish clarity on consumer healthcare data privacy gaps, research policy to enforce more effective consumer healthcare privacy practices, and create a roadmap to demonstrate how policy will influence the management of cybersecurity technology to better protect consumer healthcare data privacy. The technology roadmap outlines the evolution of cybersecurity technology through the key drivers, technology, product features, and resources that will impact consumer healthcare data privacy protection. The interactions of these factors are a key element of the roadmap. Policy will be a key catalyst in technology adoption. The resource element of our roadmap consists of local policy, federal policy, funding, and organizations. Local policy will begin driving change in product features, enabled by technology management in year one. Federal policy should expand these changes in year two.

More strict enforcement of these policies regarding data breach response is projected by year three, as well as the mass adoption of data security technologies propelling greater innovation in the industry. Data management and visibility technology features will interact with the economic, government, and technology drivers as explained through the policy proposals above.

Due to the dynamic nature of consumer data privacy public policy, our reference to academic literature in this area was limited. The lack of experience on the research team in specific cybersecurity technology management presented an additional limitation in the work, which was constrained to 8 weeks. Public policy is essential to enforce the implementation of this technology by corporations, but the research team recommends the development of a more comprehensive technology and product feature roadmap to provide clear guidelines for corporations striving to do the right thing to best protect their consumers before these policies are passed and enforced. The team recommends further research displaying the current privacy standards of corporations, to compare where these organizations stand on the spectrum of cybersecurity technology implementation and best practices – such as the "Easy-to-read Privacy Nutrition Labels" available on the Apple App Store (Privacy, n.d.). Mapping major data breach response cases that have established precedent to enforce consumer data privacy policy, such as the $5 billion Facebook penalty, would provide more context on the trend of policy enforcement. Lastly, a more thorough economic analysis of the revenue streams driven by these policy changes would ensure the proposed standards are sustainable for nonprofit organizations, government agencies, and the corporations which they will guide.

References

Apple (n.d.). https://www.apple.com/privacy (accessed 15 August 2022).

Ariu, D., Didaci, L., Fumera, G. et al. (2015a). Yet another cybersecurity roadmapping methodology. *2015 10th International Conference on Availability, Reliability and Security*, 719–726. https://doi.org/10.1109/ARES.2015.87

Ariu, D. et al. (2015b). Yet another cybersecurity roadmapping methodology, 10th international conference on availability. *Reliability and Security* 2015: 719–726. https://doi.org/10.1109/ARES.2015.87.

Drees, J. (2020). The future of health data privacy: 6 things to know. *Becker's Health IT* https://www.beckershospitalreview.com/cybersecurity/the-future-of-health-data-privacy-6-things-to-know.html.

Dutcher and Zatkowsky (2021). Big tech's foray into healthcare and your privacy. https://www.rochesterelderlaw.com/big-techs-foray-into-health-care-and-your-privacy

Federal Trade Commission (2019). FTC imposes $5 billion penalty and sweeping new privacy restrictions on Facebook. https://www.ftc.gov/news-events/news/press-releases/2019/07/ftc-imposes-5-billion-penalty-sweeping-new-privacy-restrictions-facebook.

Gershman, J. (2020). Health data after Covid-19: more laws, less privacy. *The Wall Street Journal* https://www.wsj.com/articles/protecting-health-data-after-covid-19-more-laws-less-privacy-11599750100.

Grant Thornton (2021). Build your roadmap to 2023 US privacy compliance. https://www.grantthornton.com/insights/articles/advisory/2021/build-your-roadmap-to-2023-us-privacy-compliance (accessed 14 August 2022).

HHS (2009). Breach notification rule. https://www.hhs.gov/hipaa/for-professionals/breach-notification/index.html

HHS Office for Civil Rights (2013). Summary of the HIPAA Privacy Rule. https://www.hhs.gov/hipaa/for-professionals/privacy/laws-regulations/index.html#:~:text=The%20Privacy%20Rule%20protects%20all,health%20information%20(PHI).%22

Hill, K. (2022). Deleting your period tracker won't protect you. *The New York Times*. https://www.nytimes.com/2022/06/30/technology/period-tracker-privacy-abortion.html

Klosowski, T. (2021). The state of consumer data privacy laws in the US (and why it matters). *New York Times, Wirecutter,* September 6, https://www.nytimes.com/wirecutter/blog/state-of-privacy-laws-in-us

Koebler, J. and Merlan, A. (2022). This is the data Facebook gave police to prosecute a teenager for abortion. Vice. https://www.vice.com/en/article/n7zevd/this-is-the-data-facebook-gave-police-to-prosecute-a-teenager-for-abortion

Landi, H. (2022). Healthcare data breaches hit all time high in 2021, impacting 45m people. Fierce Healthcare. https://www.fiercehealthcare.com/health-tech/healthcare-data-breaches-hit-all-time-high-2021-impacting-45m-people

McGuire Woods (2021). Healthcare data breaches hit all time high in 2021, impacting 45m people. mcguirewoods.com https://www.mcguirewoods.com/client-resources/Alerts/2021/10/ftc-issues-reminder-breach-notification-requirements-health-apps-connected-devices-service-providers#:~:text=Research%20from%20the%20Organization%20for,average%20of%20250%20per%20day.

McKeon J. (2022). How new federal, state laws impact healthcare data privacy. https://healthitsecurity.com/features/how-new-federal-state-laws-impact-healthcare-data-privacy

Privacy Rights Clearinghouse. (n.d.). Data breach notification in the united states and territories. https://privacyrights.org/resources/data-breach-notification-united-states-and-territories (accessed 14 August 2022).

Rahnama, H. and Pentland, A. (2022). The new rules of data privacy. *Harvard Business Review*. https://hbr.org/2022/02/the-new-rules-of-data-privacy

Ravindranath M. (2020). What big tech companies aren't saying about HHS data rules. Politico. https://www.politico.com/news/2020/01/10/big-tech-hhs-data-rules-097356

Reicin E. (2022), Protecting consumer health data privacy beyond HIPAA. *Forbes*. https://www.forbes.com/sites/forbesnonprofitcouncil/2022/05/10/protecting-consumer-health-data-privacy-beyond-hipaa/?sh=38bca3177b4e

Riggi J. (2022).The importance of cybersecurity in protecting patient safety. AHA Center for Health Innovation, American Hospital Association. https://www.aha.org/center/cybersecurity-and-risk-advisory-services/importance-cybersecurity-protecting-patient-safety

Schmidt, H. (2022). HIPAA Protection Slides Post-Roe, Healthcare Cybersecurity Under Siege. *Healthcare Exec Intelligence* (July 12): https://healthcareexec-intelligence.healthitanalytics.com/news/hipaa-protection-slides-post-roe-healthcare-cybersecurity-under-siege

Strielkina, A., Illiashenko, O., Zhydenko, M., and Uzun, D. (2018). Cybersecurity of healthcare IoT-based systems: Regulation and case-oriented assessment. 2018 IEEE 9th International Conference on Dependable Systems, Services and Technologies (DESSERT). IEEE. 67–73. https://doi.org/10.1109/DESSERT.2018.8409101

Taylor, P., Allpress, S., Carr, M. et al. (2018). *Internet of Things: Realizing the Potential of a Trusted Smart World*. Royal Academy of Engineering.

Thomason, J. (2021). Big tech, big data and the new world of digital health. *Global Health Journal* 5 (4): 165–168. ISSN 2414-6447, https://doi.org/10.1016/j.glohj.2021.11.003. https://www.sciencedirect.com/science/article/pii/S2414644721000890.

Torchinsky, R. (2022). How period tracking apps and data privacy fit into a post-*Roe v. Wade* climate. *NPR*, June 24. https://www.npr.org/2022/05/10/1097482967/roe-v-wade-supreme-court-abortion-period-apps

Zhou, Y. et al. (2011). A policy dimension required for technology roadmapping: learning from the development of emerging wind energy industry in China. In: *2011 Proceedings of PICMET '11: Technology Management in the Energy Smart World*, 1–9. PICMET.

13

Technology Roadmapping: Monitoring the Status of a Technology Roadmap with Data-driven Roadmapping Approach

Nathasit Gerdsri[1], Ummaraporn Pora[2], Sudatip Puengrusme[1], and Ronald Vatananan[1]

[1] *College of Management, Mahidol University, Bangkok, Thailand*
[2] *Technopreneurship and Innovation Management Program, Chulalongkorn University, Bangkok, Thailand*

13.1 Introduction

Technology roadmapping (TRM) is a strategic approach that allows organizations to link their technology strategies with their business strategies (Groenveld 1997; Nauda and Hall 1991; Wells et al. 2004). Some organizations implement TRM to support the visualization of their strategies and guide their operations or activities. Since 2000, several studies have been taken to systemize the TRM development process as well as compare the cases of TRM implementation from different organizations to see whether the implementation process can be standardized (Gerdsri et al. 2009, 2013; Hirose et al. 2020, 2021).

Under today's fast-changing business environment and the emergence of breakthrough technologies, keeping a roadmap alive seems to be a major challenge for many organizations as the roadmap itself reflects only a snapshot of an organization's strategic plan based on the information available at the point when the roadmap was developed (Brown and O'Hare 2001; Gerdsri et al. 2019; Phaal et al. 2001, 2005; Phaal and Muller 2007, 2009; Vatananan and Gerdsri 2012; Wells et al. 2004). It is crucial that an organization monitors changes to its business environment as well as the development progress.

The current practice is to set the schedule for roadmap review and update it periodically (Brown and O'Hare 2001; Gerdsri et al. 2019; Holmes and Ferrill 2008; Phaal et al. 2001; Phaal and Muller 2009; Wells et al. 2004). However, this approach may not be sufficiently proactive in monitoring the status of a roadmap and allowing management to make a timely decision on its revision (Gerdsri et al. 2019).

Future-Oriented Technology Assessment: A Manager's Guide with Case Applications, First Edition.
Edited by Haydar Yalçın and Tugrul U. Daim.

Furthermore, a roadmap updating schedule may lead to extra costs for an organization if it is unnecessary (Hirose et al. 2022).

A roadmap can automatically learn about the relevant changes needed in order to react properly when assumptions on which the roadmap was built are no longer applicable to the current environment (Pora et al. 2022). The learning process can be designed to continuously analyze and interpret data based on empirical evidence. This is known as the "data-driven" approach.

This chapter presents the framework to monitor and assess the status of a technology roadmap by evaluating the impacts of changes from both internal and external environment with a semiautomated decision support system to reflect any changes and recommend whether a roadmap needs to be slightly adjusted, totally revised, or simply remain unchanged.

With the proposed framework for monitoring the status of a roadmap with data-driven roadmapping (DDRM) in place, the management can take proactive actions in monitoring the status of a roadmap and calling for an immediate action to review or revise a roadmap rather than waiting for the next periodical review schedule.

13.2 Literature Review

13.2.1 Impacts of Changes on a Roadmap

As the current global environment becomes more complex, organizational managers have to face many complicated challenges. Challenges may result from several causes, including rapid changes from internal factors (e.g. technology development, research and development programs) and/or external factors (e.g. business drivers, market opportunities). Once the changes in any of these factors (internal/external factors) occur, it may affect the strategic targets of a roadmap. If the impacts are significant enough, a roadmap should then be adjusted or totally revised to align with the changing situations. Otherwise, a roadmap could become obsolete.

As shown in Figure 13.1, the changes of external drivers; D_1 and/or D_2, would affect the market opportunity; M_1. If the changes of D_1 and/or D_2 turn out to be more severe than the initial anticipation, then it would cause market opportunity M_1 to be materialized sooner. Therefore, an organization should consider revising its roadmap to set the target of M_1 sooner; otherwise, an organization will miss this opportunity. The same way of thinking is also applied if the changes of D_1 and/or D_2 turn out to be less severe than the initial anticipation. An organization can consider pushing out the target of M_1 so that an organization would not end up in a premature market.

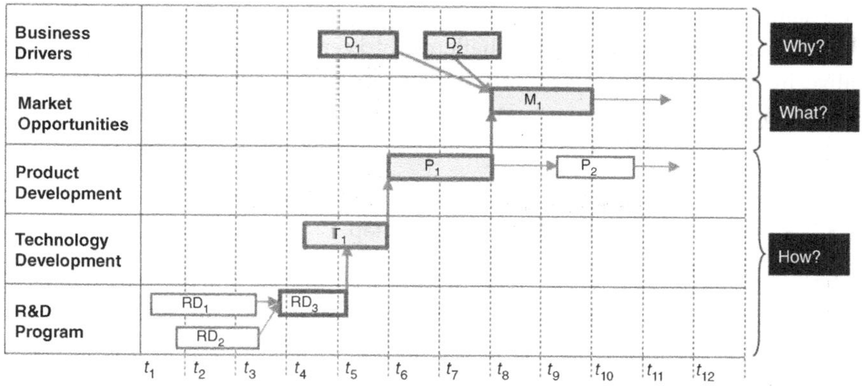

Figure 13.1 The example structure of a product-technology roadmap (Gerdsri et al., 2019/with permission of Elsevier).

The changes of internal factors also affect a roadmap. For example, product P_1 is a product that an organization initially plans to capture the market opportunity M_1 at the time of t_8. And technology T_1 is a new technology that is required to develop product P_1. Technology T_1 can be developed either from within or outside a firm; however, the development of T_1 has to be finished by the time of t_6 in order to start the development of product P_1. If technology T_1 is not ready by the time of t_6, the delay in the development of product P_1 may occur. As a result, P_1 may be ready and able to launch to the market later than the initial schedule of t_8. For instance, if product development of P_1 finishes at the time of t_{10}, this leads to the loss of market opportunities M_1 which P_1 intends to capture. In this situation, the delay of the technology development impacts the strategic focal point of a roadmap. Therefore, the strategy and roadmap need to be constantly reviewed and updated to reflect the actual situation.

On the other hand, if the development of technology T_1 finishes sooner than the targeted schedule of t_6, the development of product P_1 can start earlier than the specified time of t_6. The product development of P_1 may finish prior to the targeted strategic time of t_8. As a result, P_1 would end up being too early for the market opportunity M_1. The managers should, in this case, review and revise their strategies either trying to slow down the product development to wait for the market opportunity to arise or trying to put more efforts in developing the new market opportunity so that the market can be materialized sooner. The updated version of a roadmap should be communicated to all related stakeholders. This is to ensure that the updated plan on a roadmap is acknowledged by every related team.

In brief, the changes in both external drivers and internal factors can impact the status of a roadmap. It is important for a firm to assess whether the changes are significant enough that an updated version of a roadmap is required. Some researchers have addressed the challenge of keeping the roadmapping process alive by proposing the operation guideline to periodically update and review technology roadmaps (Brown and O'Hare 2001; Gerdsri 2013; Holmes and Ferrill 2008; Phaal et al. 2001, 2005; Phaal and Muller 2009; Wells et al. 2004).

13.2.2 Monitoring the Status of a Roadmap

The status of a roadmap is determined based on the impact assessment of changes from both external and internal factors on a roadmap. The status signal of a roadmap indicates whether a roadmap is still usable or it is required to be updated. If the assessment results indicate that the roadmap is not usable, a major revision of a roadmap must be implemented. If the roadmap is still usable but needs to be adjusted, a minor adjustment of a roadmap is then required. If the roadmap is not impacted by the changes, the roadmap can maintain as it is. Therefore, the definition of the status of a technology roadmap is referred as following table.

An organization should monitor the changes of these factors and then evaluate whether the changing situation has impacted the targets/strategic focal points of a roadmap. It is important for an organization to have a certain managerial procedure for monitoring the status of a roadmap. The mechanisms for data collection and analysis should also be defined to track changes and evaluate the impacts of changes from both external and internal factors on a roadmap. Once the impacts from the changes have gone beyond the acceptable level of an organization, a roadmap should then be considered to make an amendment. In this study, the managerial procedure is proposed in five steps, as shown in Figure 13.2.

The first step is to implement activities as planned on a roadmap. Thus, a roadmap should be clearly communicated to all stakeholders. This would help ensure the achievement of the strategic targets within a specified timeline identified on a roadmap.

The second step is to monitor the changes from both internal and external factors as specified on a roadmap. The components of external factors can be considered from social, technology, economic, environmental, policy, whereas internal factors can be considered from the progress of technology development.

The third step is to assess the impacts of changes on a roadmap. As the changes from the external and/or internal factors have already been notified, the identified changes will be used in this step to evaluate the impacts of the changes on a roadmap. Different organizations have different perception of the impacts of the changes. Some organizations are sensitive to the changes while others are not. Therefore, this step is defined based on the perception of the organization.

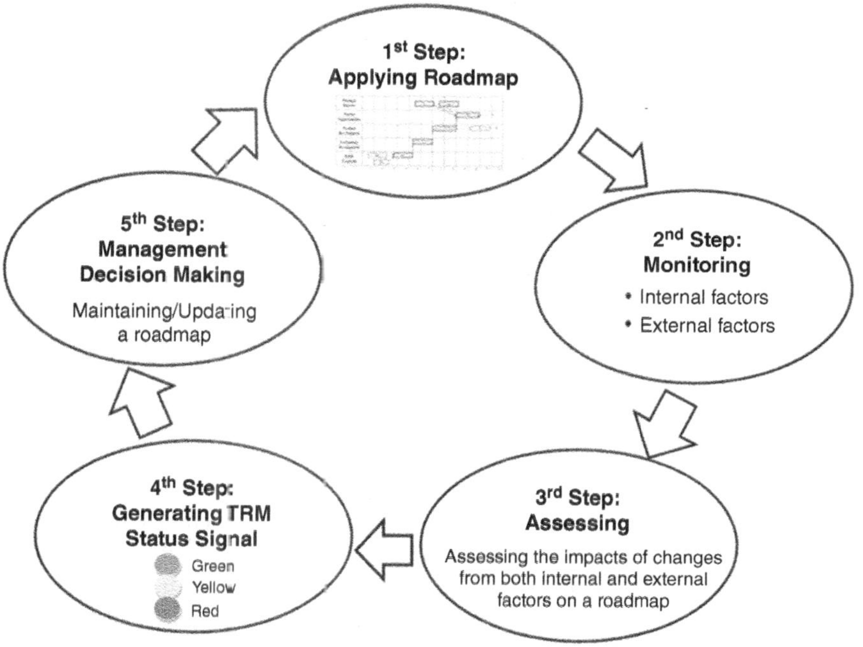

Figure 13.2 The conceptual framework to assess the TRM status signal by considering the impacts of changes from both internal and external factors (Gerdsri et al., 2019/with permission of Elsevier).

The fourth step is to generate the TRM status signal. After evaluating the impacts of changes from both internal and external factors, the TRM status signal can be generated and visualized in the form of color-coded signals; red, yellow, and green. The meaning of each status signal is already discussed in Table 13.1.

In the fifth step, management makes a final decision about whether a roadmap needs to be slightly adjusted, totally revised, or kept unchanged.

13.2.3 Assessing the Status of a Roadmap

The degree of impacts on changes from both external and internal factors is predetermined to reflect the management's perception on how much changes can be accepted. The tolerance intervals associated with each level of impact are also determined. The status assessment of a roadmap is executed by analyzing the deviation between the actual and the predetermined baseline values at the measurement level of the evaluation models. Then, the changes are compared with the tolerance intervals to determine the degree of impact.

Figure 13.3 demonstrates the framework for assessing the impact of changes on a roadmap. The upper evaluation model is used to determine the impacts

Table 13.1 Description of a roadmap status and managerial implications.

Status	Color code	Managerial implications
"Maintain"	Green	Changes from both external and internal factors are still within the acceptable range and there is no impact or minor impact on a roadmap. Therefore, a roadmap can still stay as it is without revision.
"Adjust"	Yellow	Changes from both external and internal factors have impacts to a certain extent on a roadmap. Therefore, there should be some minor amendments on a roadmap.
"Revise"	Red	Changes from both external and internal factors are severe and have critical impacts on a roadmap. It causes a roadmap to be invalid. Therefore, there should be some major amendments on a roadmap.

from changes in external factors on a roadmap status; particularly the changes in key drivers related to social, technology, economic, environmental, policy. The lower evaluation model is used to determine the impacts from changes in internal factors, particularly the changes in technology development progress, on the status of a roadmap. The result obtained from the evaluation models will be calculated to determine the status of a roadmap. The TRM status signal will then be generated in green, yellow, or red status.

13.2.3.1 Assessing the Impacts on the Changes of External Drivers

This section presents the conceptual framework using an analytical approach to construct and operate an evaluation model that can assess the impact of changes from the external business environment on the current status of a roadmap. The objective of the model is to *determine the status of the organization's roadmap by analyzing changes in key drivers and their collective effect on the roadmap.*

Development of an Evaluation Model The key drivers affecting the changes on the status of a roadmap are identified. Each key driver is also broken down into sub-drivers in which their changes can be measured directly. Figure 13.4 shows the hierarchical structure of key drivers and sub-drivers analyzed in the evaluation model.

Level 1 – Model objective: The top level of the evaluation model represents the overall objective of the model, which is to analyze the effect of changes in key drivers on an organization's roadmap to determine its current status.

Level 2 – Key drivers: The key drivers (KD_i) are identified to reflect the organization's external business environment. The relative importance (w_i) of each

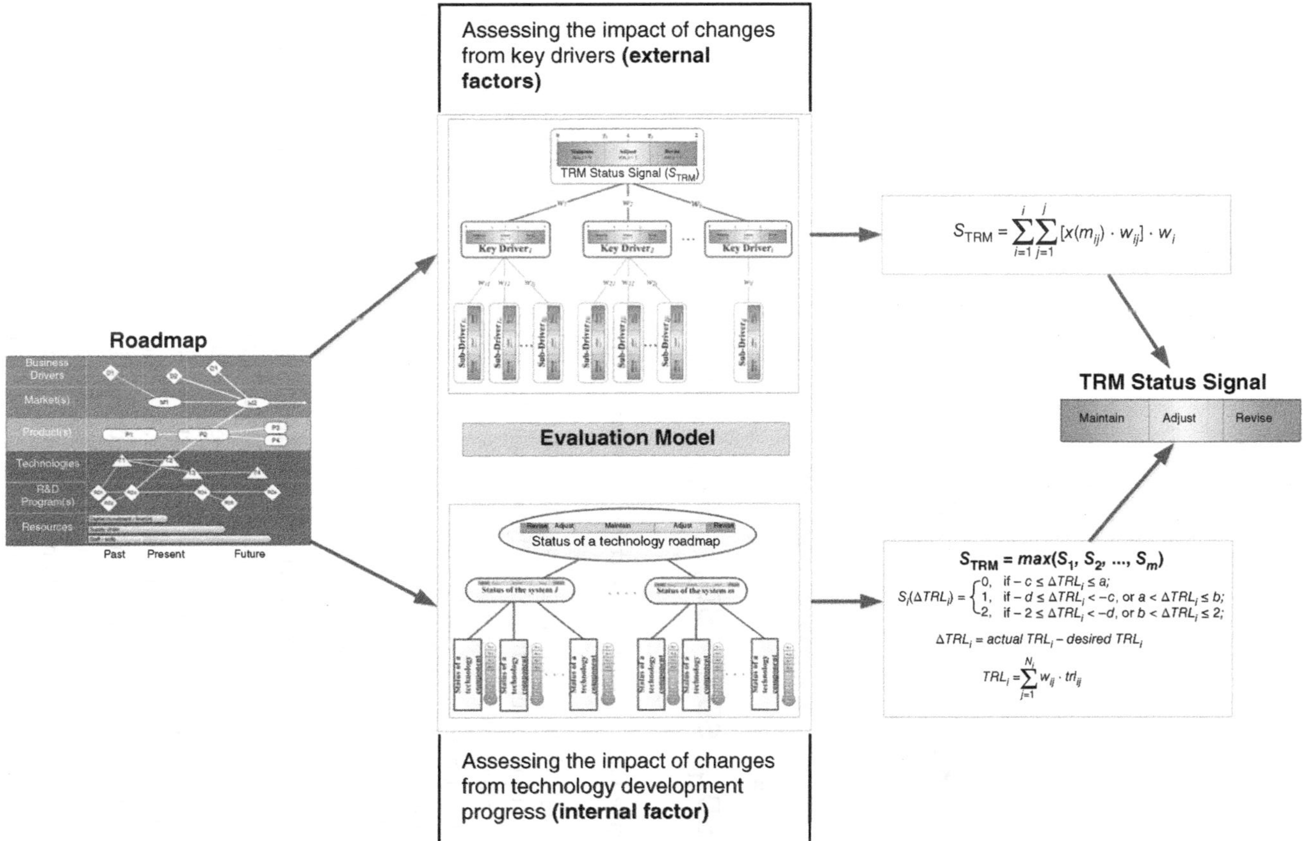

Figure 13.3 Conceptual framework to assess the impacts of changes on the status of a roadmap (Gerdsri et al. 2019/with permission of Elsevier).

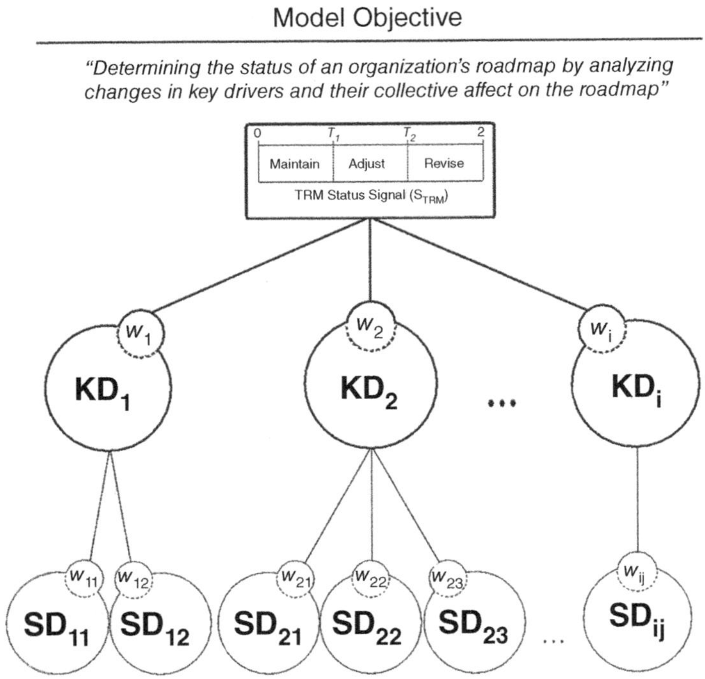

Figure 13.4 Hierarchical structure of an evaluation model.

key driver with respect to an organization's objective is determined. The value of w_i represents the intensity, with which a key driver affects an organization's strategy. However, an organization's environment is composed of numerous key drivers that can influence its operations and strategic plan. As a part of the roadmap development, it might be impractical if not impossible to include all of the organization's key drivers. Therefore, only the most influential key drivers of an organization's environment are considered for the evaluation model.

Level 3 – Sub-drivers: A set of sub-drivers (SD_{ij}) is identified and their relative importance (w_{ij}) with respect to the key driver (i) is determined. Next, a unit of measurement is set for each sub-driver measurement value (m_{ij}) together with a data source to support the measurement. Now, for each SD_{ij} experts need to define the tolerance limits and intervals of the measurement values, to specify the various degree of impact from a change in these measurements as shown in Figure 13.5. The baseline value (m_{base}) of each sub-driver is specified as the assumption in the roadmap development. The tolerance intervals are defined by the management to classify for the level of impact $x(m_{ij})$ on any changes (m_{ij}) from the baseline value (m_{base}).

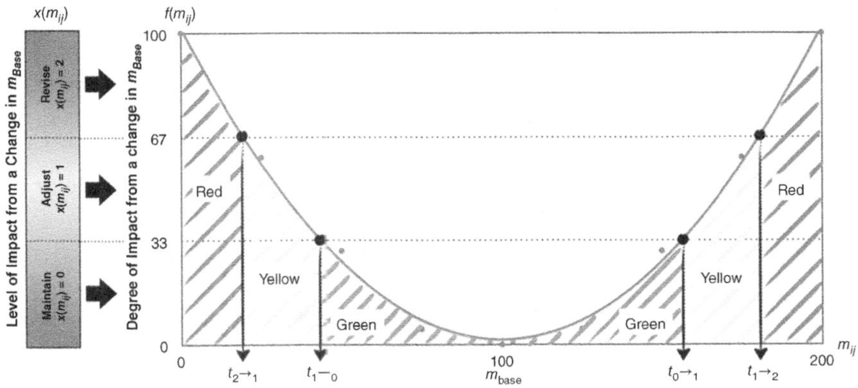

Figure 13.5 Generic pattern of a perception curve on the tolerance interval for a sub-driver measurement.

Computational Model As a part of assessing the current state of a roadmap, changes in the organization's external factors; key drivers (KD_i) and sub-drivers (SD_{ij}), are tracked. The baseline values (m_{base}) and tolerance limits for each measurable sub-driver are determined. The level of impact $x(m_{ij})$ from such a change is assessed by analyzing the deviation of the sub-driver measurement values (m_{ij}) from their baseline values (m_{base}).

As a result, the computational model produces the TRM status signal (S_{TRM}) as shown in Eq. (13.1), which serves as an indicator for the current state of a roadmap.

$$S_{TRM} = \sum_{i=1}^{i}\sum_{j=1}^{j}[x(m_{ij}) \cdot w_{ij}] \cdot w_i \tag{13.1}$$

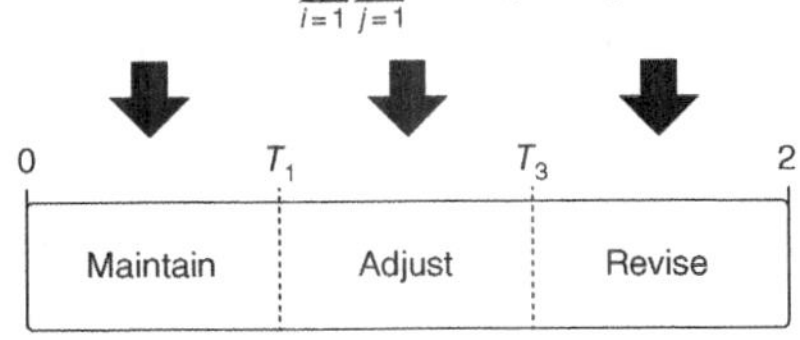

where:

w_i	: Relative importance of key driver (i) with respect to the objective.
w_{ij}	: Relative importance of sub-driver (j) associated with key driver (i).
m_{ij}	: Measurement value of sub-driver (j) associated with key driver (i).

$x(m_{ij})$: A signal representing an integer value ($x[m_{ij}] = [0, 1, 2]$) to indicate the level of impact from a change in the measurement value m_{ij} on a roadmap for the sub-driver (j) under key driver (i).

- $x(m_{ij}) = 0$: Changes in m_{ij} have little or no impact.
- $x(m_{ij}) = 1$: Changes in m_{ij} have a moderate impact
- $x(m_{ij}) = 2$: Changes in m_{ij} have a major impact

$KD_i = \sum_{j=1}^{j} x\left(m_{ij}\right) \cdot w_{ij}$: A collective signal (S_{KDi}) indicating the status of measurement values for all sub-criteria j (i.e. sub-driver [SD_j]) under the criterion i (i.e. key driver [KD_i]).

S_{KDi} : Key driver status signal

$TRM = \sum_{i=1}^{I}\sum_{j=1}^{J}[x\left(m_{ij}\right) \cdot w_{ij}] \cdot w_i$: A collective signal indicating the overall status of a roadmap.

S_{TRM} : TRM status signal

13.2.3.2 Assessing the Impacts on the Changes of Internal Factors

This section presents the conceptual framework using technology readiness level (TRL) to construct and operate an evaluation model that can assess the impact of changes from internal factors, particularly from the technology development progress. As aforementioned, the changes in technology development progress either behind or ahead of the desired progress can impact the timing of strategic targets specified on a roadmap. The objective of the model is to *determine the status of the organization's roadmap by analyzing the impact of changes in technology development progress on the roadmap.*

Development of an Evaluation Model The development of one product can be broken down into the level of systems and technology components. The status of product development at any given time is depended on the completion of technological components of each system. Therefore, the status of product development specified on a roadmap is determined as an aggregate on the development progress of all technology components. Figure 13.6 shows the hierarchical structure of systems and technology components required for the completion of product development specified on a roadmap.

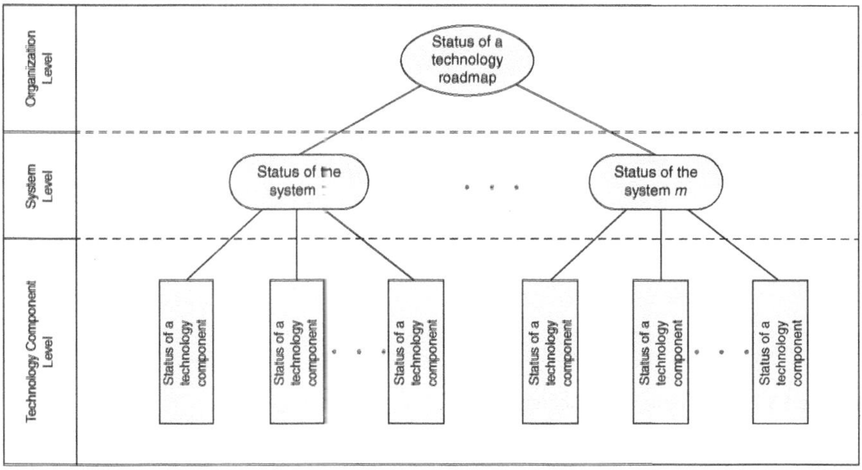

Figure 13.6 Hierarchical structu·e of an evaluation model for assessing TRM status signal.

Level 1 – Model objective: The top level of the evaluation model represents the overall objective of the model, which is to analyze the effect of changes in technology development progress on an organization's roadmap to determine its current status.

Level 2 – Systems: This level represents all systems required as a part of the targeted product. The development of these systems is set to complete within the specified timeline on a roadmap.

Level 3 – Technology components: This level represents technology components needed to be developed as a part of each system specified at Level 2.

The development of each technological component is not always equally important. The level of technology readiness of some technology components may be close to the desired level as targeted on a roadmap. In this case, the development of these components may not as critical compared to the development of some other technology components in which more development are needed to be done before meeting the desired level.

Computational Model A unit of measurement on the actual development progress for each technology component is considered in TRL scales ranging from 1 to 9. TRL1 is the level of pure research in which the basic principles are observed and reported, while the TRL9 is the level that an actual system is successfully operated. A tool used to measure Technology Readiness Levels, called "TRL calculator," is designed in the form of Microsoft Excel Spreadsheet and can be used to evaluate

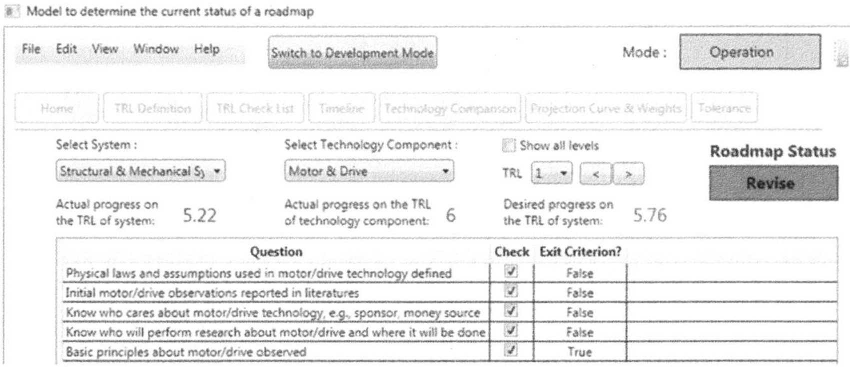

Figure 13.7 Example of TRL calculator.

the development of both software and hardware technology. The calculator would provide sets of questions for the user and employ the answer gained to calculate the result on the level of TRL (Altunok and Cakmak 2010; Nolte et al. 2003; Technology Readiness Assessment Guide 2011). Figure 13.7 provides an example of a TRL calculator used in this study.

The TRL calculator measures the actual development progress of a technology component at current milestone. The level of importance on the development of each technology component is taken into account to consolidate the actual development progress of all technology components in order to calculate the actual development progress for a system. The actual development progress of a system gives management an overview on the status of the development progress at a current milestone.

The desired development progress of each technology component at different milestones is preidentified during the planning stage. To calculate the desired development progress for a system, the desired development progress of all technology components are consolidated with the consideration of the level of importance on the development of those technology components to the completion of the system. The desired development progress of a system gives management an overview on the progress projection of a system at each milestone.

Changes in the development progress of a system are calculated based on the deviation between the actual and desired development progress of a system at the time of assessment. The degree of impacts from changes is predetermined to reflect how much the management can tolerate to changes in development progress before considering whether to adjust or completely revise the status of a roadmap. Four equations are used in the computational part of the evaluation model to assess the TRM status signal:

1. Equation to compute the progress (actual and desired development progress) of the system i (TRL_i)
2. Equation to compute the changes in development progress of the system i (ΔTRL_i)
3. Equation to compute the development status of the system i (S_i)
4. Equation to compute the status of a technology roadmap (S_{TRM})

To calculate the development progress (actual/desired development progress) of a system by using TRL scales, the development progress of all technology components is consolidated with consideration to the level of importance on the development of those technology components to the completion of a system. Therefore, the level of technology readiness on the development progress of the system i (TRL_i) can be computed from:

$$TRL_i = \sum_{j=1}^{N_i} w_{i,j}\, trl_{i,j} \tag{13.2}$$

where:

i = index of systems
j = index of technology components
$trl_{i,j}$ = the level of technology readiness on the development progress of technology component j belong to the system i
N_i = the number of technology components belong to the system i
$w_{i,j}$ = the relative importance on the development of technology component j with respect to the completion of system i

The Eq. (13.2) is used at the *model development stage* to construct a *progress projection curve* for a system by consolidating the progress projection curve of all technology components with the consideration of the level of importance on the development of those technology components to the completion of the system. The progress projection curve for a system represents the projection on the development progress of a system at different milestone.

In addition, the Eq. (13.2) will also be applied at the *model operation stage* to evaluate the actual level of technology readiness on the development progress of a system by consolidating the actual level of technology readiness on the development progress of all technology components with the consideration of the level of importance on the development of those technology components to the completion of the system.

In brief, the Eq. (13.2) can be applied for the computation of the desired development progress for a system at the model development (planning) stage as well as the actual development progress for a system at the model operation (monitoring) stage. Therefore, the desired progress and actual progress on the development of a system i, which represented by TRL can be computed from the Eqs. (13.2.1) and (13.2.2), respectively.

$$desired\ TRL_i = \sum_{j=1}^{N_i} w_{i,j} \cdot desired\ trl_{i,j} \tag{13.2.1}$$

$$actual\ TRL_i = \sum_{j=1}^{N_i} w_{i,j} \cdot actual\ trl_{i,j} \tag{13.2.2}$$

where:

i = index of systems
j = index of technology components
$desired\ TRL_i$ = the desired level of technology readiness on the development progress of system i
$desired\ trl_{i,j}$ = the desired level of technology readiness on the development progress of technology component j belong to the system i
$actual\ TRL_i$ = the actual level of technology readiness on the development progress of system i
$actual\ trl_{i,j}$ = the actual level of technology readiness on the development progress of technology component j belong to the system i
N_i = the number of technology components belong to the system i
$w_{i,j}$ = the relative importance on the development of technology component j with respect to the completion of system i

To determine the changes in development progress of a system, the difference between the actual development progress and desired development progress of the system at the measuring milestone is considered. The comparison between the actual progress ($actual\ TRL_i$) and the desired progress ($desired\ TRL_i$) reveals whether the current development progress of a system i is delayed or ahead of the strategic timeline identified on the technology roadmap by how far, as shown in Eq. (13.3). The changes in development progress of the system i (ΔTRL_i) can be computed from:

$$\Delta TRL_i = actual\ TRL_i - desired\ TRL_i \tag{13.3}$$

where:

$actual\ TRL_i$ = the actual level of technology readiness on the development progress of system i

$desired\ TRL_i$ = the desired level of technology readiness on the development progress of system i

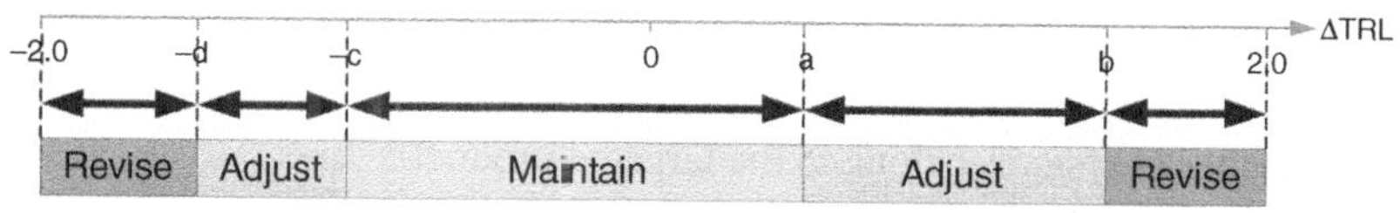

Figure 13.8 Tolerance intervals indicating how much an organization can tolerate to changes in development progress.

Note that the ΔTRL_i with negative values indicate that the current development progress of the system i is *less than* the desired development progress which means that the development of the system i is currently delayed. While the ΔTRL_i with positive values indicate the current development progress of the system is *more than* the desired development progress which means that the development of the system i is currently ahead of the targeted timeline. Eq. (13.3) is used at the *model operation stage* to determine the level of changes in development progress of a system at current milestone. The development status of the system i (S_i) would be represented by integer value as following:

Maintain status (green signal) = 0
Adjust status (yellow signal) = 1
Revise status (red signal) = 2

The development **status of the system *i* (S_i)** can be computed from:

$$S_i(\Delta TRL_i) = \{0, if - c \leq \Delta TRL_i \leq a; 1, if - d \leq \Delta TRL_i < -c, \;\; or \;\; a < \Delta TRL_i \leq b; 2, if - 2 \leq \Delta TRL_i < -d, \;\; or \;\; b < \Delta TRL_i \leq 2\} \tag{13.4}$$

where ΔTRL_i is the changes in development progress of system i, and $S_i = 0$ indicates a "maintain" status, $S_i = 1$ indicates an "adjust" status, and $S_i = 2$ indicates a "totally revise" status. In Figure 13.8, *a*, *b*, *-c*, and *-d* are the points of tolerance intervals that represent how much an organization can tolerate changes in development progress before the development status affects be final status.

Since a targeted product identified on a technology roadmap can be comprised of more than one system, status of a technology roadmap (S_{TRM}) is derived from the development status of the system (S_i, $i = 1, 2, \ldots, m$), which has *maximum* integer value of the development status. This is because all systems are part of the targeted product. The maximum integer value on the development status of a system reflects the most severe development situation of the targeted product for the management to be aware of. However, the management can also track back to understand the development status of any system belong to the targeted product if they are interested to.

With the rationale previously explained, the **status of a technology roadmap (S_{TRM})** can therefore be computed from:

$$S_{TRM} = (s_1, s_2, \ldots, s_m) \tag{13.5}$$

where:

$s_1, s_2, \ldots, s_m$ are the statuses of the system $1, 2, \ldots, m$, respectively.
m = The number of systems belong to the targeted product

Note that Eq. (13.5) would be applied in the final step of *model operation stage* in order to determine the current status of a technology roadmap once the development status of all systems have already been determined.

Model Measurement and Operation Technology manager is collaborated with technology development team to monitor the actual development progress of each technology component. The technology developers use the TRL calculator to assess the actual development progress of a technology component. Experts of an organization then determine the level of importance on the development of each technology component to the completion of a system.

To assess the changes in technology development progress at a system level, the difference between the actual and desired development progress of a system is calculated. For example, the desired development progress of a system should achieve TRL7 at current milestone; however, the actual development progress of a system at current milestone is only at TRL 6. The development progress of a system is currently delayed by 1 level of TRL. Therefore, the changes in development progress would indicate not only whether the current development progress is delay or ahead of the desired plan but also by how far. Once the changes in development progress of a system are determined, the impact of changes on the status of a roadmap is then needed to be evaluated. The experts of an organization would determine the *tolerance intervals*, which represent how much an organization can tolerate changes in development progress before the status must be changed to adjust or revise status. The tolerance intervals are separated into three intervals of *maintain (green)*, *slightly adjust (yellow)*, and *totally revise (red)* interval. If the changes in development progress are little and occur within the maintain (green) tolerance interval, the changes may not affect the roadmap. Therefore, the roadmap can probably stay as it is. However, if the changes are significant and go beyond the maintain (green) tolerance interval, then the roadmap requires the updating to reflect the changing situation. Failure to update the roadmap at the appropriate time could result in inefficient management of resources and budget and affect the company's competitiveness in the market. Figure 13.9 demonstrates the evaluation model to determine the current status of a technology roadmap with the consideration of the impacts of changes in technology development progress.

13.2.4 Data-driven Roadmapping Approach

Data is considered to be one of the key components in the success of roadmap development and implementation. Data can be classified by data type (structured

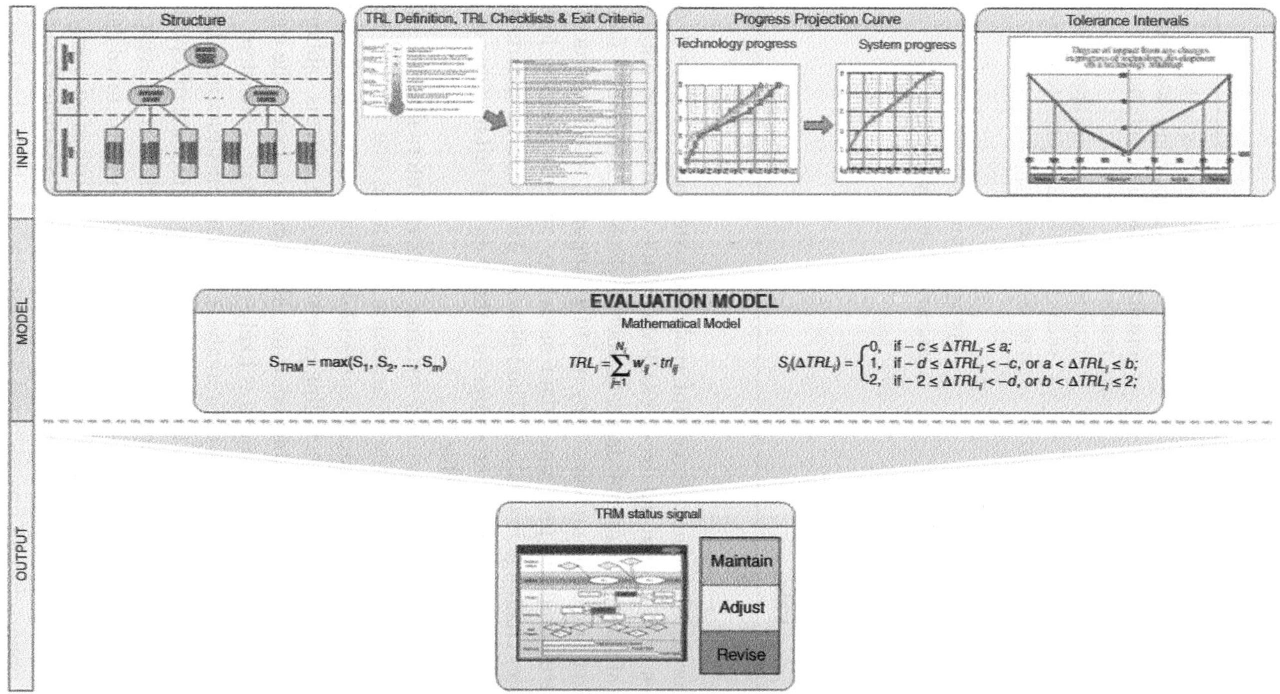

Figure 13.9 Evaluation model to determine the current status of a technology roadmap with the consideration of the impacts of changes in technology development progress (i.e. internal factor).

Table 13.2 Summary of type and source of data.

	Unstructured data	Structured data
Internal source	Email CRM Customer survey Employee survey Product catalog Technical specification Annual report	Data from ERP systems such as financial and accounting, sales and distribution, supply chain, human resources Data from other business applications such as point-of-sale (POS) terminals and e-commerce applications
External source	Industrial report Patents Research Bibliometric News Social media Video	Open government data Statistical subscription data

and unstructured) and data source (internal and external). Structured data usually resides in relational databases (RDBMS) while unstructured data may be stored within a nonrelational database such as NoSQL. Structured data is highly organized and easily understood by machine language, while unstructured data is not structured via predefined data models or schema and the format creates a greater challenge in analysis. The following 2 × 2 quadrant charts help to distinguish between the type and source of data. Data can fall into one of the four quadrants, as presented in Table 13.2.

Data can come from either internal or external sources. Internal sources represent data under the control of the organization, including that relating to all business applications. External sources represent data that could be generated or obtained outside an organization, including government data, patents, bibliometric, social media, and external syndicated. Internal structured data is the easiest to access. Data from internal operations is likely to be accurate, consistent, and structured in a way that is conducive for analysis. Internal unstructured data is available and scattered throughout a company but not set up for ease of reporting and analysis. External structured data is usually organized in the form of aggregate totals and can be easily adapted for further analysis. External unstructured data is mostly available in the form of content but not under the control of business, less likely to be reliable, and not easily aggregated. The organization should consider the analytical opportunity of unstructured data and the possibility that it can be integrated into roadmapping.

There are many different methods and techniques involved in data science from transforming raw data into valuable insight. In this study, four data science methodologies are discussed: supervised learning, unsupervised learning, time series, and text mining.

Supervised (or predictive) learning is the main approach to machine learning; a subspecialty of computer science that gives computers the ability to learn based on data to discover models, patterns, and other regularities (Mitchell 1997; Murphy 2012; Samuel 2000). The system learns to approximate a function from a collection of labeled training data for mapping between a set of input objects (variable x) and desired output value (variable y, also called a *supervisory signal*) and apply this mapping to predict the output of unseen data (Cunningham et al. 2008; Han et al. 2011; Qiu et al. 2016). The objective of supervised learning or inferred functions can be divided mainly into two problem types: regression (when the output is real value) and classification (when the output variable is a category). Examples of regression and classification problems are how to predict future energy consumption in manufacturing; and how to predict the response of customers to new technology, respectively.

Unsupervised learning does not require labeled training data (unlabeled or unclassified examples) and the algorithms learn the inherent structure from the input data (variable X) without any desired targets (no corresponding output variables) (Qiu et al. 2016). The objective of unsupervised learning is to model the underlying structure or distribution of data to learn more about it. It can be divided mainly into two types of problems: clustering (to discover the inherent groupings or classes within the data) and association (to discover rules that describe large portions of data). Two examples of clustering and association problems are how to cluster users into groups based on their purchasing behavior to develop product innovations; and how to analyze the retail baskets, respectively.

A time series is a sequence of observations made through different points in time (Chatfield 2000). The values must be measured at equal time intervals, such as every hour, every day, or every month (Azevedo et al. 2012). The aim of time series forecasting is to develop an appropriate model for predicting future values (or behavior) based on the historical values (or past observations) (Stepchenko et al. 2017). Time series has gained extensive popularity in predicting future trends and is used for many industrial applications and in various scenarios. Examples of time series applications include economic and financial forecasting, technology forecasting, sales planning, production planning, inventory management and control, and exchange rate forecasting. To analyze the time series, it is necessary to understand the pattern of data during a particular period of time. The pattern contains different components such as trends, cyclical variations, seasonal variables, and irregular fluctuations. Simple moving average, exponential smoothing, auto-regressive integration moving average (ARIMA), and neural network

(NN) are commonly recognized as effective statistical techniques for time series forecasting.

Text mining aims to perform knowledge discovery using the structured method for analysis of unstructured databases or textual data (Feldman and Sanger 2007). Text mining provides a more efficient approach for browsing and searching a huge volume of documents (Atkinson 2000). The text-mining approach retrieves and extracts key information like keywords and terminology from data sources such as patent documents, social media, emails, and product manuals. Using a predefined pattern, this approach can automatically classify documents into categories (Sebastiani 2002) or cluster a collection of documents according to similarity (Dhillon and Modha 2001). Different algorithms of text mining are based on computational linguistics and statistical methods (Feldman and Sanger 2007; Weiss et al. 2010). Natural language processing (NLP) is a component of text mining with a variety of techniques for performing linguistic analysis such as automatic summarization, word stemming, and relation extraction.

13.3 Monitoring the Status of a Roadmap with Data-driven Roadmapping Approach

13.3.1 Analysis Steps

Any business organization needs to continually monitor changes from key drivers in order to evaluate the impact of a changing situation on the strategic roadmap targets. This includes monitoring the social, technological, economic, environmental, and political aspects as well as competitors, suppliers, and customers. To support its development, the roadmapping team may use several techniques to understand the business environment and predict the future.

To continually monitor the status of roadmap, an automated procedure should be implemented to assess the impact of changes on key drivers, using dashboard to track the current situation. Table 13.3 shows the five analysis steps proposed to assess the impact of changes in key drivers.

13.3.2 Case Demonstration: Managerial Procedure for Data-driven Roadmapping (DDRM) Approach

The analysis steps and computational details for assessing the impacts of changes from both external and internal drivers on the status of a roadmap are demonstrated through two case examples:

- The development of exoskeleton published in *Journal of Engineering and Technology Management* (Gerdsri et al. 2019).

Table 13.3 Analysis steps for data-driven roadmapping to assess the impact of changes in key drivers.

Step 1: develop an evaluation model for roadmap status

Level 1: model objective

1.1 Define the objectives.

Level 2: key drivers

1.2 Identify the key drivers (KD_i).

1.3 Determine the relative importance (w_i) of each key driver with respect to the objective.

Level 3: sub-drivers

1.4 Identify the sub-drivers (SD_{ij}).

1.5 Determine the relative importance (w_{ij}) of each sub-driver with respect to the key driver (KD_i).

1.6 Define a unit of measurement (m_{ij}) for each sub-driver.

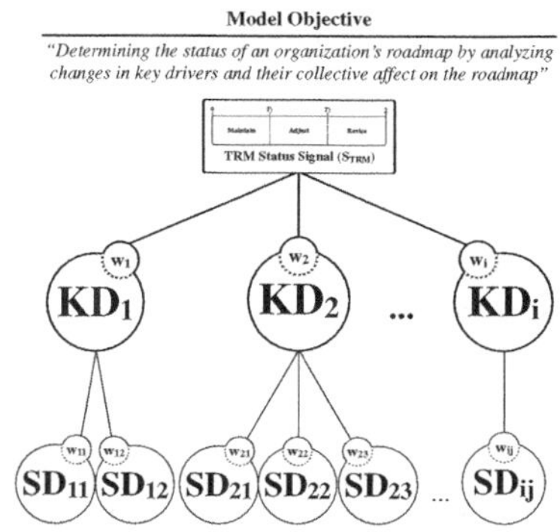

(Gerdsri et al., 2019 / with permission of Elsevier)

Step 2: design of data-driven roadmapping architecture and develop the approach

2.1 Design the layered architecture for data-driven roadmapping to support the maintenance and updating processes.

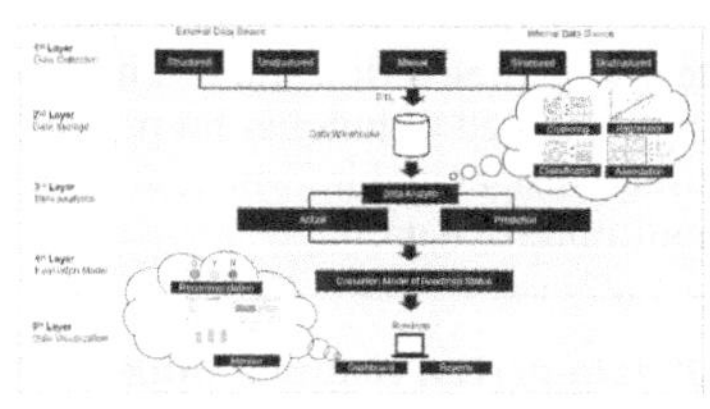

Data collection (first layer of data-driven roadmapping architecture as shown in Figure 13.1)

2.2 Identify the data sources of all units of measurement (m_{ij}) for each sub-driver. Data can be separated by type (structure and unstructured) of data and source (internal and external).

	Unstructured Data	Structured Data
Internal Source	Email CRM Customer survey Employee survey Product catalog Technical specification Annual report	Data from ERP system such as financial and accounting, sales and distribution, supply chain, human resources Data from other business application such as point-of-sale (POS) terminals and e- commerce applications
External Source	Industrial report Patents Research Bibliometric News Social media Video	Open government data Statistical subscription data

2.3 Design a conceptual schema by creating an entity-relationship diagram (ERD) and collecting data from identified sources.

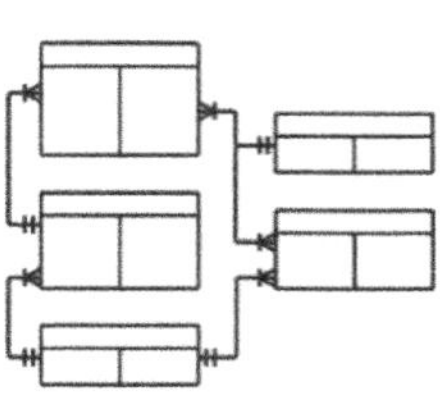

(Continued)

Table 13.3 (Continued)

Data storage (second layer of data-driven roadmapping architecture as shown in Figure 13.1)

2.4 Extract and convert data on the unit of measurement (m_{ji}) for each sub-driver from multiple formats and structures into a standardized system (computer-readable) before inserting into data warehouse (ETL process).

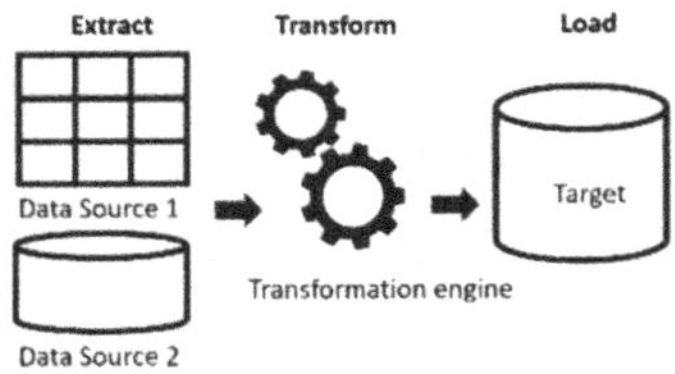

Data analytics (third layer of data-driven roadmapping architecture as shown in Figure 13.1)

2.5 Retrieve the data from the data warehouse and develop the algorism for analysis:

- Track the actual value (recent data) for the unit of measurement (m_{ij}) of each sub-driver.
- Predict future trends for the unit of measurement (m_{ij}) of each sub-driver.
- Gain insight into the particular situation or phenomena to identify the sub-drivers impacting on the business environment.

Supervised learning, unsupervised learning, time series, and text mining are suggested techniques for providing useful insight and information from a multitude of data.

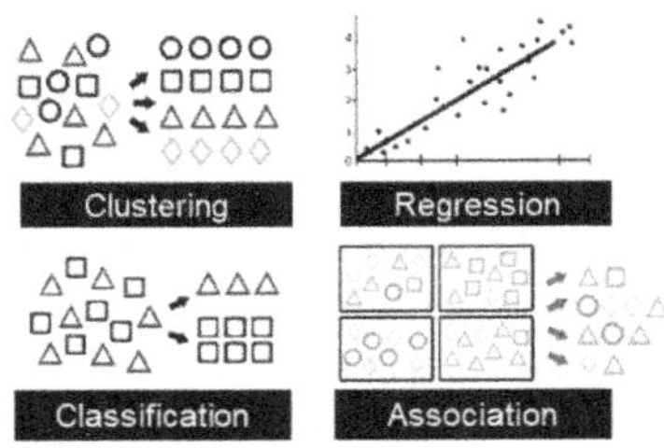

Step 3: data-driven roadmapping (DDRM) operation

Evaluation model of roadmap status (fourth layer of data-driven roadmapping architecture as shown in Figure 13.1)

3.1 Define the tolerance limits and intervals of the measurement values for each sub-driver.

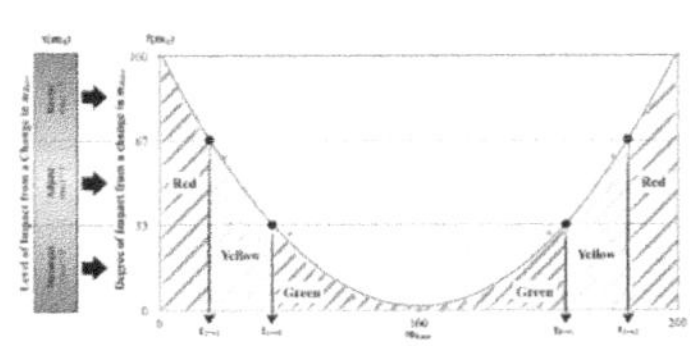

(Gerdsri et al., 2019 / with permission of Elsevier)

3.2 Continuously monitor the unit of measurement value (m_{ij}) for each sub-driver by scheduling an automatic scan for deviations in the measurement value (m_{ij}).

$$x(m_{ij}) = \begin{cases} 0 \text{ if } t_{1\to 0} \le m_{ij} \le t_{0\to 1} \\ 1 \text{ if } t_{2\to 1} \le m_{ij} < t_{1\to 0} \text{ or } \\ \quad t_{0\to 1} < m_{ij} \le t_{1\to 2} \\ 2 \text{ if } m_{ij} < t_{2\to 1} \text{ or } t_{1\to 2} < m_{ij} \end{cases}$$

(Continued)

Table 13.3 (Continued)

3.3	Analyze the impact of change from deviations in the sub-driver measurement values (m_{ij}) compared to baseline values (m_{base}) using Eq. (13.1).	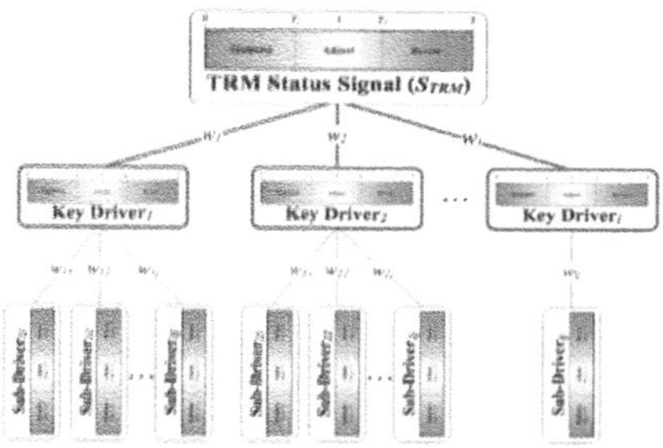
3.4	Calculate the TRM status signal (S_{TRM}) using Eq. (13.1) and display the status using color coding; maintain (green), adjust (yellow), and revise (red). The roadmap status is determined based on an impact assessment of changes in key drivers (evaluate S_{TRM}) and submit the result.	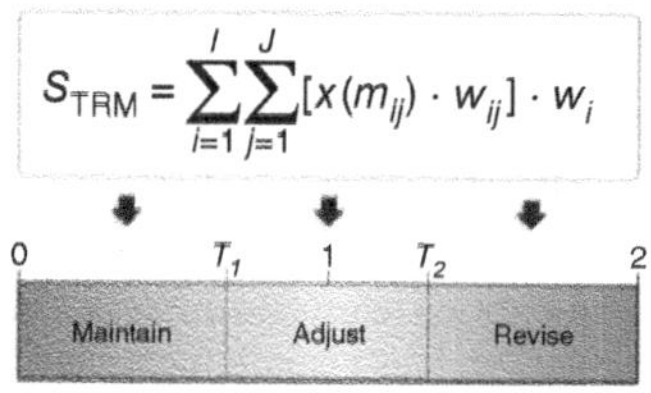
Data visualization (fifth layer of data-driven roadmapping architecture as shown in Figure 13.1)		
3.5	Data visualization for the actual value (recent data), trend prediction, and insight in relation to the unit of measurement value (m_{ij}) for each sub-driver of the roadmap through the dashboard.	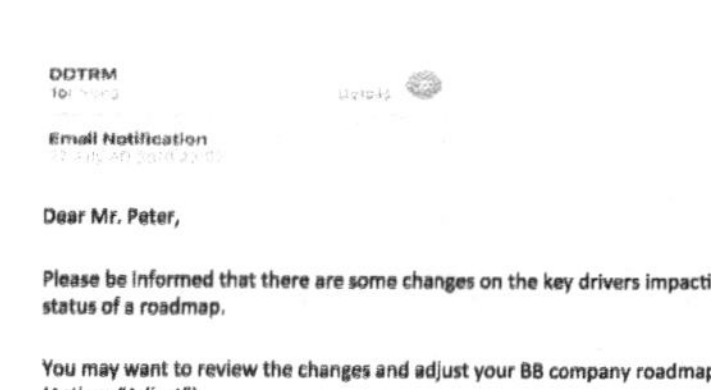
3.6	Report the roadmap status to top management resulting from the impact assessment of changes in key drivers. Notify when changes in key drivers are affecting the roadmap status according to the predefined conditions and timelines.	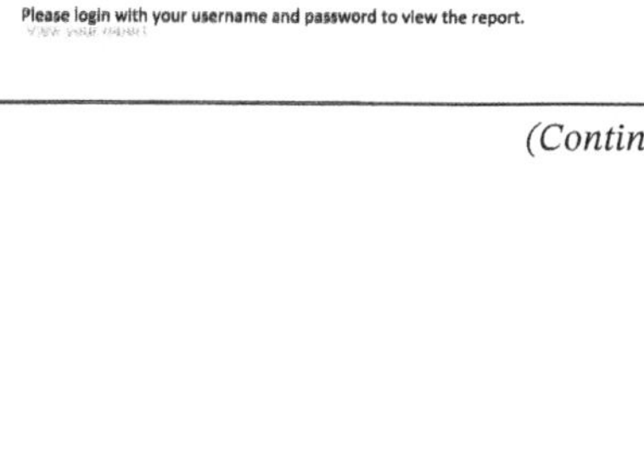 DDTRM Email Notification Dear Mr. Peter, Please be informed that there are some changes on the key drivers impacting the status of a roadmap. You may want to review the changes and adjust your BB company roadmap. (Action: "Adjust") Please login with your username and password to view the report.

(Continued)

Table 13.3 (Continued)

3.7	Data visualization of the current roadmap status can be classified and visualized using three distinct colors: ● **Maintain** – There is no impact or a minor impact on the roadmap. ● **Adjust** – Some minor amendments should be made to the roadmap. ● **Revise** – Some major amendments should be made to the roadmap.	
Step 4: monitor the impact of change on the roadmap status		
The TRM operations team continues to monitor changes in the business environment as well as visualizing recent information, trends, and insights as specified in the roadmap via the dashboard provided. The system will trigger the notification to indicate that changes are impacting key drivers and causing the roadmap to be "adjusted" or "revised." The TRM operations team will take action to inform the management team of any changes.		
Step 5: managerial decisions suggested by the TRM status report		
Management makes a final decision on whether to review the roadmap based on the TRM status recommendation. If adjustment or revision is required, an updated version of the roadmap should be communicated to all related stakeholders.		

13.4 The Development of Integrative Restaurant Services and Intelligent Management Published in IEEE Transactions on Engineering Management (Pora et al. 2022)

The managerial procedure in applying DDRM approach is proposed with five steps, as shown in Figure 13.10. The roles between the management, TRM operations and data science teams through the DDRM approach are also described according to the dynamic interaction among key players involving in roadmapping implementation (Gerdsri et al. 2009).

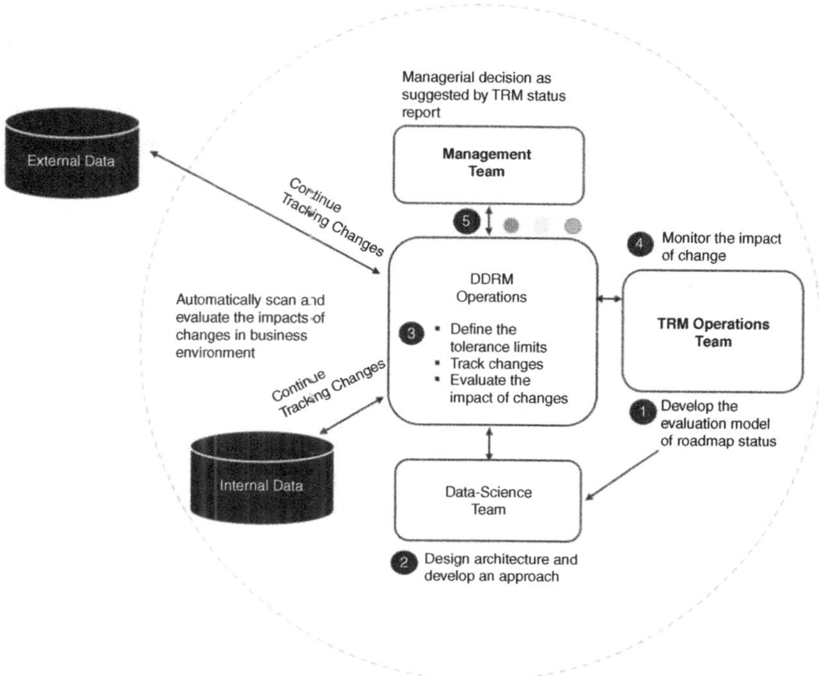

Figure 13.10 Managerial procedure for the data-driven roadmapping approach (Pora et al. 2022/with permission of IEEE).

The first step is to develop an evaluation model of roadmap status by identifying the key drivers and sub-drivers that may create a critical impact. Such factors can be social, technological, economic, environmental, or political, and may include internal factors such as customer insight. This step is carried out by the TRM operations team.

The second step is to design the DDRM architecture (including identifying different types and sources, determining how to collect, store, and track the data, how to analyze the data and predict trends, and how to visualize the data) and develop a data-driven support tool for roadmap status monitoring. This step would be carried out by the data science team.

The third step is to define the tolerance limits and intervals of the measurement values for each sub-driver after tracking an actual value (recent data) per unit of measurement (m_{ij}) for each sub-driver. DDRM involves creating a schedule to automatically scan and evaluate the impact of changes in the business environment according to predefined conditions. The roadmap status can be generated and visualized in the form of color-coded signals: red, yellow, and green. The system provides recent data, trend prediction, and key insight through a web application, and is displayed on a dashboard for both the TRM operations team and management team. This step would be performed by the data science team and operations team.

The fourth step is to monitor the impact of changes to the business environment on the roadmap status. Such changes can be continuously monitored and visualized using recent information, trends, and insight into the business environment as specified on the roadmap dashboard. The roadmap status report automatically delivers in accordance with the defined time period and condition. In the event that any changes impact on the key drivers, causing the roadmap to "adjust" or "revise," the system will trigger a notification to indicate that the roadmap needs to be updated or totally revised. This step would be carried out by the TRM operations team.

The final step involves making a decision as to whether the roadmap should be adjusted or revised according to the TRM status recommendation. If the roadmap requires adjustment or revision based on the perception of the organization, an updated version should be communicated again to all related stakeholders. This step would be performed by the management team.

13.5 Conclusion

The proposed framework for monitoring and assessing the status of a technology roadmap with DDRM is suitable for companies that routinely integrate roadmapping into an ongoing business process and manage their operations on the basis of roadmap evidence. The system will provide notification of significant changes with the insight to alert the roadmap operation team; therefore, companies could quickly respond to any changes of external and/or internal environment.

The systematic approach of DDRM transforms a traditional roadmap into a data-driven, semi-automated system. During the roadmap development stage, an organization can exploit data and create insight to reflect any changes in the business environment and then assess their impact on the roadmap. With this approach, the management team can effectively make decisions based on evidence from the analysis rather than management's judgment alone. A computerized decision support with data-driven system could dynamically strengthen roadmapping process as the status of a roadmap can be proactively monitored and revised as needed in a timely manner.

The data can be both unstructured and structured, ranging from the voice of customers (VOC) to ERP data. The type of data to be tracked is determined by the roadmapping team in order to gain insight into the changes of sub-drivers as indicated on a roadmap.

Five analysis steps are proposed to assess the impact of changes in key drivers including: (i) developing an evaluation model for roadmap status, (ii) designing DDRM architecture and developing the approach, (iii) developing a DDRM operation, (iv) monitoring the impact of change on the roadmap status, and (v) implementing the managerial decisions suggested by the TRM status report.

The effectiveness of the proposed framework can be evaluated by how well the status of a roadmap could reflect the impact of changes so that the management can make a timely decision on its revision. To effectively implement the proposed framework, practitioners or researchers engaging in the initiative have to consider the following actions:

- Clearly articulate to the top management why the status of a roadmap needed to be actively monitored in order to reflect the impact of changes in business competition, customer preference, technology trend, etc.
- Communicate with all stakeholders that implementing the DDRM system is not just setting up computer programs or data analytic tools, it requires to adjust the managerial procedure and the roles of the management.
- Beware that not all data needed would be available or could be directly used. The management and data science team have to work together to identify which data to be used, how to collect them, or whether the data transformation is required before applying into the DDRM system.
- Setup the DDRM system is not a one-time effort. The responsible team has to constantly review and calibrate the system. Overtime, the business condition may change; therefore, the weight of key drivers and sub-drivers along with their tolerance limits may have to be adjusted.

References

Altunok, T. and Cakmak, T. (2010). A technology readiness levels (TRLs) calculator software for systems engineering and technology management tool. *Advances in Engineering Software* 41 (5): 769–778.

Atkinson, J. (2000). Text mining: principles and applications. *Revista Facultad de Ingenieria* 7: 57–62.

Azevedo, J.M., Almeida, R., and Almeida, P. (2012). Using data mining with time series data in short-term stocks prediction. *International Journal of Intelligence Science* 2 (4A): 177–181.

Brown, R. and O'Hare, S. (2001). The use of technology roadmapping as an enabler of knowledge management. In: *IEEE Seminar Managing Knowledge for Competitive Advantage*, 7/1–7/6. IEEE.

Chatfield, C. (2000). *Time-series Forecasting*. CRC Press.

Cunningham, P., Cord, M., and Delany, S.J. (2008). Supervised learning. In: *Machine Learning Techniques for Multimedia: Case Studies on Organization and Retrieval* (ed. M. Cord and P. Cunningham), 21–49. Springer.

Dhillon, I.S. and Modha, D.S. (2001). Concept decompositions for large sparse text data using clustering. *Machine Learning* 42 (1–2): 143–175.

Feldman, R. and Sanger, J. (2007). *The Text Mining Handbook: Advanced Approaches in Analyzing Unstructured Data*. Cambridge University Press.

Gerdsri, N. (2013). Implementing technology roadmapping in an organization. In: *Technology Roadmapping for Strategy and Innovation: Charting the Route to Success*, 191–210. Springer.

Gerdsri, N., Vatananan, R., and Dansamasatid, S. (2009). Dealing with the dynamics of technology roadmapping implementation: a case example. *Technological Forecasting and Social Change* 76 (1): 50–60.

Gerdsri, N., Kongthon, A., and Vatananan, R.S. (2013). Mapping the knowledge evolution and professional network in the field of technology roadmapping: a bibliometric analysis. *Technology Analysis & Strategic Management* 25 (4): 403–422.

Gerdsri, N., Puengrusme, S., Vatananan, R., and Tansurat, P. (2019). Conceptual framework to assess the impacts of changes on the status of a roadmap. *Journal of Engineering and Technology Management* 52: 16–31.

Groenveld, P. (1997). Roadmapping integrates business and technology. *Research-Technology Management* 40 (5): 48–55.

Han, J., Pei, J., and Kamber, M. (2011). *Data Mining: Concepts and Techniques*. Elsevier.

Hirose, Y., Phaal, R., Farrukh, C. et al. (2020). Characterizing maturity levels for organization-wide roadmapping implementation. *IEEE Engineering Management Review* 48 (4): 133–143.

Hirose, Y., Phaal, R., Farrukh, C. et al. (2021). Practical roadmapping implementation: what we learned from QinetiQ group. *IEEE Engineering Management Review* 49 (4): 108–114.

Hirose, Y., Phaal, R., Farrukh, C. et al. (2022). Sustaining organizational roadmapping implementation – lessons learned from subsea 7. *Research-Technology Management* 65 (3): 50–57.

Holmes, C. and Ferrill, M. (2008). A process for the update and review of operation and technology roadmaps. *International Journal of Innovation and Technology Management* 5 (02): 247–258.

Mitchell, T.M. (1997). *Machine Learning*, vol. 45(37), 870–877. Burr Ridge, IL: McGraw Hill.

Murphy, K.P. (2012). *Machine Learning: A Probabilistic Perspective*. MIT Press.

Nauda, A. and Hall, D. (1991). Strategic technology planning-developing roadmaps for competitive advantage. In: *Technology Management: The New International Language*. IEEE.

Nolte, W.L., Kennedy, B.C., and Dziegiel, R.J. (2003). Technology readiness level calculator. In: *NDIA Systems Engineering Conference*. NDIA.

Phaal, R. and Muller, G. (2007). Towards visual strategy: an architectural framework for roadmapping. In: *Portland International Conference on Management of Engineering & Technology*, 1584–1592. IEEE.

Phaal, R. and Muller, G. (2009). An architectural framework for roadmapping: towards visual strategy. *Technological Forecasting and Social Change* 76 (1): 39–49.

Phaal, R., Farrukh, C.J., and Probert, D.R. (2001). Characterisation of technology roadmaps: purpose and format. In: *Portland International Conference on Management of Engineering & Technology*, 367–374. IEEE.

Phaal, R., Farrukh, C.J., and Probert, D.R. (2005). Developing a technology roadmapping system. *Technology Management: A Unifying Discipline for Melting the Boundaries* 31: 99–111.

Pora, U., Gerdsri, N., Thawesaengskulthai, N., and Triukose, S. (2022). Data-driven roadmapping approach and case demonstration. *IEEE Transactions on Engineering Management* 69 (1): 209–227.

Qiu, J., Wu, Q., Ding, G. et al. (2016). A survey of machine learning for big data processing. *EURASIP Journal on Advances in Signal Processing* 2016 (1): 67.

Samuel, A.L. (2000). Some studies in machine learning using the game of checkers. *IBM Journal of Research and Development* 44 (1.2): 206–226.

Sebastiani, F. (2002). Machine learning in automated text categorization. *ACM Computing Surveys* 34 (1): 1–47.

Stepchenko, A., Chizhov, J., Aleksejeva, L., and Tolujew, J. (2017). Nonlinear, non-stationary and seasonal time series forecasting using different methods coupled with data preprocessing. *Procedia Computer Science* 104: 578–585.

US Department of Energy (2011). *Technology Readiness Assessment Guide*. Washington, D.C.: U.S. Department of Energy https://www.directives.doe.gov/directives-documents/400-series/0413.3-EGuide-04/@@images/file.

Vatananan, R.S. and Gerdsri, N. (2012). The current state of technology roadmapping (TRM) research and practice. *International Journal of Innovation and Technology Management* 9 (04): 1250032.

Weiss, S.M., Indurkhya, N., Zhang, T., and Damerau, F. (2010). *Text Mining: Predictive Methods for Analyzing Unstructured Information*. Springer Science & Business Media.

Wells, R., Phaal, R., Farrukh, C., and Probert, D. (2004). Technology roadmapping for a service organization. *Research-Technology Management* 47 (2): 46–51.

Printed and bound by CPI Group (UK) Ltd, Croydon, CR0 4YY
17/09/2026
14964255-0004